T0155673

Off-Grid Electrical Systems in Developing Countries

Henry Louie

Off-Grid Electrical Systems in Developing Countries

 Springer

Henry Louie
Electrical and Computer Engineering
Seattle University
Seattle, WA, USA

ISBN 978-3-030-06322-1   ISBN 978-3-319-91890-7   (eBook)
https://doi.org/10.1007/978-3-319-91890-7

This Springer imprint is published by the registered company Springer International Publishing AG part
of Springer Nature.
The registered company address is: Gewerbestrasse 11, 6330 Cham, Switzerland

To Kristine

Preface

Over one billion people do not have access to electricity. The majority live in developing countries in Sub-Saharan Africa and South Asia. The consequences of this form of energy poverty can be severe. Most activities end at sunset, unless dangerous and expensive kerosene lamps or candles are used. Children breathe air polluted by smoke from open fires. Women give birth in darkness or without the aid of life-saving electronic medical devices. Rural communities become even more isolated without access to news and information by radio or television.

Now, perhaps more than ever, electricity access has caught the attention of the global community. Access to affordable and sustainable energy is one of the United Nation's Sustainable Development Goals. Large philanthropic and development organizations are prioritizing electricity access. Even multi-billion-dollar technology companies have started electricity access initiatives.

An estimated US$50 billion per year is needed through 2030 to achieve universal electricity access. It is estimated that over 100,000 mini-grids will be needed, and that one in three households presently without electricity access will have an off-grid system of some form. This will not happen without a workforce of engineers well prepared to innovate and design systems in the unique context of off-grid communities in developing countries.

The purpose of this book is to provide engineers with the essential foundational knowledge of designing and operating off-grid electricity systems in developing countries. This is a first-of-its-kind book that brings together the electrical engineering concepts relevant to off-grid systems. The scope is broad. Throughout the book, examples, design approaches, and practical considerations especially relevant to off-grid systems in developing countries are provided.

The book focuses on electrical aspects of off-grid systems. It assumes the reader has basic proficiency in DC and AC circuit analysis, including phasor and steady-state power analysis. Previous exposure to balanced three-phase circuit analysis is helpful, but not required. Wherever possible, the single-phase (per-phase) model is used. It is appropriate for third or fourth year undergraduate students, or first year graduate students. Although aspects of renewable energy engineering and power electronics are covered, it does not replace a course or book dedicated to these topics. Practitioners may find the book as a useful reference.

The book is arranged in four parts. The first part, Chaps. 1 to 4, is focused on electricity access in general. Chapters 1 and 2 describe the state of energy consumption in the world in general, and in off-grid communities in developing countries in particular. Chapter 3 describes electricity access through grid extension, whereas Chap. 4 introduces off-grid systems.

The second part of this book, Chaps. 5 through 7, is focused on the energy conversion technologies used in off-grid systems. Readers with a background in renewable energy and electromechanical energy conversion may find some of this material familiar. However, the information is presented considering small-scale off-grid applications, which most readers will find fresh.

The third part of this book, Chaps. 8 and 9, covers energy storage and electronic converters and controllers. Chemical batteries are incredibly important in many off-grid systems, yet most engineers do not have a firm understanding of the underlying electrochemistry. Chapter 9 discusses converters. Readers with a background in power electronics will find some concepts familiar.

The fourth and final part of the book ties the concepts presented in the previous chapters. In Chap. 10 we see how the components discussed in the second and third part of the book operate together in an off-grid system. Chapter 11 begins a two-chapter description of how off-grid systems are designed, beginning with load and resource assessment. Off-grid system design is the focus of Chap. 12. Here a realistic example of a solar-based mini-grid is used to illustrate the design of the energy production and distribution systems. Chapter 13 is focused on solar home systems and solar lanterns. The book concludes with a short chapter on practical considerations.

In writing this book, I drew heavily upon my experience in off-grid electrical systems. My work with the nonprofit organization KiloWatts for Humanity, IEEE Smart Village, and time living in Zambia as a Fulbright Scholar is especially formative. Most of the text is oriented toward electricity access in the Sub-Saharan African experience.

It is with some hesitation that the title of this book includes the term "developing country." To some, this is a degrading term. Of course, it is not intended to be interpreted in this way. Rather, the term is used to connote the general circumstances that the off-grid systems discussed in this book exist: in at-risk, underserved, and/or impoverished communities. These conditions also exist in countries not classically considered developing.

I am especially grateful to the many reviewers, organizations, and individuals whose insight and feedback helped shape this book. In particular, Dr. Paul Neudorfer and Dr. Eric Watson, S.J. of Seattle University; Steve Szablya, P.E., and Daniel Nausner of KiloWatts for Humanity; Peter Dauenhauer from the University of Strathclyde; Dr. Pritpal Singh of Villanova University; Brett Bauer of Canyon Industries; Frank Bergh, P.E. of Sigora Haiti; Ifeanyi Orajaka of GVE Projects; and Isaiah Lyons-Galante and Sam Slaughter of Power Gen.

I am thankful for the willingness of so many individuals and organizations for allowing their images to be used in this book: Canyon Industries, BBOXX, d.light, Energy Sector Management Assistance Program, HOMER Energy, Itek Energy, KTH University, Ella Louie, Eli Patten, Outback Power, PowerGen, Robert Ngoma, and World Bank Group. Several Seattle University students assisted in proofreading and developing figures: Yahya Alyami, Greg Hirose, and Mahekdeep Singh.

This book was made possible by the resources provided by Fr. Francis Wood Chair at Seattle University.

Lastly, I am immeasurably grateful for the early mentorship and inspiration from Dr. Bert Otten, S.J., of the Chikuni mission in Zambia.

Seattle, WA, USA   Henry Louie
March 2018

Contents

Acronyms

AAAC	All aluminum alloy conductor
ABC	Aerial bundled conductor
ABS	Acrylonitrile butadiene styrene
ACSR	Aluminum conductor steel reinforced
AGM	Absorbed glass mat
AGR	Annual growth rate
Ah	Amperehour
AM	Air Mass
ARPU	Average revenue per user
AVR	Automatic voltage regulator
AWEA	American Wind Energy Association
BTU	British Thermal Unit
CAR	Central African Republic
CF	Coincidence Factor
CFL	Compact florescent light
CHP	Combined heat and power
CI	Compression ignition
DF	Demand factor
DoD	Depth-of-discharge
DRC	Democratic Republic of Congo
ELC	Electronic load controller
GOGLA	Global Off-Grid Lighting Association
GPRS	General Packet Radio Service
GPS	Global Positioning System
HDI	Human Development Index
IC	Incremental conductance
ICE	Internal combustion engine
IEC	International Electrotechnical Commission
IEEE	Institute of Electrical and Electronic Engineers
ITC	Internet, telecommunications, computers
KCL	Kirchhoff's Current Law

KVL	Kirchhoff's Voltage Law
koe	kilogram of oil equivalent
LCOE	Levelized cost of energy
LED	Light Emitting Diode
LF	Load factor
LI	Lithium ion
LPG	Liquefied petroleum gas
LVD	Low voltage disconnect
MHP	Micro hydro power
MOSFET	Metal–oxide–semiconductor field-effect transistor
MPP	Maximum power point
MPPT	Maximum power point tracker
MTF	Multi-Tier Framework
NOCT	Normal Operating Cell Temperature
OPzV	Ortsfest PanZerplatte Verschlossen
PCD	Pitch Circle Diameter
PDF	Probability density function
PF	Power factor
PMSG	Permanent magnet synchronous generator
PO	Perturb & observe
PV	Photovoltaic
PVC	Polyvinyl chloride
PWM	Pulse width modulation
REAs	Rural Electrification Authorities
REMP	Rural electrification master plan
RMS	Root mean square
RPM	Revolutions per minute
SA	South Asia
SHS	Solar home system
SI	Spark ignition
SL	Solar lantern
SLA	Sealed lead–acid
sLCOE	Simplified LCOE
SLI	Starting, lighting and ignition
SPWM	Sinusoidal pulse width modulation
SOC	Standard operating conditions
SoC	State-of-charge
SSA	Sub-Saharan Africa
STC	Standard test conditions
SWER	Single Wire Earth Return
toe	Tonne of oil equivalent
THD	Total harmonic distortion
TSR	Tip speed ratio
UPE	User premise equipment

UPS	Uninterruptible Power Supply
VRLA	Valve regulated lead–acid
WECS	Wind energy conversion system

Part I
Electricity Access

Chapter 1
Energy and Development

1.1 Introduction

Off-grid electrification refers to providing electricity to an unserved population by a means other than a connection to an existing centralized power grid. This book is concerned in particular with off-grid electrification in the context of so-called developing countries.[1]

Worldwide, 1.1 billion people—approximately one in seven—do not have access to electricity [12]. This form of energy poverty disproportionately afflicts those living in at-risk rural communities in Sub-Saharan Africa[2] (SSA) and South Asia, as shown in Fig. 1.1.

For many, the prime motivation for studying off-grid electrification stems from the idea that access to electricity improves people's lives. While the linkage between electricity access and quality of life surely exits, this perspective misses an important aspect of off-grid electrification: it need not be strictly a humanitarian effort. For-profit companies are increasingly active in this area. For example, over 130 million off-grid solar lighting units have been sold since 2010 with little or no subsidy, and the market for off-grid appliances—TVs, radios, refrigerators and the like designed to operate in off-grid conditions—exceeded US$500 million in 2015 [4, 7]. Even still, a substantial market remains. In 2015, people without electricity spent an estimated US$27 billion on candles, kerosene, and other stopgap fuels that could be replaced by off-grid solutions [10]. Off-grid electrification is a growing industry—over 100 companies are actively involved in electricity-access products and services—and there is a growing need for well-prepared engineers, technicians, managers, and entrepreneurs to support this effort [10].

---

[1] See the Preface for a discussion on the definition and usage of this terminology.

[2] Sub-Saharan Africa refers to the 48 African countries south of the Sahara Dessert as defined by the World Bank.

© Springer International Publishing AG, part of Springer Nature 2018
H. Louie, *Off-Grid Electrical Systems in Developing Countries*,
https://doi.org/10.1007/978-3-319-91890-7_1

Fig. 1.1 The majority of the world's population without electricity access live in Sub-Saharan Africa (SSA) or South Asia (SA)[11, 15]

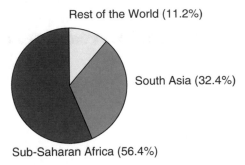

Rest of the World (11.2%)

South Asia (32.4%)

Sub-Saharan Africa (56.4%)

Fig. 1.2 A cluster of houses in rural Zambia (courtesy of the author)

Much of this book is dedicated to off-grid electrification in a rural setting like that shown in Fig. 1.2. Although over one hundred million people live in urban areas without electricity access, electrification rates are much lower in rural areas, as shown in Fig. 1.3. Off-grid electrification is associated with rural areas because it is usually economically unfavorable to connect remote, sparsely populated communities to the grid. This is true even in developed countries. While there is no strict definition of a rural community, they tend to exhibit these common characteristics [9]:

- decentralized population;
- geographic isolation;
- underserved in terms of health care, education, clean water, sanitation, and other infrastructure;
- unable to participate in regional and national markets.

Under certain conditions, rural communities can be better served by off-grid solutions, including solar lanterns, solar home systems, and mini-grids (see Fig. 1.4) rather than the national grid. Indeed, some projections estimate that as many as one-in-three people presently without a grid connection will one day be served by some form of off-grid solution.

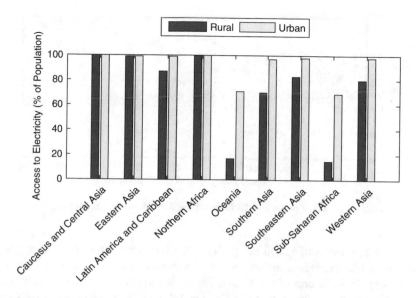

Fig. 1.3 People living in rural communities are less likely to have access to electricity than people in urban areas [11]

Fig. 1.4 Solar-powered mini-grid in Tanzania (courtesy of PowerGen)

   This chapter discusses the important linkage between energy use and human development, with a focus on electrical energy. Basic descriptions of the various types of off-grid electrical systems—the primary focus of this book—are given. The chapter concludes with an overview of grid-connected power systems.

1.2 Energy and Human Development

Energy in its many forms underwrites all human endeavors. Our most basic needs—growing and harvesting food, accessing potable water, and transporting goods and people—and our most complex undertakings, from robotics to space exploration, require access to inexpensive and abundant energy sources.

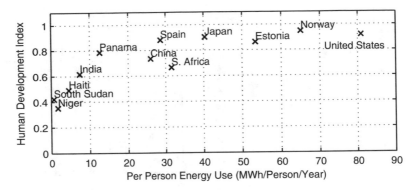

Fig. 1.5 Total primary energy use and the Human Development Index (2014) [15]

Access to energy, of which electricity is an important form, correlates closely with human development. Indeed, universal access to clean and affordable energy by the year 2030 is one of the United Nation's Sustainable Development Goals. This notion is succinctly captured by Fig. 1.5, which plots the Human Development Index (HDI) versus per person energy use[3] for various countries. The HDI is a commonly used metric that attempts to measure the development or well-being of a country, accounting for income, health, and education [13]. Higher HDI scores are associated with more developed countries. Countries with high HDI scores tend to use more energy per person than those with lower HDI scores [1]. However, a saturation effect is apparent: there is little sensitivity of HDI to variation in energy consumption for countries with high HDI scores, but for countries with HDI scores below approximately 0.7, modest increases in energy use are associated with sharp rises in HDI. A similar relationship can be shown for HDI and per person electricity consumption, suggesting that electricity is an enabling and critical infrastructure for high economic, health, and educational attainment.

Electricity provides lighting, which is important for work, study, socializing, and safety. It is supportive of improved health care at hospitals and clinics, and education at schools; it can reduce the toil of manual labor by powering pumps and motors for agriculture and milling; it can support income generation by allowing work to be done more efficiently and extending business hours[4]; and to many people, having electricity brings with it a newfound sense of dignity and modernity. In some countries, for example, Brazil and South Africa, access to electricity has been declared a fundamental right, in recognition of its importance to wellness and prosperity.

---

[3] Here "energy use" refers to primary energy before transformation to other end-use fuels. For example, it accounts for the energy used as input to a power plant, not just the electrical output of the power plant.

[4] It must be noted that evidence suggests that electricity alone does not necessarily increase productivity, but other factors, for example, access to machinery and education, are also required.

1.3   Units of Energy

It is more common to discuss energy use on an average annual basis rather than daily basis and to exclude the energy in the food people eat and the work animals they might use. The International System unit of energy is the joule (J). A joule is a derived unit, which is defined as

$$1\,\text{J} = 1\,\text{N m} = \text{Pa m}^3 = 1\frac{\text{kg m}^2}{\text{s}} \tag{1.1}$$

where N is the unit for force (newton) and Pa is the unit for pressure (pascal). Energy can also be defined with respect to electrical quantities as

$$1\,\text{J} = 1\,\text{W s} \tag{1.2}$$

where W is the unit watt. It is not uncommon in popular media to confuse energy and power. The difference is extremely important in electrical systems. Power, expressed in watts, describes the rate in which energy is supplied or consumed. More generally

$$P = \frac{dE}{dt} \tag{1.3}$$

where P is power, E is energy, and t is time. The relationship between power and energy is analogous to the relationship between speed and distance.

Example 1.1 A small generator linearly increases its power output from zero watts to 100 watts over a 10-min period. Compute the total energy output by the generator during this period.

Solution Equation (1.3) can be rearranged to solve for the energy output, noting that there are 600 s in 10 min:

$$E = \int P(t)dt = \int \frac{100}{600}t\,dt.$$

This yields

$$E = \int_0^{600} \frac{100}{600}t\,dt = \frac{1}{6}\frac{t^2}{2}\Big|_0^{600} = 30,000\,\text{J}.$$

Depending on the context and scale considered, different units of energy are in common use. For example, electrical engineers often use units based on the watthour, whereas mechanical engineers might favor BTUs (British Thermal Unit) or joules. Economists or those describing energy characteristics of an entire country

Table 1.1 Conversion of
energy units

Unit	Joules
Joule (J)	1
Calorie (cal)	4.1868
British Thermal Unit (BTU)	1055.87
Watthour (Wh)	3600
Kilocalorie (C, kcal)	4186.8
Kilowatthour (kWh)	3.6×10^6
Kilogram of oil equivalent (koe)	41.868×10^6
Megawatthour (MWh)	3.6×10^9
Tonne of oil equivalent (toe)	41.868×10^9
Quad (quad)	1055.87×10^{15}
Gigajoule (GJ)	1×10^9
Terawatthour (TWh)	3.6×10^{15}

might favor tonne[5] of oil equivalent (toe) or quad (quadrillion BTU). Several of these are defined with reference to the joule in Table 1.1. For context, the average house in the United States consumes 30 kWh (108 MJ) of electricity each day, an automotive battery can supply approximately 0.5 kWh (1.8 MJ), and a typical LED bulb consumes 0.009 kWh (0.0324 MJ) over the course of 1 h. We will most often use gigajoules to refer to energy in general and kilowatthours to refer to electrical energy in particular.

Example 1.2 The 2013 average annual per person energy consumption in Zambia was 26.6 GJ. Compute the average daily consumption in kilowatthours per day and kilocalories per day.

Solution Consumption of 26.6 GJ per year translates into $\frac{26.6 \text{ GJ/yr}}{365 \text{ days/yr}} =$ 72.87 MJ/day. Converting to kilowatthours:

$$\frac{72.87 \text{ MJ}}{1 \text{ day}} \times \frac{1 \text{ kWh}}{3.6 \text{ MJ}} = 20.24 \text{ kWh/day}$$

and to kilocalories

$$\frac{72.87 \text{ MJ}}{1 \text{ day}} \times \frac{1 \text{ kcal}}{4.1868 \times 10^{-3} \text{ MJ}} = 17,406 \text{ kcal/day}$$

which is less than half of the world average.

---

[5]Here "tonne" refers to a metric ton, equal to 1000 kg.

1.4 World Energy System

The global energy system is a complex network of energy sources, bulk transportation, storage, distribution, and end use. The energy system is constantly evolving to meet growing demand and responding to economic, social, and environmental constraints. The existing world energy system does not adequately meet the needs of all people, nor is it sustainable in the long-term. Forty percent of the world lives in energy poverty. Greater innovation in technology, business models, and policy is needed by engineers, scientists, business managers, and government administrators across several decades to transform the energy system into one that provides sustainable energy for all.

1.4.1 Human Energy Use Throughout History

Human beings require approximately 2.8 kWh (2400 kcal) in the form of food each day to be able to function and do work. However, of this only about 500 Wh is available for motor function. The rest is used for resting metabolic function, including regulating our temperature, and digestion. A life restricted to this amount of energy was quite limiting and uncomfortable. It was the reality for humans until approximately 400,000 years ago when people began using fire [14].

Fire liberated energy stored in plant matter, and so its use increased our energy consumption. Along with this increased consumption came improvements in our lives: food could be cooked, broadening our diet and reducing spoilage, warmth was provided, evening lighting provided heat and some protection from predators, and basic tools could be made.

Agriculture was developed approximately 10,000 years ago and animal domestication began. The use of animal power for agricultural and travel increased our energy consumption further. Although small amounts of coal were used for heating during the middle ages, its consumption soared during the Industrial Revolution. The Industrial Revolution, ushered in part by the invention of the steam engine in the 1700s, began a rapid increase in mankind's energy consumption. By 1860, the average daily consumption of non-food and work animal energy was 2.4 kWh, nearly equal to the energy consumed in the form of food.

Electricity access, beginning in the late 1800s, provided an even more convenient form of energy which could be used for safe, high-quality lighting and to power motors which drove the industrial base. The automobile and airplane further spurred energy consumption. Nearly 20% of our energy is now devoted to transportation. By 1950, the average daily inanimate energy consumption was 20.9 kWh (18,000 kcal).

In more recent times, the daily energy consumption rose from 42.5 kWh (36,540 kcal) in 1971 to 61.4 kWh (52,790 kcal) in 2014. Whereas our ancestors consumed 2.8 kWh in food a day, we now average 22 times this amount in non-food energy consumption. The story of mankind's energy appetite is one of continued growth as our population increases and as we employ greater amounts of energy to improve our quality of life.

1.4.2 Total Energy Consumption

Figure 1.6 shows that the per person consumption of energy has been increasing over the last several decades. The trend, however, is not universal. Several countries with developed economies have seen a stabilization or even a decline of per person consumption. Germany, for example, uses the same annual per person energy (159 GJ) today as it did in 1970, as shown in Fig. 1.7. More astonishingly, the link between energy use and standard of living has been at least weakened as some countries are able grow their economies while using less energy. This is largely due to increases in efficiency associated with energy production, distribution, and end use.

However, these reductions in per person consumption are offset globally by rapidly increasing consumption, primarily in China, India, and developing countries. Another important factor is population growth, which compounds the increase in per person consumption. The global population grew by 1.5% per year from 1980 to 2014, increasing the population 66% from 4.4 billion to 7.3 billion. Over this same period, the total energy consumed each year increased 93%, from 282 quad to 545 quad.

Presently, over 80% of the worldwide energy is supplied by fossil fuels, which are depletable and whose carbon dioxide emissions contribute to global warming. Fossil-fuel resources tend to be concentrated in certain regions. Countries without

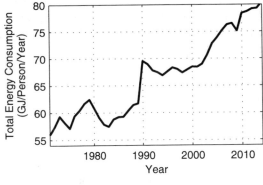

Fig. 1.6 Worldwide per person total energy consumption has increased rapidly since the 1970s [15]

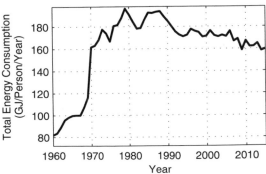

Fig. 1.7 Per person total energy consumption in Germany has started to decline [15]

natural supplies of fossil fuels must import energy. This can pose energy supply challenges in developing countries with weak currencies, poor credit, and without adequate infrastructure such as ports, pipelines, and railways.

1.4.3 Energy Inequality

Throughout history global per person consumption of energy has trended upward, with the world average annual energy consumption reaching approximately 80 GJ per person per year in 2014. However, there is vast inequality in energy consumption. Per person consumption is higher in developed countries, particularly in those whose climate necessitates use of space heating and air conditioning. The per person consumption in Canada, for example, is 318 GJ. On the other hand, the African continent—with a population of nearly 1 billion—consumed less than 4 % of the world's total energy in 2012. The average annual per person consumption in Sub-Saharan Africa is just 29 GJ.

Referring back to Fig. 1.5, it has been noted that no country with an HDI above 0.7—a reasonable target for a comfortable standard of living—has a per person consumption below 33.5 GJ (9.35 MWh). This suggests that, given the average consumption of 80 GJ per person, the present energy supply is sufficient to achieve a universal comfortable standard of living and that the problem is in the equitable distribution of energy.

1.4.4 Electricity Supply and End Use

Electricity is a convenient form of energy because it is easy to distribute with minimal losses. Electricity is versatile. It powers everything from motors, heaters, and chemical reactions to computers. The worldwide gross electricity production in 2014 was 23,815 TWh. Electricity is generated from power plants using a variety of input fuels, as summarized in Fig. 1.8.

Fig. 1.8 Electricity generation by fuel source

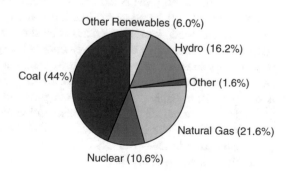

The general reliance on fossil fuels in the electric power sector leads to a question of the environmental impact, namely, carbon dioxide emissions, of increasing electrification rates and consumption in developing countries. Under reasonable scenarios, the impact of increasing access to electricity on global carbon dioxide emissions is marginal, less than 1%, as shown in the following example:

Example 1.3 The global carbon dioxide emissions due to human activity in 2012 were 35 billion tonnes, approximately 40% of which were related to the electric power sector. The total electricity consumption in 2012 was 23,815 TWh. If the 1.1 billion people presently without electricity access were given a grid connection and each consumed a modest 100 kWh per year, what is the percent increase in global carbon dioxide emissions? Assume the electricity supplied produced the same amount of carbon dioxide per unit energy as the global average.

Solution Using the global average, the amount of carbon dioxide emitted for every terawatthour of consumption is

$$\frac{0.40 \times 35 \text{ billion tonnes}}{23,815 \text{ TWh}} = 587,865 \text{ tonnes/TWh}$$

The annual consumption of electricity would increase by

$$1.1 \text{ billion} \times 100 \text{ kWh} = 110 \text{ TWh}$$

so that the total carbon dioxide emissions rise by 64.7 million tonnes. This only represents a 0.18% increase in global carbon dioxide emissions. Note that this does not count the reduction in carbon dioxide emissions associated with the *reduction* of kerosene and biomass usually associated with access to electricity.

As with total energy, per person consumption of electricity has rapidly increased, as shown in Fig. 1.9. In countries like the United States, the more recent trend of per person electricity consumption has been stagnant—having peaked in 2005 with 2015 levels similar to 1995. Improvements in energy efficiency and the transition from energy-intensive industries such as manufacturing to information, finance, and service industries are likely causes. Keep in mind that some of this decrease is really a just a displacement to other regions. In other countries, notably China and India, the per person electricity consumption has gone up rapidly, increasing by a factor of seven since 1971, as shown in Fig. 1.10. Because these countries have large population bases—over 2.5 billion people total—their total consumption of electricity has dramatically increased.

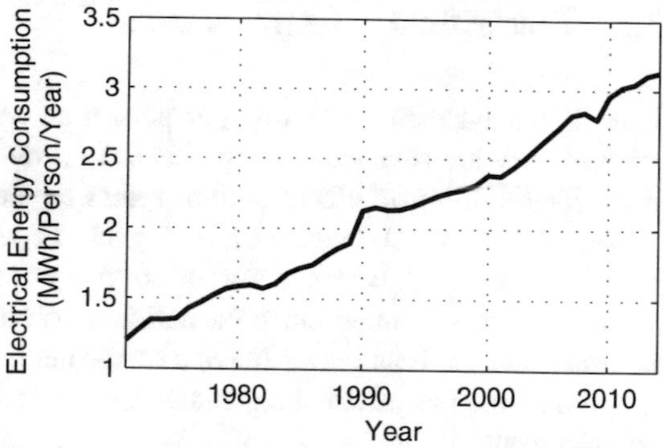

Fig. 1.9 Trend in worldwide per person electricity use. Note the dip in 2008, corresponding to an economic downturn, highlighting the linkage between electricity use and economic activity

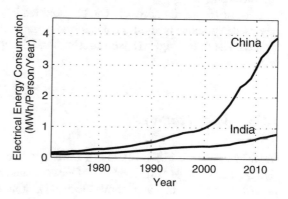

Fig. 1.10 Trend in per person electricity use in China and India [15]

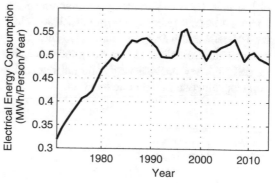

Fig. 1.11 Trend in per person electricity use in Sub-Saharan Africa [15]

In Sub-Saharan Africa, the story is different. Incredibly, in Sub-Saharan Africa as a whole, a region of 1 billion people, the per person consumption has recently regressed to its level in the 1980s, as shown in Fig. 1.11. This is largely attributed to a rapidly expanding population base without a similar expansion of electrical infrastructure. Consumption per person, therefore, went down.

1.5 Electrification Approaches

One significant factor affecting per person energy consumption is access to electricity. There are two general approaches to providing access to electricity: on-grid and off-grid. On-grid electrification refers to the provision of electricity through a connection to a large centralized grid. In most countries, there is a single interconnected grid, which is also referred to as the "national grid."

In rural areas, connection to the national grid usually requires the construction of infrastructure such as power lines, transformers, and substations; in urban areas, it might only require establishing a low-voltage connection and modest infrastructure enhancements.

Off-grid electricity access to can be provided in several ways [8]. The terminology used in practice is not strict, and in some cases the characteristics are not sharply defined. The following is a basic description of the off-grid approaches to electricity access.

1.5.1 Solar Lanterns

Solar lanterns (see Fig. 1.12), also known as "pico solar," are designed to provide lighting and perhaps USB charging of devices such as mobile phones. They rely on photovoltaic (PV) cells to generate electricity. Solar lanterns have peak PV capacity of less than 10 W—with many solar lanterns sized at less than 1 W. They typically contain a battery using lithium-ion or lead-acid chemistry. In most solar lanterns, the PV cells and battery are integrated into the lantern, rather than being separate components. Solar lanterns are designed to be portable and low-cost, typically US$5 to US$20. They provide very minimal but still quite useful electrical power to a home or someone in a remote location.

Fig. 1.12 A solar lantern includes a small solar panel and LED light (courtesy of d.Light)

Fig. 1.13 Solar home
systems are often capable of
powering several LED lights
and small appliances
(courtesy of BBOXX Ltd.)

Fig. 1.14 A solar-powered energy kiosk in Zambia (courtesy of KiloWatts for Humanity)

1.5.2 Solar Home Systems

Solar home systems (see Fig. 1.13) refers to products with PV modules rated
between 10 to 350 W. The battery is in a ruggedized case, often supporting several
LED lights and USB ports. Some can power televisions and fans. These are not
portable devices. They are installed in a semipermanent fashion.

1.5.3 Energy Kiosks

Energy kiosks (see Fig. 1.14) provide centralized access to electricity to a community [6]. They use a "walk-up" retail model of electricity access, rather than
direct wired connections to households. Community members can have batteries

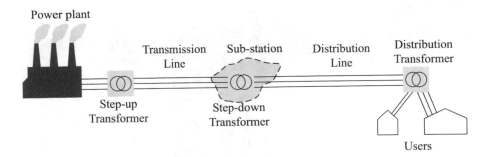

Fig. 1.15 Traditional flow of power in a national grid

recharged at the kiosk, and some kiosks provide services for higher-powered appliances such as refrigeration. Energy kiosks typically have higher power output capacities than solar home systems. The electrical system is often housed in a small building and requires on-site staff to interact with customers.

1.5.4   Micro-Grids and Mini-Grids

Micro- and mini-grids generally refer to stand-alone electrical systems that serve multiple customers through wired connections. They can be powered by multiple sources such as PV modules, fossil-fueled generators, and wind turbines. There is no strict distinction between a "micro-grid" and a "mini-grid." However, a micro-grid has the connotation that it is smaller in peak power rating than a mini-grid. Hereafter, we shall use the term "mini-grid" for the sake of clarity. We shall consider mini-grids whose peak power rating is less than 1 MW. While this serves as an upper limit, most mini-grid systems are in the 1 to 10 kW range.

1.5.5   National Grid Electrification

Although this book focuses on off-grid systems, we will briefly discuss how larger national grids are configured. This will equip us with the background needed to understand the role of off-grid solutions in the larger electrification context.

Providing access to electricity is more complex than simply running a wire to a house or factory. The interconnected electric power system is one of the most complex creations ever engineered by mankind. Figure 1.15 shows the traditional flow of power in a national grid.

Electric power systems can be separated into three subsystems: generation, transmission, and distribution. Electricity is generated by power plants fueled by sources such as coal, nuclear, natural gas, hydro, wind, and solar. The bulk of the energy generated occurs at large power plants often with capacities exceeding 1000 MW. More recently, distributed generation—small-scale power plants that are located near or in cities and neighborhoods—has become popular and may supplement generation from large power plants. Rooftop solar power is one example of distributed generation. In order to reliably supply power, it is necessary for a country to have enough generation capacity to serve the maximum anticipated simultaneous load. Otherwise, the country will be forced to import power, often at high prices, in order to avoid a complete or partial blackout. In many developing countries, there routinely is insufficient generation capacity to meet the demand, and short- or long-term blackouts are common.

Most power plants produce electricity using three-phase generators at a frequency of either 50 Hz or 60 Hz, depending on national standards. The voltage at which the generators output the electricity, typically 10 kV to 30 kV, is too low for efficient transmission across long distances. Transformers are used to increase the voltage and thus decrease the losses for long-distance transmission.

Transmission voltages typically range between 110 kV and 750 kV. Transmission lines often run hundreds of kilometers atop metal lattice or guyed wooden or composite poles. Reliable national grids have meshed networks, meaning there are several parallel transmission paths between the power plants and the load. They are also interconnected, allowing neighboring countries or regions to share generation resources.

Once the transmission lines reach their destination—typically an industrial or population center—the voltages are reduced at a substation. Transformers step down the voltage to the local distribution lines. Typical distribution voltage levels are 11 kV, 22 kV, and 33 kV. These lower voltages are safer and require less clearance than transmission lines and so are more suitable to populated areas. Transformers located near customers again step the voltage down to levels that are safe for consumption, typically 120 or 230 V. Due to losses, the power can only be transmitted a short distance at this low-voltage level.

Increasing electrification rates often requires investment not only in power lines and connections but also power plants and other equipment that keep the grid operating. In the context of rural electrification, the following must be kept in mind:

- higher voltages are required for efficient transmission;
- transmission and distribution lines are expensive to construct;
- energy supplied by the grid is nearly always less expensive than from mini- or micro-grids.

Additional information on power systems design and analysis can be found in the numerous textbooks on the subject [2, 3, 5]. Finally, we note that even in countries with expansive national grids, there may be some portions of the population without access, even in the United States, as shown in Fig. 1.16.

Fig. 1.16 An off-grid system in the Navajo Reservation in the United States (courtesy D. Terry, Navajo Tribal Utility Authority)

1.6 Summary

This chapter established the motivation for the study of off-grid electrification. Access to electricity is a widely recognized humanitarian challenge; more recently, it is being viewed as an emerging market for commercial enterprise. Energy consumption is closely tied to human development and has increased on a per person and absolute basis. Worldwide average energy consumption stands at 80 GJ per person per year, but there is wide variation between countries. Developing countries use substantially less total energy and electrical energy than developed countries. Electricity access can be increased through connection to the national grid, or by off-grid solutions, including solar lanterns, solar home systems, energy kiosks, and mini-grids.

Problems

1.1 Place the following quantities of energy in order from highest energy content to lowest: 1 Calorie, 1 BTU, 1 MWh, 100 kWh, 5000 J, 10,000 calories, 1 toe.

1.2 A hamburger at a popular fast-food restaurant contains 600 kcal. How many of these hamburgers are needed to equal the energy consumption by the average Canadian each day, assuming their average consumption is 318 GJ per year?

1.3 One liter of gasoline contains 34.2 MJ of energy. If humans could drink gasoline to satisfy our daily caloric need of 2400 kcal, how many days could a human survive on one liter of gasoline? Assume the cost of gasoline is US\$1 per liter. What is the daily cost of such a diet?

1.4 Assume that a kilowatthour from an outlet costs US\$0.15. If humans could consume electricity to satisfy our daily caloric need of 2400 kcal, what would be the cost of such a diet?

1.5 Research three companies that manufacture or distribute either solar lanterns or solar home systems. Select a product from each company and describe its features and rating of the PV module (in watts) and battery (in amp-hours or watthours).

1.6 The power produced by a PV module with a 200 W rating on a certain day is described by the equation:

$$P(t) = \begin{cases} P_{PV} \frac{900}{1000} \cos\left(\frac{t\pi}{12} - \pi\right) : 6 < t < 18 \\ 0 \qquad\qquad\qquad : \text{else} \end{cases}$$

where P_{PV} is the rating of the module and t is in hours. Compute the energy produced by the PV module, in kilowatthours, on this day.

1.7 A certain off-grid system is supplied by a diesel generator set. The generator produces a constant 1 kW of power. How many 200 W PV panels are needed to replace the energy supplied by the generator, if the power output by each panel is described by the equation in the previous problem?

1.8 Compute and compare the total electricity consumption of Sub-Saharan Africa to the United States in 2012. The population of Sub-Saharan Africa was 926 million, and the United States was 314 million. The per person consumption in the United States in 2012 was 12.96 MWh per year. Consult Fig. 1.11 for the per person consumption in Sub-Saharan Africa.

1.9 A proposed mathematical model estimating HDI based on per person energy consumption is

$$\hat{HDI} = 0.1185 + 0.1412 \ln(E)$$

where E is the total primary energy demand per person measured in gigajoules. Plot this function and compute the percent error for the following countries: United States (HDI: 0.912, Energy: 289 GJ/person), Gabon (HDI: 0.697, Energy: 60.1 GJ/person), Niger (HDI: 0.380, Energy: 6.36 GJ/person), Thailand (0.743 Energy: 83.0 GJ/person).

1.10 Compute and plot the derivative of the HDI model presented in the previous problem, and describe what it suggests in terms of the relationship between HDI and per person energy consumption.

1.11 In 2015, the United Nations set 17 Sustainable Development Goals to be achieved by the year 2030. Research and select three of these goals (other than Number 7), and describe the role of electricity access in achieving them.

References

1. Arto, I., Capellan-Perez, I., Lago, R., Bueno, G., Bermejo, R.: The energy requirements of a developed world. Energy Sustain. Dev. **33**, 1–13 (2016). doi:http://dx.doi.org/10.1016/j.esd.2016.04.001
2. Berge, A., Vittal, V.: Power Systems Analysis, 2nd edn. Prentice-Hall (2000)
3. El-Sharkawi, M.: Electric Energy: An Introduction, 3rd edn. CRC Press (2004)
4. Global LEAP: The state of the global off-grid appliance market (2015). URL http://globalleap.org/resources/
5. Gönen, T.: Electrical Power Transmission System Engineering, 3rd edn. CRC Press (2014)
6. Kemeny, P., Munro, P., Schiavone, N., van der Horst, G., Willans, S.: Community charging stations in rural sub-saharan Africa: Commercial success, positive externalities, and growing supply chains. Energy Sustain. Dev. **23**, 228–236 (2014). doi:https://doi.org/10.1016/j.esd.2014.09.005. URL http://www.sciencedirect.com/science/article/pii/S0973082614000921
7. Lighting Global: Off-grid solar market trends (2018). URL https://www.lightingglobal.org/2018-global-off-grid-solar-market-trends-report/
8. Louie, H., O'Grady, E., Acker, V.V., Szablya, S., Kumar, N.P., Podmore, R.: Rural off-grid electricity service in sub-saharan Africa [technology leaders]. IEEE Electrification Mag. **3**(1), 7–15 (2015). doi:10.1109/MELE.2014.2380111
9. Mandelli, S., Barbieri, J., Mereu, R., Colombo, E.: Off-grid systems for rural electrification in developing countries: Definitions, classification and a comprehensive literature review. Renew. Sustain. Energy Rev. **58**, 1621–1646 (2016). doi:https://doi.org/10.1016/j.rser.2015.12.338
10. Sturm, R., Njagi, A., Blyth, L., Bruck, N., Slaibi, A., Alstone, P., Jacobson, A., Murphy, D., Elahi, R., Hasselsten, J., Melnyk, M., Peters, K., Appleyard, E., Orlandi, I., Tyabji, N., Chase, J., Wilshire, M., Vickers, B.: Off-grid solar market trends (2016). URL http://documents.worldbank.org/curated/en/197271494913864880/Off-grid-solar-market-trends-report-2016
11. Sustainable Energy for All: Global tracking framework (2015). URL http://www.seforall.org/global-tracking-framework
12. Sustainable Energy for All: Global tracking framework (2017). URL http://www.seforall.org/global-tracking-framework
13. United Nations Development Programme: Human development report (2016). URL http://hdr.undp.org/en/2016-report
14. Weissenbacher, M.: Sources of power: how energy forges human history. ABC-Clio/Praeger/Greenwood (2009)
15. World Bank: World bank open data (2017). URL https://data.worldbank.org/

Chapter 2
Energy Poverty

2.1 Introduction

Energy poverty is *the lack of access to modern fuels*. It is estimated that 2.9 billion people, approximately 40 % of the world's population, are energy-impoverished in some way [28]. Energy poverty is associated with economic poverty, and people living in rural areas are more likely to be energy-impoverished than those in urban areas.

The energy-impoverished rely on solid fuels such as fuel wood, animal dung, and crop residue, rather than more convenient fuel sources such as electricity, natural gas, or liquefied petroleum gas (LPG). The energy-impoverished tend to:

- consume low amounts of energy overall;
- rely on human and animal power for mechanical tasks;
- devote considerable time to procuring and processing fuel;
- spend a relatively large portion of their income on fuel.

Energy poverty detracts from income, educational attainment, and health, and so it impedes human development. Energy poverty is also seen as a gender issue. Women tend to disproportionately bear the burden of fuel collection and are more exposed to the health and safety risks associated with energy poverty. One approach to combating energy poverty is to implement local mini-grids that do not require connection to the national grid.

An off-grid electrical system that is successful and appropriate requires careful, holistic consideration of the energy resources and constraints of the community it serves. As such, it is important to understand how off-grid communities presently meet their energy needs: how much energy is used, from what sources, and what are the advantages and challenges of the status quo? This chapter presents this information, with particular attention to electricity use in energy-impoverished communities and how electricity access is quantified.

© Springer International Publishing AG, part of Springer Nature 2018
H. Louie, *Off-Grid Electrical Systems in Developing Countries*,
https://doi.org/10.1007/978-3-319-91890-7_2

2.2   Rural Community Energy Needs

Although this book focuses on off-grid electricity access, it is important to understand the complete energy needs and characteristics of off-grid communities. Since the majority of those without electricity access live in rural areas, we focus in particular on these communities.

The energy needs of a rural community can be divided into three categories [20]:

1. Energy for households
2. Energy for community services
3. Energy for productive uses

Each of these categories has different technical and economic requirements, and each provides different benefits.

2.2.1   Energy Use of Households

Household (domestic) consumption constitutes the majority of the energy used in rural areas. The energy is primarily used for cooking, water heating, and spacing heating, with the balance used for lighting and powering electronic devices and appliances such as mobile phones, radios, and in some cases fans and televisions. Rural households tend to rely on several fuel sources: fuel wood, charcoal, crop residue, animal dung, kerosene, candles, batteries, and in some cases small diesel- or petrol-powered generator sets and solar-powered systems. The per-person annual energy consumption for domestic purposes is often between 5 and 15 GJ per year [1, 3, 9, 15, 24]. However, consumption above this range does occur. For example, consumption of 50 GJ per person per year is not uncommon in colder climates. Even still, this is far below the global average of 80 GJ.

2.2.2   Energy Use of Community Services

Community services include schools (see Fig. 2.1), medical facilities (Fig. 2.2), churches, and community centers, among others. The typical energy uses in these facilities are lighting, ITC (Internet, telecommunications, computers), health-care equipment (refrigeration for medicine and vaccines, suction machines, diagnostic equipment), and clean and hot water. In addition, street or pathway lighting is desirable for protection against crime, wildlife, and tripping hazards. This can increase school attendance, especially for girls, as their commute to and from school is less dangerous [25].

Electricity is required for effective community services. In addition to direct benefits, electricity access helps health-care facilities and teachers attract and retain

Fig. 2.1 A government primary school in Zambia (courtesy of author)

Fig. 2.2 A rural health post in Malawi (courtesy of author)

qualified workers. The importance of this should not be overlooked [10]. Studies in Ghana, for example, have shown that 85% of teachers cited "lack of access to potable (drinkable) water and electricity" for the reason they turned down positions in rural areas [26].

2.2.2.1   Health-Care Facilities

Electricity access at health-care facilities is critical. Over 1 billion people do not have adequate access to health services due to energy poverty. The typical electricity requirements of health-care facilities are listed in Table 2.1. Health-care facilities have special electrical requirements. The electricity must be reliable, vaccines must

Table 2.1 Health facility energy use in rural areas [13]

Type	Energy (kWh/day)	Peak power (kW)
Hospital (>120 beds)	15–35	9
Health center (60–120 beds)	10–20	5
Health clinic (<60 beds)	4–10	2.4
Health post (intermittent use and storage)	0	0

be kept at tightly regulated temperatures, and of high quality—voltage spikes can damage sensitive medical equipment, which often cannot be easily or expeditiously repaired or replaced. As a result, medical clinics connected to the national grid often require backup generator sets or Uninterruptible Power Supplies (UPS). A UPS is a battery-based backup electricity source.

2.2.2.2   Education Facilities

Four out of five schools in Africa lack access to electricity [29]. Over 200 million children attend a school that has no electricity [26]. These schools are unable to make use of enhanced teaching technologies such as projectors, radios, computers, printers, digital cameras, and the Internet. Instruction times are limited to daylight hours. Electricity access in schools generally reduces absenteeism and increases enrollment, graduation rates, and test scores [17, 26]. However, these results are not universal. Some studies have found decreased educational performance when electricity is also available at home, as children watched television instead of completing homework and stayed awake later in the evening, decreasing their ability to concentrate in school [6, 26]. Others have noted that the increased economic opportunities in communities with electricity access may result in children joining the workforce at an early age.

The electricity requirements vary with the level of services provided and size of the school. This can range from less than 1 kWh per day for basic lighting of small schools to much more than 10 kWh per day if computer labs and other enhanced teaching technologies are used.

2.2.3   Energy Use for Productivity

Productive uses of energy typically refer to income-generating activities. These are often related to agriculture in the rural context. Typical uses include land preparation and cultivation, irrigation pumping, harvesting, milling of grain and other agricultural processing, brick and charcoal manufacturing, and small appliances such as shavers for barbershops and televisions for video halls. Solid and liquid (petroleum for generation sets) fuels and electricity are important in increasing or expanding productivity. The electricity requirements vary but often exceed those of households and social service facilities. The power of an irrigation pump depends on the pump type, the vertical distance pumped, and the flow rate. For example, a submersible pump lifting water 65 m and at a flow rate of 2.1 m^3/h requires approximately 1.2 kW. Solar pumping systems commonly range from less than 1 kilowatt peak to perhaps 10 kW. Hammer mills, used for processing maize, are rated from several kilowatts to perhaps 20 kW.

2.3   Fuel Choice

A variety of sources can be used to meet the needs of a rural community. The spectrum of fuels includes:

- Electricity
- Liquid Petroleum Gas (LPG)
- Kerosene
- Charcoal
- Fuel wood
- Animal dung and crop residue

This list can be further subdivided. Electricity can be provided from the national grid, mini-grid, energy kiosks, solar lanterns, solar home systems, and disposable batteries. LPG can be expanded to include natural gas (methane) in addition to propane and butane. We will focus our discussion of fuel choice to households, as energy use in households is better defined than in social institutions and industry.

2.3.1   Energy Content

The energy content by fuel source, in megajoules (1×10^6 J), is listed in Table 2.2. There is wide variability in the specific energy (energy per unit mass) of biomass fuels (wood, charcoal, dung, and straw) depending on moisture and ash content, so the values provided for these sources are not exact. Unprocessed biomass fuels such as green (freshly cut) wood tend to have low specific energy due to the added moisture.

Table 2.2 Energy content by
fuel source [4]

Fuel	Unit	MJ/Unit
Charcoal	kg	30.8
Dung	kg	14.5
Electricity	kWh	3.6
Kerosene	liter	35.0
LPG (Gas)	kg	45.0
Wood (Dry)	kg	16.0
Straw	kg	13.5

Example 2.1 A household uses 100 GJ of fuel wood, 10 GJ of kerosene, and 10 GJ of LPG each year. Compute the quantity, in kilograms, of fuel wood and LPG, and the quantity, in liters, of kerosene required each year.

Solution Using the values in Table 2.2, the quantities of each fuel source are

$$\text{Fuel wood} = \frac{100\,\text{GJ}}{0.016\,\text{GJ/kg}} = 6250.00\,\text{kg}$$

$$\text{Kerosene} = \frac{10\,\text{GJ}}{0.035\,\text{GJ/liter}} = 285.71\,\text{liter}$$

$$\text{LPG} = \frac{10\,\text{GJ}}{0.045\,\text{GJ/kg}} = 222.22\,\text{kg}.$$

2.3.2 Fuel Attributes

Three attributes influence a household's decision to use a particular fuel [23]:

- Quality
- Convenience
- Cost

The desirable characteristics of these attributes are summarized in Table 2.3. Electricity is considered a desirable fuel source. It can be used in a variety of high-efficiency appliances. It is consistent, is easy to control, and has zero local emissions. It is therefore a high-quality energy source. If an electrical connection exists, electricity is extremely convenient, provided the connection is reliable. Electricity can be low cost, especially if it is subsidized; however, appliances must be purchased, which discourages its use. Note that a fuel that is desirable in one context might not be in another.

Table 2.3 Fuel source attributes

Attribute	Desirable characteristics
Quality	High efficiency, controllable output, consistent performance, low or zero emissions
Convenience	Delivered to home or locally-available, reliable supply, little or no processing, familiar/easy to use, versatile, safe, does not require storage
Cost	Low cost of fuel, low cost of equipment

Fig. 2.3 The energy ladder shows the transition from fuel sources with low desirability to those with high desirability as income increases. The overlap represents the simultaneous use of multiple fuels

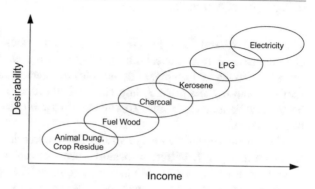

2.3.3 Energy Ladder

The different fuel sources can be arranged from least desirable to most desirable as seen in Fig. 2.3. Households tend to transition from least desirable fuels to more desirable fuels as their income increases. This concept was first articulated in the 1980s and is known as the "energy ladder"[19]. It remains an important starting point in understanding the dynamics of fuel transition.

The notion that a household "climbs the energy ladder," transitioning from one fuel source to the next as income increases, is somewhat simplistic. Many other factors, particularly local conditions, influence the transition. These include in particular access to the fuel, the cost of the equipment to use the fuel—such as burners and appliances—and the "lumpiness" of the fuel payments. A fuel with a lumpy payment is one whose minimum purchasable amount is relatively large, requiring a sizable up-front payment. LPG, whose minimum canister size might be 2 to 5 kg, is one example. Even if this bulk-buying behavior results in a substantial reduction in energy costs in the long-term, vulnerable households with variable income often prefer purchasing fuels available in smaller minimum quantities, such as fuel wood and kerosene. Other factors that tend to influence fuel transition are wealth, income, and educational attainment.

Most often, lower-quality fuels are usually not completely abandoned as income increases [23]. Lower-quality fuels might be used for cultural reasons and to guard against price increases or shortages of other fuels. Thus, households might climb or descend the energy ladder and will simultaneously span more than one rung.

In general, as household income increases, so does total energy consumption. However, the increase is not proportional. For example, increasing income by a factor of ten might only increase the household energy consumption by 25%. This is in part due to smaller household sizes associated with higher income but also the greater efficiency that can be obtained from higher-quality fuels and appliances.

2.3.4 Fuel Stacking

The use of several fuels is referred to as "fuel stacking" and is an important characteristic of rural household energy use. Fuel stacking adds flexibility and robustness to the fuel supply, as households can switch between sources based on a fuel's availability and price. A household in an urban city on the other hand might be solely reliant on electricity. This adds vulnerability to their fuel supply should a blackout occur.

The annual household energy use by fuel source in several different locations is shown in Fig. 2.4. Different locations use different fuel stacks (also known as "energy matrices"), even within the same country. The types of fuel used and their quantity depend on the local conditions. The annual household consumption ranges between approximately 70 and 130 GJ per household per year for these locations. This generally agrees with an estimate of between 5 and 15 GJ per person per year, given that rural households tend to be large, for example, between 5 and 15 people.

2.3.5 Fuel Expenditure

The amount of money spent on each fuel source also varies within and across countries based on local fuel prices, climate, household size, and income. Figure 2.5 shows the average annual household fuel expenditure in several villages across

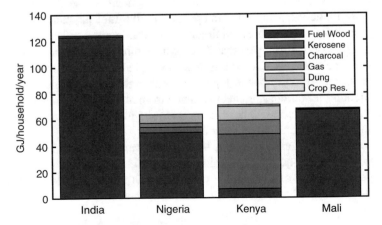

Fig. 2.4 Rural household energy use by fuel type, compiled from various sources [1, 3, 15, 24]

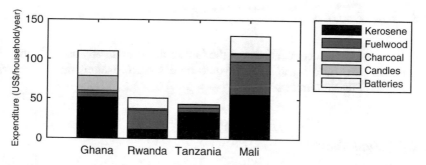

Fig. 2.5 Annual energy expenditure by fuel source for households in villages in Ghana, Rwanda, Tanzania and Mali. Adapted from [2]

Africa. Money spent on grid-connected electricity, which only a small number of households had access to, are omitted. There is a wide range in household expenditure, from approximately US$40 to US$140 per year, with some variation in the fuels used. A considerable portion of income is spent on fuels primarily used for lighting—kerosene and batteries.

The cost of fuel is a contributing factor to energy poverty. A desirable goal is that fuel expenses should be no more than 5 to 10% of a household's income [8, 12]. In some households, however, it exceeds 30%. This limits the resources available to purchase food, medicine, agriculture inputs and pay school tuition.

Example 2.2 Compute the annual fuel expenditure of the household in the previous example. Assume the cost of fuel wood is US$0.05 per kilogram, kerosene is US$1.10/liter, and LPG is US$1.60/kg. What must the household's annual income be so that no more than 10% of their income is spent on fuel?

Solution Using the values in Table 2.2, the expenditure on each fuel is:

$$Fuel\ wood = 6250 \times 0.05 = US\$312.50$$

$$Kerosene = 285.71 \times 1.10 = US\$314.28$$

$$LPG = 222.22 \times 1.60 = US\$355.55$$

for a total of US$982.33 per year. The household would need to earn more than US$9823.30 per year to spend no more than 10% of their income on fuel. This amount is much greater than the income of most rural households. Descending the energy ladder, by transitioning from LPG to fuel wood and reducing kerosene use, would reduce the energy expenditure.

2.4 Fuel Sources

We now consider the characteristics of the fuel sources commonly used in rural off-grid households. An off-grid system would be designed to supplement or replace some or all of these sources.

2.4.1 Fuel Wood

Fuel wood is perhaps the most important fuel in rural areas and often supplies the majority of the energy consumed [21]. It is primarily used for heating spaces and water, as well as cooking as shown in Fig. 2.6.

Fuel wood can be purchased at a low cost or be gathered from trees for free throughout the year. Fuel wood is self-replenishing; however, there is some concern about overharvesting of fuel wood, in particular, to create charcoal. Depending on the household size and location, between 5 and 25 h per week are spent gathering fuel wood. Preparation of the fuel wood, if needed, can add a few additional hours each week. Fuel wood has low specific energy and low energy density, and so large quantities are needed. The typical fuel wood consumption for rural areas ranges from 1 to 5 kilograms per person each day [1, 9, 21]. In colder climates, it can exceed 10 kilograms per person each day. Seasonal variations also influence fuel wood consumption—less wood is consumed during summer months.

Over the course of a year, a typical household will collect several tonnes of fuel wood. Walking more than 5 km to collect wood is common. The burden generally falls upon women and girls, which is cited as one reason why energy poverty disproportionately affects women.

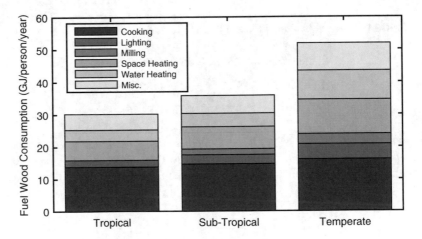

Fig. 2.6 Fuel wood use in rural communities in varying climates [9]

Fig. 2.7 Three-stone cook stoves are widely used in rural Sub-Saharan Africa (courtesy of author)

Table 2.4 Typical cook stove efficiencies [14]

Type	Efficiency (%)
Fuel wood	11
Charcoal	20
Kerosene	45
LPG	55
Electricity	75

The efficiency of cook stoves also influences the quantity of fuel wood consumed. In many communities, traditional three-stone cook stoves are used to heat water and to cook food (see Fig. 2.7). Traditional cook stoves have a low efficiency, around 11%. Improved designs can double the efficiency but, overall, they are less efficient than stoves designed for modern fuels like LPG and electricity, as shown in Table 2.4. Nonetheless, doubling the efficiency of the cook stove reduces the fuel wood demand by a factor of two. This saves both time and money.

2.4.2 Charcoal

Charcoal has about twice the specific energy of dry fuel wood and so is more convenient to transport, store, and use. It also lights easily and requires little if any preparation. Charcoal is often used in small stoves or braziers as shown in Fig. 2.8.

Charcoal is wood that has undergone a process called "pyrolysis." Pyrolysis is a thermal–chemical reaction in which wood is exposed to high temperatures—over 300°C—over many hours or days in the absence of oxygen. This prevents the wood from catching fire. At temperatures around 300°C, wood spontaneously breaks down resulting in water vapor, methanol, acetic acid, tars, and gases (primarily carbon dioxide, carbon monoxide, and methane). At these temperatures, the process

Fig. 2.8 Container of
charcoal and cook stove
("jiko" in Kenya) (courtesy of
the author)

becomes exothermic (energy is released) and continues until carbonized residue
remains. The resulting charcoal is 75 to 85% carbon, with some remaining ash and
volatile compounds. Because the volatile compounds have been reduced, charcoal
tends to burn with little or no flame. A typical yield is 1 kilogram of charcoal for
every 6 kilograms of input dry wood.

Creating charcoal is a labor- and energy-intensive process. It is not surprising
then that charcoal costs more than fuel wood. This limits its use in poor households.
Charcoal use among the urban and peri-urban (near urban), including those with
access to electricity, is high. In many developing countries, charcoal production
and distribution play a major role in the economy [16]. Over 36 million tonnes
of charcoal, valued at over US$ 11 billion, were produced in Sub-Saharan Africa
(SSA) in 2012 [14].

2.4.3 Animal and Crop Residue

Animal dung—from cows, horses, goats, llamas, yaks, camels, and even
elephants—and crop residue are often abundant in rural settings, as many
households are engaged in agriculture and animal husbandry. These fuels are
free or extremely low cost, but are usually undesirable.

As shown in Table 2.2, the energy density of these sources is low, even after
drying. The smoke from burning animal dung and crop residue is associated with
negative health outcomes, including lung cancer and respiratory infections. When
burned, cow dung releases more particulate matter and higher levels of microbial
products than when fuel wood is burned or are emitted from diesel-powered
generator sets [22].

Fig. 2.9 A kerosene lamp
made from an up-cycled
tomato paste can (courtesy of
author)

2.4.4   Kerosene

Kerosene, sometimes locally known as "paraffin," is a transparent petroleum-based
liquid. It has high specific energy but also high cost. Kerosene is used to start
fires and for lighting. It may also be used for cooking by higher-income families.
Kerosene lamps are widely used in SSA, and it is common to see the lamps
constructed using upcycled materials, as shown in Fig. 2.9. Kerosene can often be
purchased from petrol stations, even in remote areas. Kerosene is often purchased
in small quantities, perhaps 1 liter or less. Repurposed plastic water or soda bottles
are used to transport it home.

The consumption of kerosene by a lamp varies considerably depending on the
physical characteristics of the lamp. It typically ranges between 5 and 50 liters per
year based on 3.5 h of evening use per lamp. Most lamps use a simple flat wick
which is low-cost and can be manufactured locally.

Kerosene lamps pose safety and health risks. They are used indoors, where
they might leak, be knocked over, or come in contact with flammable materials.
Combustion of kerosene releases emissions including particulate matter, carbon
monoxide, formaldehyde, polycyclic aromatic hydrocarbons, sulfur dioxide, and
nitrogen oxides [30]. These are harmful to human health. Research has found an
increased risk of cancer, respiratory infection, asthma, tuberculosis, and cataracts
related to kerosene use. Kerosene is transparent and is often stored in reused
beverage bottles, increasing the risk of accidental ingestion by children.

Although energy from kerosene often contributes to less than 10 % of the total
energy of a household, it can comprise about half of total energy expenditure.
Kerosene use is affected by government subsidies and its availability in rural
areas. In some regions such as Zambia, candles or disposable battery flashlights
("torches") are preferred to kerosene [18]. The high cost coupled with the low
quality and efficacy of kerosene lamps presents an opportunity to greatly reduce
fuel expenditure by replacing kerosene with LED lighting.

2.4.5 Batteries

Some households are transiting to electric lighting solutions. Although solar lanterns are becoming more common, disposable dry cell battery flashlights (torches) and lamps using highly efficient LED bulbs are common in some areas [7].

A challenge with the proliferation of flashlights in rural areas is the disposal of the dry cell batteries. Professionally managed disposal or recycling facilities are uncommon. The batteries are depleted relatively quickly—a rural household uses between 4 and 15 per month. The batteries, which contain hazardous materials, are commonly discarded in nature or in latrines [7]. Batteries discarded in latrines can concentrate the hazardous chemicals from the batteries, which could possibly cause greater harm then if discarded in nature.

2.5 Household Electricity Needs

Electricity can replace or complement existing fuels used in rural households. Complete replacement is not likely unless a subsidized grid connection is available due to the high cost of high-quality electricity service, appliances, and electrical energy, as explored in the following examples.

Example 2.3 Consider a household whose consumption is 60 GJ (16,667 kWh) per year using fuel wood and kerosene. Of this, 36 GJ from fuel wood is used for cooking and water heating on a cook stove with an efficiency of 12%, and 20 GJ from fuel wood is used for space heating. Kerosene is used for lighting, with an annual consumption of 4 GJ. Compute the annual energy required, in kilowatthours, to replace the fuel with electricity from the national grid. Assume the electric cook stove is 75% efficient and the electric lamps consume 45 kWh per year.

Solution The use of an electric cook stove reduces the consumption from water heating and cooking to:

$$36 \times 0.12/0.75 = 5.76 \, \text{GJ}.$$

The electrical energy consumption excluding lighting is

$$(60 - 36 - 4) + 5.76 = 25.76 \, \text{GJ}.$$

Converting to kilowatthours and adding the energy from the electric lights:

$$25.76 \times 10^9 \times \frac{1}{3.6 \times 10^6} + 45 = 7201 \, \text{kWh}.$$

The transition to electricity reduced the energy consumption by 57%.

Example 2.4 Consider the energy consumption described in the previous example. Assume that each liter of kerosene costs US$1.10, and that 75% of the fuel wood is gathered for free, and the rest is purchased at US$0.05 per kilogram. Compute the annual energy expenditure of the household and compare it to the annual expenditure if electricity is used, assuming the electricity price is unsubsidized at US$0.15/kWh.

Solution Consulting Table 2.2, the household consumes

$$\frac{36\,\text{GJ} + 20\,\text{GJ}}{0.016\,\text{GJ/kg}} = 3500 \text{ kg of fuel wood}$$

$$\frac{4\,\text{GJ}}{0.035\,\text{GJ/liter}} = 114.3 \text{ liters of kerosene.}$$

The total annual expenditure is

$$3500\,\text{kg} \times (1 - 0.75) \times 0.05\,\text{US\$/kg} + 114.3\,\text{liter}$$

$$\times 1.10\,\text{US\$/liter} = \text{US\$169.48.}$$

If electricity replaces kerosene and fuel wood, the annual expenditure becomes

$$7201\,\text{kWh} \times 0.15\,\text{US\$/kWh} = \text{US\$1080.15}$$

which is likely cost-prohibitive for this household.

Example 2.4 showed that switching entirely to electricity is not likely to be economically viable. Instead of complete replacement, electricity can be used to complement the existing fuel sources by being used for services that traditional fuels cannot provide. For example, fuel wood cannot be used to power fans and televisions. Although traditional fuels can be used for lighting, it is inferior in quality to electric light. The electricity needs of a household therefore is a continuum, based upon the electrical services required.

There are seven broad categories of household electricity services, as shown in Table 2.5. We can estimate the electrical energy required for these services, or a combination of these services, based on the duration of use and electrical characteristics of the appliances. For example, we can estimate the energy requirement of providing 6 h of single-room electric lighting and 2 h of television per day. Using typical appliance ratings of 9 W per LED light bulb and 30 W for a television, the daily requirement is $(6 \times 9 + 2 \times 30) = 0.114$ kWh per household per day. This is far less than if the entire energy needs were replaced by electricity.

Estimates of the electricity need range from about 20 to 100 kWh per person per year, depending on the services provided. Some organizations target 365 kWh per

Table 2.5 Household services and power requirements [8]

Service	Very-low power	Low-power	Medium-power	High-power	Very high-power
1. Lighting	Task	General	–	–	–
2. Communication & entertainment	Phone charging, radio	TV, computer, printer	–	–	–
3. Space cooling & heating	–	Fan	Air cooler	–	Air conditioner, heater
4. Refrigeration	–	–	Refrigerator, freezer	–	–
5. Mechanical loads	–	–	Water pump, food processor	Washing machine	Vacuum cleaner
6. Product heating	–	–	–	Iron, hair dryer	Water heater
7. Cooking	–	–	Rice cooker	Toaster, microwave	Electric range (cooker)

Table 2.6 Electricity consumption of various services [11]

Services	Consumption (kWh/person/year)
Task lighting	1
Light or TV or radio	2
Light, phone, radio, small TV	22
Light, phone, radio, TV, fan, productive uses	220

year (1 kWh per day) per household [8], which is 20 to 30 times less than consumed in most households in developed countries. Always be careful to distinguish between energy use per person and per household. The assumption of appliance type also strongly influences the required energy; for example, an LED light can be up to twice as efficacious as a compact florescent light. Table 2.6 is an example of per person annual energy requirements.

2.6   Electricity Access

Electricity access of a region or country is quantified by the *electrification rate*—the percent of the population with access at home to the electricity from the national grid. The simplicity of this definition makes it easy to interpret but, as discussed later, has shortcomings as well. The 2012 global electrification rate is 85% and has generally been increasing over the last several decades as shown in Fig. 2.10. An 85% electrification rate seems high, but it means that 1.1 billion people do not have access to electricity.

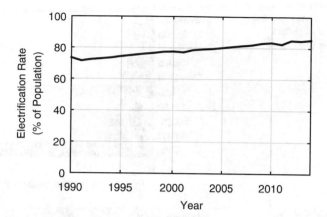

Fig. 2.10 World electrification rate

Fig. 2.11 Countries with the lowest electrification rate [28]

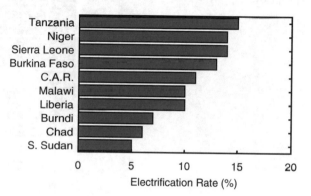

Approximately 80% of those lacking electricity access live in just 20 countries in Sub-Saharan Africa and South Asia. People without access tend to live in rural areas. The global electrification rate in urban areas is 96% but just 72% in rural areas [28]. This "rural penalty" is evident in Sub-Saharan Africa, where the urban and rural rates stand at 69% and 15%, respectively. The countries with the lowest electrification rates are all found in SSA, as shown in Fig. 2.11 [28]. In terms of total population without access, India is the unfortunate leader, with over 250 million people without electricity, as shown in Fig. 2.12.

Achieving universal electrification—a state in which the electrification rate globally is 100%—will be expensive and take time. Although challenging to estimate with high accuracy, the cost of universal access will be between US$0.5–0.9 trillion, approximately US$45 billion per year through 2030. Present investment levels are far under this target, at approximately US$10 billion per year [28].

Fig. 2.12 Countries with the
largest number of people
without access to electricity
[28]

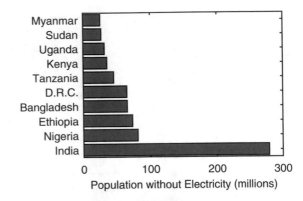

Fig. 2.13 Public primary
school electrification rates
(various years) [27]

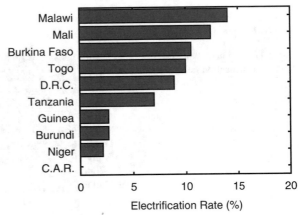

2.6.1 Electricity Access of Community Service Institutions

As discussed in Sect. 2.2.2, social institutions, schools, health-care facilities, and
government offices require electricity to effectively provide their services. The
electrification rate of schools and medical clinics in various countries are provided in
Figs. 2.13 and 2.14. Social institutions are often given high priority by governments
for electrification because they support improved education, health care, and other
development outcomes.

2.6.2 Annual Growth Rate of Electricity Access

The (AGR) of electricity access is used to quantify the change in electricity access
over a period of time. The use of the AGR is preferred over reporting the change
in the electrification rate or the change in the absolute number of people with or
without access. These indicators alone without accounting for population dynamics
can be misleading.

Fig. 2.14 Health clinic electrification rates (various years) [27]

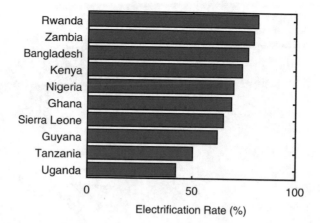

Consider the following. A government points to an increasing electrification rate within the country as a sign of progress. At the same time, the opposition party claims that the total number of people without access to electricity in the country has increased. Both the government and the opposition can be correct. How is it possible for the electrification rate to increase, yet there is also an increase in the number of people without electricity?

Consider a brief case study of Malawi, a country in southern Africa. The population of Malawi in 2010 was 14.8 million; by 2012 it had grown to 15.7 million. The electrification rate during the same period increased from 9% to 10%— a small sign of progress. Based on the 2010 population, the number of people without access to electricity can be calculated as $14.8 \times (1.00 - 0.09) = 13.47$ million people. Repeating the calculation using the 2012 data shows that 14.13 million people do not have access to electricity, an increase of 660,000 people— a step backward.

The cause of these seemingly incompatible results is that Malawi's population increased faster than the number of people that gained access. The number of people with access increased but so did the number of people without access, because the population as a whole increased.

Sub-Saharan Africa as a whole is another example of a region where electrification rates are increasing but so is the number of people without access to electricity. Here, the electrification rate increased from 32 to 35% from 2010 to 2012, but the population grew from 874 million to 923 million over the same period, leading to an additional 5 million people without electricity. Whether or not this represents progress depends on one's perspective.

The AGR captures the effects of changes in the electrification rate and population dynamics. The AGR considers the changes in number of people with access $A[y]$, in a given year y, and the population, $P[y]$, over a period of years t. The AGR is the change in the number of people with access between the years considered, minus the change in population during the same span, divided by the population. It is divided by the number of years considered to determine the annual rate.

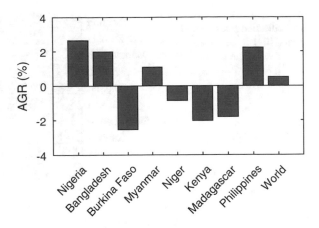

Fig. 2.15 Annual Growth Rate of electrification in various countries

$$AGR = \frac{(A[y] - A[y-t]) - (P[y] - P[y-t])}{P[y]} \times \frac{1}{t} \tag{2.1}$$

The AGR is often expressed as a percentage, in which case (2.1) is multiplied by 100.

The AGR is positive if the number of people with access increased faster than the population grew over the same period. A country's AGR is typically between ±3%. Some countries have obtained much greater AGRs, for example, Cambodia at 7% and Afghanistan at 10%. Encouragingly, the worldwide AGR is positive. Progress is attributed to a few countries with both a high AGR of electricity access and a large population. India, Nigeria, and Bangladesh are examples of this. The AGR for select countries between 2010 and 2012 are shown in Fig. 2.15.

Example 2.5 Compute the AGR of electricity access for Malawi between 2010 and 2012.

Solution The AGR is found by applying (2.1) to the data for 2010 and 2012 presented in the text. The population with access is found by subtracting the population without access from the total population each year:

$$14.8 - 13.47 = 1.33 \text{ million people with access (2010)}$$

$$15.7 - 14.13 = 1.57 \text{ million people with access (2012)}.$$

The AGR is therefore:

$$AGR = \frac{(1.57 - 1.33) - (15.7 - 14.8)}{15.7} \times \frac{1}{2} = -0.0211.$$

The AGR is negative indicating that the number of people gaining access is outpaced by the increase in population.

Example 2.6 Compute the AGR for the world between 2010 and 2012. The population of the world was 6.92 billion in 2010 and 7.09 billion in 2012. The electrification rates were 83% and 85% for 2010 and 2012, respectively.

Solution The population with access is found by multiplying the population by the electrification rate for each year

$$6.92 \times 0.83 = 5.744 \text{ billion people with access (2010)}$$

$$7.09 \times 0.85 = 6.027 \text{ billion people with access (2012)}.$$

Applying (2.1), the AGR is computed as:

$$AGR = \frac{(6.027 - 5.744) - (7.09 - 6.92)}{7.09} \times \frac{1}{2} = 0.0080.$$

From the above example, we see that the world AGR is positive so that the electrification rate is increasing and the number of people without access is decreasing. The AGR, however, is very small and below what is needed to achieve universal access by the United Nations' target year of 2030.

2.7   Multi-Tier Electricity Access Framework

The electrification rate by itself suffers from several criticisms:

- it ignores the quality of the electricity supply;
- it treats all electricity supplies the same;
- it ignores the reliability and availability of the electricity supply.

For example, a household with a connection that can supply at most 1 kVA of apparent power—not enough for cooking—is only available a few hours a day and at a tariff[1] that prohibits regular use is treated the same as a 20 kVA connection, available 24 h a day at an affordable tariff. The method also ignores the modest electricity supplied by solar lanterns, solar home systems, and low-capacity mini-grids. These criticisms have led to the development of several alternative measures [5, 8]. In general, these approaches consider the *quality* of the electricity supply in defining electricity access. We will consider the *Multi-Tier Framework* (MTF) developed by ESMAP [8].

---

[1]The term "tariff" refers to the pricing structure for electricity. It typically consists of a fixed monthly "connection" fee and a variable "energy" fee.

The MTF is part of a comprehensive approach to assessing energy access in general. It considers energy access for household uses (electricity, cooking, and heating), for productive uses, and for community uses (street lighting, health and education facilities, and community and public buildings). We will focus on household electricity use.

There are three different MTFs for assessing household electricity access based upon attributes of the supply, the services such as lighting and cooking made possible by the supply, and the consumption level. These MTFs provide a more holistic accounting of a household's access to electricity than the electrification rate. The MTFs are also technology neutral. They can equally be applied to on- and off-grid electricity supplies. However, more time and resources are required to collect the data needed to apply the MTFs than to determine the electrification rate.

2.7.1 Supply Multi-Tier Framework

The Supply Multi-Tier Framework assigns a household's electricity access to one of the six tiers based on *attributes* of the electrical supply. A household with higher-tier access is served by a higher-quality electricity supply. There are seven attributes considered:

1. Capacity
2. Availability
3. Reliability
4. Affordability
5. Legality
6. Health and Safety
7. Quality of Voltage and Frequency

The seven attributes are separately evaluated according to a predefined scale and each assigned a tier. The tier of the supply is the minimum tier achieved when all attributes are considered.

2.7.1.1 Capacity Attribute

The capacity attribute of an electricity supply is assessed using one of two scales. The first scale is well-suited for grid or mini-grid connections. It is based on two indicators: the peak power capacity and the daily total energy that can be provided by the supply. This scale is provided in Table 2.7. The minimum peak power and daily energy escalate with the tier level and are related to the type of appliances that could likely be operated at each tier. Although peak power and daily energy are often related, their tiers can be different. The overall tier of the capacity attribute is set to the lower of the peak power tier or daily energy tier. For example, if a supply is capable of providing 9000 Wh (Tier 5) of energy per day, but the peak capacity

Table 2.7 Capacity attribute scale: peak and daily energy

Indicators	Tier 0	Tier 1	Tier 2	Tier 3	Tier 4	Tier 5
Power capacity rating	–	≥3W	≥50W	≥200W	≥800W	≥2000W
Daily energy potential	–	≥12Wh	≥200Wh	≥1000Wh	≥3400Wh	≥8200Wh

Table 2.8 Capacity attribute scale: service

Indicator	Tier 0	Tier 1	Tier 2	Tier 3	Tier 4	Tier 5
Service	–	≥ 1,000 lm/hr of lighting/day	Lighting, air circulation, television, and phone charging possible	–	–	–

rating is only 1000 W (Tier 4), then the capacity attribute of the supply is Tier 4. Note that the supply is evaluated based on its potential capability, not whether or not a particular amount of peak power or daily energy is actually consumed by the household.

The second scale is based on the services that can be supported by the supply, according to the scale in Table 2.8. Tier 2 is the maximum tier that can be achieved using this scale. This scale can more readily be applied to households that do not have a grid or mini-grid connection but rather are served by solar lanterns or solar home systems. The quantity of lighting required for Tier 2 service capacity is 1000 lumen-hours (lm hr) per day. A lumen is a unit of luminous flux, which can be thought of as the light output of a lamp. Note that either the scale in Table 2.7 or Table 2.8 can be applied, but not both.

2.7.1.2 Availability Attribute

The availability attribute is based on the length of time that the supply is available each day. Supply availability can be limited when countries face electricity shortages and must intentionally disconnect customers; in off-grid systems, the availability might be reduced, for example, if generator sets are only run during certain hours of the day to reduce fuel costs, or in a solar-powered system if there is insufficient battery capacity to supply power throughout the evening. Availability is important for loads that need to be operated continuously or during certain hours, such as lighting, climate control, and refrigeration. The availability attribute scale is shown in Table 2.9.

Two indicators are used to determine the tier of the availability attribute of a supply: the total number of hours per day the supply is available and the number of evening hours in particular that the supply is available. Here, evening hours is defined as the 4 h after sunset. The tier of the availability attribute is set to the lowest of the two indicators.

Table 2.9 Availability attribute scale

Indicator	Tier 0	Tier 1	Tier 2	Tier 3	Tier 4	Tier 5
Hours available per 24 h period	–	≥4 h		≥8 h	≥16 h	≥23 h
Hours available per evening	–	≥1 h	≥2 h	≥3 h	≥4 h	

Table 2.10 Reliability attribute scale

Indicator	Tier 0	Tier 1	Tier 2	Tier 3	Tier 4	Tier 5
Number of disruptions/week	–	–	–	–	≤ 14	≤ 4 AND aggregate duration < 2 hrs/week

Table 2.11 Affordability attribute scale

Indicator	Tier 0	Tier 1	Tier 2	Tier 3	Tier 4	Tier 5
Cost of consuming 365 kWh/year	–	–	–	< 5% of annual income		

2.7.1.3   Reliability Attribute

The reliability attribute refers to the frequency of unplanned interruptions of service. Although similar to availability, reliability is a distinct concept, and the consequences of poor reliability are different from poor availability. The reliability attribute scale is shown in Table 2.10. If more than 14 disruptions occur each week, the supply is assigned Tier 3 for its reliability attribute.

2.7.1.4   Affordability Attribute

An electricity supply does little to improve the quality of life if it is not affordable. A household's expenditure on electricity is dependent on the price of electrical energy and the level of consumption. However, the amount of the expenditure alone does not fully encapsulate the affordability of the supply—the income of the household must be considered. The affordability attribute scale is shown in Table 2.11.

The affordability attribute of a supply is determined from whether or not an annual consumption of 365 kWh is, or would be, less than of 5% of the household's income. Note this calculation is independent of whether or not the household actually consumes 365 kWh each year. If this is less than 5% of the household's income, then the affordability attribute is Tier 5; otherwise, it is Tier 2.

2.7.1.5   Legality Attribute

In many developing countries, theft of electricity is a pervasive problem. Electricity theft threatens personal safety (through illegal connections) and the financial viability of the supplier (the utility or off-grid system owner). It is also associated with corruption. The legality attribute scale is shown in Table 2.12.

Table 2.12 Legality attribute scale

Indicator	Tier 0	Tier 1	Tier 2	Tier 3	Tier 4	Tier 5
Bill Payment	–	–	–	–	Paid to supplier or authorized agent	

Table 2.13 Health and safety attribute scale

Indicator	Tier 0	Tier 1	Tier 2	Tier 3	Tier 4	Tier 5
Wiring installed per national standards	–	–	–	–	No past accidents and no perception of high shock/electrocution risk	

Table 2.14 Quality attribute scale

Indicator	Tier 0	Tier 1	Tier 2	Tier 3	Tier 4	Tier 5
Voltage	–	–	–	–	Voltage is within the parameters specified by the grid code	

Determining the legality of a supply is challenging as people might not disclose this information, and utilities might not have an accurate or precise estimate of illegal connections. Like affordability, the legality attribute is based on a binary indicator: if the bill for electricity is paid to the supplier—or one of its agents—then it is Tier 5; otherwise, it is Tier 3.

2.7.1.6 Health and Safety Attribute

An electricity supply can pose a serious threat to safety if improperly installed. For a supply to be safe, it should meet the country's electric code. The indicator for health and safety is whether or not there has been past accidents or perceived future risk of electrocution, as shown in Table 2.13.

2.7.1.7 Quality Attribute

Designers of consumer appliances and electronics assume the electricity supply meets certain quality standards. Voltages and frequencies above or below the range specified in the standards can cause devices to malfunction, become inoperable, or fail. The quality attribute scale is shown in Table 2.14.

2.7.1.8 Overall Supply Tier

To compute the overall tier of a supply, each attribute is assessed independently. The overall tier of the supply is the lowest tier assigned to any attribute. For example, if a supply that is Tier 5 in six of the attributes but Tier 3 in the seventh, then it is a

Tier 3 supply. While this might seem strange, the MTF is useful in that it identifies how this supply can be increased to Tier 5—by improving the deficient attribute.

Example 2.7 Determine the energy access tier of a customer whose supply has the following characteristics:

- Power Capacity: 700 W
- Daily Capacity: 15,400 Wh
- 22 h of availability per day, 4 h of which are after sunset
- 7 disruptions per week
- Consumption of 1 kWh per day costs 4% of daily income
- Bill for electricity supply is paid to the supplier
- No past accidents and no perceptions of future risk; supply meets local electrical code
- Voltage spikes and sags are common, occasionally damaging equipment

What would the energy access index be if the peak power capacity was improved to 1000 W and the voltage quality was within the parameters specified by the grid code?

Solution We begin by assessing each attribute separately.

Capacity: Tier 3 (Tier 3 power capacity rating and Tier 5 daily energy potential)
Availability: Tier 4 (Tier 4 h per 24-h period and Tier 5 h per evening)
Reliability: Tier 4
Affordability: Tier 5
Legality: Tier 5
Health and Safety: Tier 5
Quality: Tier 3

The energy access tier of the connection is the lowest tier of any attribute. In this case it is Tier 3. If capacity and quality were improved as indicated in the problem statement, then the household would have a Tier 4 supply.

2.7.2   Aggregate Index

The indices of a set of households can be aggregated and combined into a single Aggregate Index (*Aggregate Index*) using the following formula:

$$Aggregate\ Index = \sum_{T=0}^{5} 20 \times P_T \times T \qquad (2.2)$$

where P_T is the proportion of the population with tier T supply.

Table 2.15 Portion of population with each access tier (Example 2.8)

	Tier 0	Tier 1	Tier 2	Tier 3	Tier 4	Tier 5
Portion of population	30%	20%	10%	20%	10%	10%

Table 2.16 Service-based tiers

	Service
Tier 0	–
Tier 1	Phone charging and task lighting
Tier 2	General lighting, television, and air circulation (if needed)
Tier 3	Medium-power appliances
Tier 4	High-power appliances
Tier 5	Very high-power appliances

Example 2.8 Compute the Aggregate Index of a country with the following energy access tiers in Table 2.15.

Solution The Aggregate Index is found using (2.2):

$$Aggregate\ Index = 20(0.30 \times 0 + 0.20 \times 1 +$$
$$0.10 \times 2 + 0.20 \times 3 + 0.10 \times 4 + 0.10 \times 5) = 38.0.$$

The Aggregate Index gives a general sense of the electrical system quality, but it mixes so many concepts that it cannot be interpreted in a strictly quantitative manner.

2.7.3   Services Multi-Tier Framework

The second framework is the Service MTF. It assesses electricity access from the electricity-based services available to the household. Application of the Supply MTF and the Service MTF to a household can yield different results. For example, a household might have a high-tier supply, but if the household is unable to acquire or use basic appliances, their access tier as determined by the Service MTF will be low.

The service-based tier framework is shown in Table 2.16. The listed appliance power descriptions is in reference to Table 2.5.

Table 2.17
Consumption-based tiers

	Daily (Wh)	Annual (kWh)
Tier 0	–	–
Tier 1	≥ 12	≥ 4.5
Tier 2	≥ 200	≥ 73
Tier 3	≥ 1000	≥ 365
Tier 4	≥ 3425	≥ 1250
Tier 5	$\geq 8,219$	$\geq 3,000$

2.7.4 Consumption Multi-Tier Framework

The third MTF that is used to assess a household's access to electricity is the Consumption MTF. The Consumption MTF is based on the daily or equivalent annual energy used by the household according to the scale in Table 2.17. It generally mirrors the Supply MTF and the Service MTF.

The MTF is particularly useful when planning an off-grid project. Not all systems need to provide Tier 5 quality in order to be useful and beneficial to the community. For example, an organization might target Tier 3 access and design their system accordingly.

2.8 Summary

Energy poverty is the lack of access to modern fuels such as electricity and LPG. Approximately 40% of the world, primarily those living in rural areas in developing countries, are energy impoverished. They often rely on fuels such as animal dung, crop waste, fuel wood, charcoal, and kerosene. The household energy consumption in rural areas ranges between 5 and 15 GJ (1389 to 4167 kWh) per person year, with additional energy needed for community services and to enhance productivity. Procuring fuel is often an economic burden and requires considerable time.

Energy-impoverished households tend to fuel stack, relying on several fuel sources, and the transition to fuels with higher desirability depends on income but also access to the fuel and other local factors. The electricity requirements of a household depend on the services rendered and range from 1 to over 3000 kWh per year per person.

Electricity access is often quantified by the electrification rate, which is the proportion of the population with grid-supplied electricity access. When assessing electrification progress, the Annual Growth Rate should be used as it captures the dynamics of a population. More recently, a Multi-Tier Framework, which provides a more holistic picture of electricity access, is used. The Supply Multi-Tier Framework considers capacity, availability, reliability, affordability, legality, health and safety, and voltage and frequency quality of the electricity supply.

Problems

2.1 A health clinic in a rural area requires 3.5 kWh of electricity each day. The diesel generator set that serves the health clinic has an average efficiency of 15%. Compute the annual fuel requirements of the gen set assuming the energy content of diesel is 36 MJ per liter.

2.2 A rural school is to be supplied with five desktop computers (100 W each), five CFL lightbulbs (16 W each), and two radios (15 W each). The desktop computers will be used for 3 h per day, 5 days a week. Three of the CFLs are used indoors to hold classes in the evening. These bulbs are used 2 h each day, 5 days per week. The other bulbs are used for overnight security, for 10 h each day. The radios are used 1 h per day, 5 days per week. Estimate the average daily consumption and the total yearly consumption.

2.3 The average annual per person energy use in Cameroon is 14.3 GJ. Assume that 75% of this energy is used for cooking and heating. Compute the required mass of dry wood needed for cooking and heating. Repeat the calculation assuming only charcoal is used and again assuming only dried animal dung is used.

2.4 Evaluate and compare the desirability of fuel wood and charcoal considering its quality, convenience, and cost attributes.

2.5 A kerosene lamp consumes 0.15 liters of kerosene when used for 3.5 h per day. The kerosene lamp can be replaced by an electric LED lamp that consumes 3.5 Wh each day. Compute the equivalent daily energy, in watthours, consumed by the kerosene lamp. What is the effective cost per kilowatthour of operating the kerosene lamp assuming the price of kerosene is US$1.40 per liter. How does this compare using an LED lamp, assuming the price of electricity is US$0.15/kWh?

2.6 A certain mobile phone requires 12 Wh of energy to recharge. The phone can be recharged for a fee of US$0.20. Compute the rate, in dollars per kilowatthour that is being paid to recharge the mobile phone. How does this compare to the rate for grid-connected electricity in your country? If a solar lantern capable of recharging mobile phones costs US$40, how long will it take to pay for itself if the mobile phone is recharged every 4 days?

2.7 Compute the AGR of the countries in Table 2.18.

Table 2.18 Electrification data

| Country | 2010 | | 2012 | |
	Rate (%)	Population (millions)	Rate (%)	Population (millions)
Indonesia	94.2	242.5	96.0	248.8
India	76.3	1231	79.9	1263
Nepal	67.5	27.02	75.6	27.65
Zimbabwe	35.6	14.09	36.2	14.71

2.8 The population of Zambia in 1990 was 8.03 million with an electrification rate of 13.9%. How many Zambians did not have access to electricity in 1990?

2.9 The world AGR of electricity access in 2012 was 0.51%. If this rate continues, how many people will be without access in the year 2030? Assume the population in the year 2030 is 8.5 billion, the population in 2012 is 7.10 billion and the electrification rate in 2012 is 85.3%.

2.10 What will the AGR of electricity access need to be in order to achieve universal (100%) access by the year 2030?

2.11 Compute the energy access tier for the following conditions. Determine the tier for each attribute.

- Power Capacity: 1000 W
- Daily Capacity: 2400 Wh
- 10 h of daily availability and 6 h during the night
- 4 disruptions per week with total duration of less than 2 h
- Consumption (365 kWh/year) is less than 5% of income
- Bill is paid to the supplier
- No past accidents and no perception of future risk
- Voltage spikes and sags are not common

2.12 Use the Consumption Mult-Tier Framework to evaluate the electricity service described in the previous problem.

2.13 Compute the energy access tier for the following conditions. Determine the tier for each attribute.

- Power Capacity: 5000 W
- Daily Capacity: 10,000 Wh
- 22 h of daily availability and 4 h during the night
- 1 disruption per week less than 2 h
- Consumption (365 kWh/year) is less than 5% of income
- Connection is illegal
- No past accidents and no perception of future risk
- Voltage spikes and sags are not common

References

1. Abdullah, S., Markandya, A.: Rural electrification programmes in Kenya: Policy conclusions from a valuation study. Energy Sustain. Dev. **16**(1), 103–110 (2012). doi:https://doi.org/10.1016/j.esd.2011.10.007. URL http://www.sciencedirect.com/science/article/pii/S0973082611000858
2. Adkins, E., Oppelstrup, K., Modi, V.: Rural household energy consumption in the millennium villages in sub-saharan Africa. Energy Sustain. Dev. **16**, 249–259 (2012). doi:https://doi.org/10.1016/j.esd.2012.04.003. URL http://www.sciencedirect.com/science/article/pii/S0973082612000221

3. Alabe, M.: Household energy consumption patterns in northern Nigeria. Energy Sustain. Dev. **2**(5), 42–45 (1996). doi:https://doi.org/10.1016/S0973-0826(08)60160-X. URL http://www.sciencedirect.com/science/article/pii/S097308260860160X

4. World Bank: Malawi - Household Energy Use in Malawi (English). World Bank. URL http://documents.worldbank.org/curated/en/214711468056950074/Malawi-Household-energy-use-in-Malawi

5. Baring-Gould, I., Burman, K., Singh, M., Esterly, S.: Quality assurance framework for mini-grids. Tech. Rep. NREL/TP-5000-67374, National Renewable Energy Laboratory (2016). URL https://www.nrel.gov/docs/fy17osti/67374.pdf

6. Bastakoti, B.P.: The electricity-livelihood nexus: some highlights from the Andhikhola Hydroelectric and Rural Electrification Centre (AHREC). Energy Sustain. Dev. **10**(3), 26–35 (2006). doi:https://doi.org/10.1016/S0973-0826(08)60541-4. URL http://www.sciencedirect.com/science/article/pii/S0973082608605414

7. Bensch, G., Peters, J., Sievert, M.: The lighting transition in rural Africa – from kerosene to battery-powered led and the emerging disposal problem. Energy Sustain. Dev. **39**, 13–20 (2017). doi:http://dx.doi.org/10.1016/j.esd.2017.03.004

8. Bhatia, M., Angelou, N.: Beyond connections: Energy access redefined. Tech. Rep. 008/15, ESMAP (2015)

9. Bhatt, B., Rathore, S., Lemtur, M., Sarkar, B.: Fuelwood energy pattern and biomass resources in Eastern Himalaya. Renewable Energy **94**, 410–417 (2016). doi:https://doi.org/10.1016/j.renene.2016.03.042. URL http://www.sciencedirect.com/science/article/pii/S0960148116302269

10. Cosmas, M., Sithulisiwe, B., Jenny, S.: Factors influencing trainee teachers' choice of schools on deployment after completion of training. Mediterranean J. Soc. Sci. **5**(16), 346–356 (2014). doi:10.5901/mjss.2014.v5n16p346

11. EnDev (Energising Development Progam): EnDev's understanding of access to modern energy services (2011)

12. Fankhauser, S., Tepic, S.: Can poor consumers pay for energy and water? An affordability analysis for transition countries. Energy Policy **35**(2), 1038–1049 (2007). doi:https://doi.org/10.1016/j.enpol.2006.02.003. URL http://www.sciencedirect.com/science/article/pii/S0301421506000887

13. Franco, A., Shaker, M., Kalubi, D., Hostettler, S.: A review of sustainable energy access and technologies for healthcare facilities in the Global South. Sustain. Energy Technol. Assess. **22**, 92–105 (2017). doi:https://doi.org/10.1016/j.seta.2017.02.022. URL http://www.sciencedirect.com/science/article/pii/S2213138817301376

14. IEA: Africa energy outlook: A focus on energy prospects in Sub-Saharan Africa (2014). URL https://www.iea.org/publications/freepublications/publication/WEO2014_AfricaEnergyOutlook.pdf

15. Johnson, N.G., Bryden, K.M.: Energy supply and use in a rural west African village. Energy **43**(1), 283–292 (2012). doi:https://doi.org/10.1016/j.energy.2012.04.028. URL http://www.sciencedirect.com/science/article/pii/S0360544212003088. 2nd International Meeting on Cleaner Combustion (CM0901-Detailed Chemical Models for Cleaner Combustion)

16. Jones, D., Ryan, C.M., Fisher, J.: Charcoal as a diversification strategy: The flexible role of charcoal production in the livelihoods of smallholders in central Mozambique. Energy Sustain. Dev. **32**, 14–21 (2016). doi:https://doi.org/10.1016/j.esd.2016.02.009. URL http://www.sciencedirect.com/science/article/pii/S0973082615302246

17. Kanagawa, M., Nakata, T.: Assessment of access to electricity and the socio-economic impacts in rural areas of developing countries. Energy Policy **36**(6), 2016–2029 (2008). doi:https://doi.org/10.1016/j.enpol.2008.01.041. URL http://www.sciencedirect.com/science/article/pii/S0301421508000608

18. Kornbluth, K., Pon, B., Erickson, P.: An investigation of the cost and performance of a solar-powered LED light designed as an alternative to candles in Zambia: A project case study. Renew. Sustain. Energy Rev. **16**(9), 6737–6745 (2012). doi:https://doi.org/10.1016/j.rser.2012.08.001. URL http://www.sciencedirect.com/science/article/pii/S1364032112004704

19. Leach, G.: The energy transition. Energy Policy **20**, 116–123 (1992). doi:https://doi.org/10.1016/0301-4215(92)90105-B

20. Mandelli, S., Barbieri, J., Mereu, R., Colombo, E.: Off-grid systems for rural electrification in developing countries: Definitions, classification and a comprehensive literature review. Renew. Sustain. Energy Rev. **58**, 1621–1646 (2016). doi:https://doi.org/10.1016/j.rser.2015.12.338

21. Matsika, R., Erasmus, B., Twine, W.: Double jeopardy: The dichotomy of fuelwood use in rural South Africa. Energy Policy **52**, 716–725 (2013). doi:https://doi.org/10.1016/j.enpol.2012.10.030. URL http://www.sciencedirect.com/science/article/pii/S0301421512008889. Special Section: Transition Pathways to a Low Carbon Economy

22. McCarthy, C.E., Duffney, P.F., Wyatt, J.D., Thatcher, T.H., Phipps, R.P., Sime, P.J.: Comparison of in vitro toxicological effects of biomass smoke from different sources of animal dung. Toxicol. in Vitro **43**, 76–86 (2017). doi:https://doi.org/10.1016/j.tiv.2017.05.021. URL http://www.sciencedirect.com/science/article/pii/S0887233317301467

23. Rahut, D.B., Behera, B., Ali, A., Marenya, P.: A ladder within a ladder: Understanding the factors influencing a household's domestic use of electricity in four African countries. Energy Economics **66**, 167–181 (2017). doi:https://doi.org/10.1016/j.eneco.2017.05.020. URL http://www.sciencedirect.com/science/article/pii/S0140988317301846

24. Sarmah, R., Bora, M., Bhattacharjee, D.: Energy profiles of rural domestic sector in six un-electrified villages of Jorhat district of Assam. Energy **27**(1), 17–24 (2002). doi:https://doi.org/10.1016/S0360-5442(01)00040-8. URL http://www.sciencedirect.com/science/article/pii/S0360544201000408

25. Sovacool, B.K., Clarke, S., Johnson, K., Crafton, M., Eidsness, J., Zoppo, D.: The energy-enterprise-gender nexus: Lessons from the Multifunctional Platform (MFP) in Mali. Renewable Energy **50**, 115–125 (2013). doi:https://doi.org/10.1016/j.renene.2012.06.024. URL http://www.sciencedirect.com/science/article/pii/S0960148112003758

26. Sovacool, B.K., Ryan, S.E.: The geography of energy and education: Leaders, laggards, and lessons for achieving primary and secondary school electrification. Renew. Sustain. Energy Rev. **58**, 107–123 (2016). doi:https://doi.org/10.1016/j.rser.2015.12.219. URL http://www.sciencedirect.com/science/article/pii/S1364032115016020

27. Sustainable Energy for All: Global tracking framework (2013). URL "http://www.seforall.org/global-tracking-framework"

28. Sustainable Energy for All: Global tracking framework (2015). URL "http://www.seforall.org/global-tracking-framework"

29. United Nations Educational, Scientific and Cultural Organization (UNESCO): Access to basic services in public schools by level of education: Percentage of primary schools with access to electricity, education and literacy: Africa (2015). URL http://data.uis.unesco.org/

30. World Health Organization: WHO indoor air quality guidelines: household fuel combustion (2014)

Chapter 3
Grid Extension and Enhancement

3.1 Introduction

There are two general approaches to providing electricity access: by extending the national grid or by implementing off-grid systems. The term "grid extension" refers to the provision of electricity access by the extension or enhancement of the existing national grid. Grid extension encompasses not only constructing medium-voltage distribution lines to reach new communities but also adding new connections to households in areas with existing but incomplete electricity access.

Although this book focuses on off-grid systems, attention must be given to the grid extension approach. We consider grid extension because:

- It is the most common approach to providing electricity access;
- It can be more expeditious and cost-effective than off-grid systems at providing high-tier electricity access for communities located near existing electrical infrastructure;
- Many of the technical and economic concepts and considerations related to grid extension also apply to mini-grid systems.

Grid extension is the default electrification practice by utilities and governments. It is estimated that 99% of electricity infrastructure investments, including those that improve the quality of existing connections, go toward grid extension projects [10]. Off-grid systems are only considered when there is a compelling reason not to extend the grid to a particular community. These reasons primarily include cases in which:

- Off-grid systems are less expensive on a per connection or per unit of energy consumed basis than grid extension;
- The wait for grid extension is too long;
- The electricity access tier (see Chap. 2) provided by the grid is insufficient for the community (e.g., the grid reliability is insufficient for a health-care facility).

© Springer International Publishing AG, part of Springer Nature 2018
H. Louie, *Off-Grid Electrical Systems in Developing Countries*,
https://doi.org/10.1007/978-3-319-91890-7_3

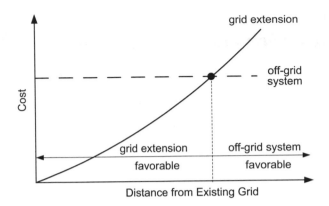

Fig. 3.1 Cost curve of grid extension and off-grid systems

Figure 3.1 shows a simplified cost curve of providing electricity access to a community as a function of distance of the community to the existing grid. The cost of grid extension and an off-grid system are shown. The cost of the off-grid system can be approximated as being constant, regardless of the distance from the grid. The cost of grid extension on the other hand rises as the distance from the grid increases. This is attributed to increased construction and material costs and losses along the line.

The two curves intersect at some distance, indicated by the dot in Fig. 3.1. To the left of the dot, grid extension is economically favorable; to the right of the dot, the cost of constructing electrical infrastructure and the associated losses make an off-grid system less expensive. This simple idea, that the distance from the grid greatly influences whether or not off-grid systems are economically favorable [2, 7, 8], is a major theme of this chapter.

3.1.1 Urban Electrification

The cost curves show that, all other considerations being equal, the most efficient use of a limited electrification budget is to connect people located near the existing electric grid through grid extension. This then biases electrification efforts to those living in urban and peri-urban areas, leaving rural, remote communities underserved. In fact, 80% of the 222 million people gaining first-time access to electricity between 2010 and 2012 lived in urban areas [9].

Increasing access in urban areas can usually be done quickly and at relatively low cost. In some regions, the per connection cost of connecting a rural household is over eight times that of an urban household [3].

Despite the focus on urban electrification, a surprisingly large number of people live within eyesight of the grid and yet have no access to it. For example, in Kenya about 50% of the 33 million people without electricity are estimated to live "under

Fig. 3.2 Millions of people live within sight of power lines but are not connected (courtesy of P. Dauenhauer)

Fig. 3.3 The basic components of grid extension include substation, three-phase medium-voltage distribution line, distribution transformer, low-voltage wiring, and user premise equipment (not shown)

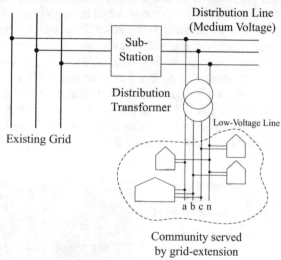

Community served by grid-extension

the grid"—within 200 m of the grid—as shown in Fig. 3.2 [6]. Across Sub-Saharan Africa (SSA), 30% of urban residents do not have access to electricity. Many of them live in slums and other informal settlements.

3.1.2 Basic Components of Grid Extension

The basic components of a grid extension project are shown in Fig. 3.3. The grid acts to connect all of the generation sources to all of the end users. These are not shown in the diagram. An existing distribution line or substation is modified so that a new distribution line can be connected to it. The distribution line is nominal voltage usually 11 kV to 33 kV. These voltage levels are commonly referred to as "medium-voltage". A transformer might be needed to change the voltage of the existing line to the nominal voltage of the new distribution line. The distribution line is typically a bare (uninsulated) overhead pole-top three-phase circuit. Once the distribution line reaches the community, a transformer reduces the voltage to the service voltage level, typically 120 or 230 V. Communities with high power consumption are served by multiple transformers. Low-voltage wiring connects the transformer secondary to the users. The home or building of each user must be internally wired and metered. In many cases, a distribution line serves several communities along its path.

3.2   Distribution Line Design

The purpose of a distribution line is to provide a pathway for power to flow from its sending end—where it is connected to the existing grid—to its receiving end, where the users are located. Distribution lines can be overhead or buried. Distribution lines to rural areas are almost always overhead as this is less expensive. A typical overhead three-phase distribution line is shown in Fig. 3.4. Overhead distribution lines are supported by poles that are 8 to 14 m tall and are made of wood, metal, concrete, or a composite material. The poles are typically spaced between 50 m and 200 m apart. The conductors are uninsulated, which increases the problems caused by falling tree branches and animals. Ceramic, glass, or polymeric insulators connect the conductors to the poles.

Fig. 3.4 A typical three-phase overhead distribution line (courtesy of R. Ngoma)

The design of a distribution line can be separated into two interrelated aspects: electrical and mechanical. The mechanical aspects include the tower or support structure, the conductor tensioning system, and the associated civil works. The electrical aspects include the selection of the conductor and insulators. Distribution line design is a rich subject and covered in other texts [4]; for our purposes, we will review the basic electrical conductor design only.

3.2.1 Power, Voltage, and Current Relationship

Before continuing with this section, the reader may wish to review the basics of three phase circuit concepts and analysis provided in the Appendix. Distribution lines consist of three phase conductors. The nominal voltage of a distribution line refers to the root mean square (RMS) magnitude of the line-to-line voltage $V_{\ell\ell}$, for example, the voltage from the a-phase conductor to the b-phase conductor. Common distribution line nominal voltages in SSA are 11 kV, 22 kV, and 33 kV. The relationship between the magnitude of the line-to-line voltage and the magnitude of the line-to-neutral voltage of a balanced three-phase system is

$$V_\phi = \frac{V_{\ell\ell}}{\sqrt{3}} \qquad (3.1)$$

where V_ϕ is the nominal line-to-neutral voltage of the distribution line. Due to real and reactive losses along the line, the power supplied to the sending end (the end connected to the grid) is different from the power at the receiving end (the end located at the community). We therefore must distinguish the sending-end quantities from the receiving-end quantities. The sending-end line-to-neutral voltage and apparent power are denoted V_s and $S_{\text{total,s}}$, respectively; the receiving-end quantities are V_r and $S_{\text{total,r}}$. Keep in mind that apparent power quantities are complex, where the real part refers to the real power P, expressed in watts, and the imaginary part refers to the reactive power, Q, expressed in voltampere reactive (VAR). Generically

$$S = P + jQ. \qquad (3.2)$$

We will arbitrarily assume that the voltages refer to the a-phase line-to-neutral voltage. Note that variables corresponding to phasors will use bold italic font. In most cases it is reasonable to assume that the magnitude of the sending-end voltage is equal to the nominal line-to-neutral voltage $|V_s| = V_\phi$.

We can assume that each phase carries an equal amount of power so that the per-phase sending- and receiving-end apparent powers S_s and S_r are

$$S_s = \frac{S_{\text{total,s}}}{3} \qquad (3.3)$$

$$S_r = \frac{S_{\text{total,r}}}{3}. \tag{3.4}$$

The per-phase real power is therefore

$$P_s = |S_s| \times PF \tag{3.5}$$

$$P_r = |S_r| \times PF \tag{3.6}$$

where PF is the power factor associated with the load. A reasonable assumption is a power factor of 0.85 lagging. The a-phase line current, also expressed in RMS, is

$$I_s = \left(\frac{S_s}{V_s} \right)^* \tag{3.7}$$

$$|I_s| = \frac{|S_s|}{|V_s|} \tag{3.8}$$

where $*$ is the complex conjugate operator. We make the approximation that the sending-end current is equal to the receiving-end current. From (3.8), we see that for a given quantity of power, there is an inverse relationship between the voltage magnitude and current magnitude. The nominal voltage of a distribution line is typically selected based upon the quantity of power it is expected to supply. For example, a utility might select 33 kV for distribution lines serving 2.5 MVA or above and 11 kV for lines serving less than 2.5 MVA. The goal of these selections is to keep the current and their resulting line losses low.

3.2.2   Distribution Line Model

A circuit model of a single phase of a distribution line is shown in Fig. 3.5. This model ignores the capacitive affects that become increasingly important as the length exceeds about 50 km. The conductor is modeled as a resistance R_{line} in series with a inductive reactance X_{line}. The impedance of a conductor in a distribution line is

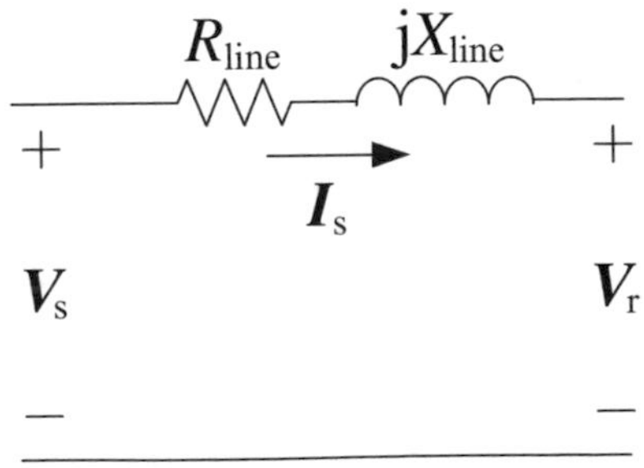

Fig. 3.5 Distribution line circuit model

$$Z_{line} = R_{line} + jX_{line}.$$ (3.9)

3.2.2.1 Resistance

The AC resistance R_{line} between the sending and receiving end of each conductor in a distribution line depends on the length, cross-sectional area, frequency, and material used:

$$R_{line} = s\rho\frac{l}{A_{line}}.$$ (3.10)

Here ρ is the resistivity in ohmmeters, l is the conductor length in meters, and A_{line} is the conductor cross-sectional area in meters squared. The coefficient s models the frequency-dependence of resistance caused by the skin effect. The skin effect increases the resistance by about 1–3% in a 50 Hz distribution line.

The resistivity is a property of the conductor material, which is typically aluminum or an aluminum alloy. Note that distribution lines are not always constructed along the shortest geographic path between two points, and so the length of a proposed line can be considerably longer than the distance between a community and the existing grid.

3.2.2.2 Inductive Reactance

In addition to resistance, distribution lines have significant inductive effects. Deriving the series inductive reactance of a conductor is beyond the scope of this book but can be found in most power system analysis textbooks [1]. It suffices to note that the inductance L of one of the phases of the distribution line depends on a variety of factors: the length of the line, the physical distance separating the phases, and the effective radius of the conductors. The inductive reactance, X_{line}, is computed from

$$X_{line} = \omega L$$ (3.11)

where ω is the frequency of the system, in radians per second.

The relationship between conductor radius and inductive reactance for three different separation distances is shown in Fig. 3.6a. Decreasing the separation distance reduces the inductive reactance. However, the separation cannot be made arbitrarily small. The conductors must be spaced far enough apart to prevent arcing or accidental contact. Increasing the conductor radius reduces the inductive reactance. It also decreases the resistance. The separation distance has no effect on resistance. The inductive reactance for a given conductor and separation distances are provided by conductor's manufacturer.

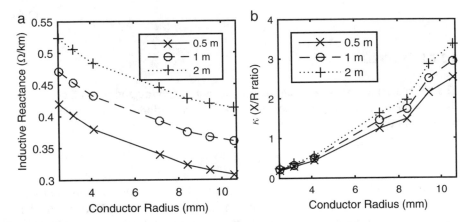

Fig. 3.6 (**a**) The effect of conductor radius on the inductive reactance per kilometer. (**b**) The X/R ratio for various conductor separation distances

It is often convenient to express the ratio of the inductive reactance to the resistance by the so-called X/R ratio κ:

$$\kappa = \frac{X_{\text{line}}}{R_{\text{line}}}. \tag{3.12}$$

The resistance is more sensitive to changes in the conductor radius than the inductance, so κ increases with the radius, as shown in Fig 3.6b.

3.2.3   Constraints

There are two important technical limitations that must be considered in designing a distribution line. The first is the voltage drop along the line, which restricts how long a distribution line of a given conductor radius can be. The second is the thermal limit of the line, which restricts the amount of current that can flow through the conductor, regardless of its length. The thermal limit is associated with the power loss along the line. The voltage drop and power loss are related, but not the same.

Selecting a conductor with a larger cross-sectional area reduces the voltage drop and increases the maximum current-carrying capability. Of course, the conductor cross-sectional area cannot be made arbitrarily large as it will be more expensive, require more mechanical support, and be more challenging to install.

3.2.3.1   Voltage Drop

The voltage drop of a distribution line is the magnitude of the difference between the voltage at the sending end and the receiving end. In general, the voltage drop should not exceed 5 to 10 % of the line's nominal line-to-neutral sending-end voltage. Large voltage drops result in low voltage at the receiving end. This can damage the users' appliances and equipment or cause them to malfunction.

The magnitude of the voltage drop depends on the amount of current and the length of the line. The maximum length that a line can be without violating a voltage drop limit for a specified amount of current is known as the "voltage reach" of the line. For example, if at a current of 180 A, the voltage along a distribution line decreases by 2.5% per kilometer, then the voltage reach of the line is four kilometers using a 10% voltage drop limit. The voltage reach would be increased if the current is decreased.

The magnitude of the voltage drop $V_{\text{drop},\phi}$ along a phase of a distribution line is

$$V_{\text{drop}} = |V_s - V_r| = |I_s| \times |Z_{\text{line}}|. \tag{3.13}$$

The voltage drop expressed as a percent is

$$V_{\text{drop},\%} = \frac{V_{\text{drop}}}{|V_s|} \times 100. \tag{3.14}$$

Example 3.1 The conductors used in a 33 kV distribution line have an impedance of $Z_{\text{line}} = 1.75 + j1.0\ \Omega$. The magnitude of the line current is 115 A. Compute the voltage drop as a percent.

Solution Each phase of the distribution line carries 115 A, and so the voltage drop is computed using (3.13) as

$$V_{\text{drop}} = 115 \times |1.75 + j1.0| = 231.79\ \text{V}.$$

We will make the standard assumption that the sending end is at the nominal voltage. The nominal voltage of the distribution line is always given as a line-to-line value. The corresponding line-to-neutral voltage is found using (3.1) to be

$$V_\phi = V_s = \frac{33,000}{\sqrt{3}} = 19052.6\ \text{V}$$

The percent drop is found from (3.14)

$$V_{\text{drop},\%} = \frac{231.79}{19052.6} \times 100 = 1.22\%.$$

This is below the typical 5–10% maximum voltage drop.

Should the length of a distribution line be limited by the voltage reach, the designer has several options:

- use conductors with larger cross-sectional area to reduce the resistance and inductive reactance;
- decrease the separation distance between the conductors to reduce the inductive reactance;
- increase the nominal voltage of the distribution line to reduce the current for a given amount of power;
- include voltage boosting equipment such as transformers or capacitor banks.

With the exception of decreasing the separation distance, these options increase the cost of the distribution line.

3.2.3.2   Thermal Limit

The current that a conductor can carry continuously is limited by its thermal characteristics. Heat is generated as current flows along the conductor due to the conductor's resistance. The generated heat is considered a power loss. The power loss along each phase and the total power for the line are computed as

$$P_{loss,\phi} = |I_s|^2 \times R_{line} \tag{3.15}$$

$$P_{loss,total} = 3P_{loss,\phi}. \tag{3.16}$$

It is possible for the heat generated by the power loss to increase the temperature of the conductor beyond the acceptable level, causing it to anneal and mechanically fail. The amount of current that can flow continuously through a line without overheating is known as the "ampacity" of the line. Note that the current through each phase has equal magnitude. The ampacity of a conductor is therefore the same as the ampacity of the line. The ampacity for a given conductor depends on the heat generated by the current in the conductor, the ambient temperature, the heat from the sun, and the cooling caused by the wind.

The ampacity of a conductor under typical conditions is provided by the manufacturer. The ampacity is not a strict limit because the conditions governing the temperature of the line, for example, the ambient temperature, vary over the course of a day and throughout the year. Some utilities will allow a distribution line to carry current in excess of the ampacity limit during the colder winter months.

The losses along a distribution line lower its efficiency. A utility might opt for a larger conductor than the one that satisfies the voltage drop and thermal limits in an effort to reduce the energy loss. A compromise must be made: increasing the cross-sectional area reduces losses but also increases the cost of the line.

3.2.4 Conductor Sizes and Types

Manufacturers produce a wide range of conductors for use in distribution lines. There are several conventions for expressing the cross-sectional area of a conductor: in square millimeters, circular mils[1], and by wire gauge.

In practice, a utility might only use certain discrete sizes of conductors, for example, 50 mm^2, 100 mm^2, or 200 m^2. Several conductors can be run in parallel for each phase to decrease the effective resistance, reduce losses, and increase the total ampacity of the line.

Overhead distribution conductors are often of the type AAAC (all aluminum alloy conductor) or ACSR (aluminum conductor steel reinforced). ACSR cables are made from several aluminum conductors that are wrapped around steel cables. The steel provides tensile strength; the aluminum provides a low-resistance conduction path.

3.2.5 Distribution Line Power Rating

The apparent power rating of a distribution line is expressed in kilovolt–amperes (kVA) or megavolt–amperes (MVA). The rating corresponds to the maximum apparent power that can be supplied by the line at the nominal voltage while remaining within the thermal limits of the line. The rating of a three-phase line S_{rated} is three times the phase capacity

$$S_{rated,total} = 3 \times I_{amp} \times V_\phi \tag{3.17}$$

where I_{amp} is the rated ampacity of the conductor. Recall that V_ϕ is the nominal line-to-neutral RMS voltage of the line. The thermal rating is independent of line length, and so it ignores voltage drop limitations of the line. Obviously, though, voltage drop is an additional consideration in sizing conductors.

Example 3.2 The conductors of a 15 km long, 22 kV distribution line have a cross-sectional area of 100 mm^2. The conductors have a resistivity of 0.274 Ω/km and are spaced so that $\kappa = 1.2$. The ampacity is 313 A. Compute the power rating of the line, the voltage drop, and the losses when operating at the rated ampacity.

(continued)

---

[1] A circular mil is the area of a circle whose diameter is one thousandth of an inch.

Solution The power rating of the line is

$$S_{rated,total} = 3 \times I_{rated} \times V_\phi = 3 \times 313 \times \frac{22\ kV}{\sqrt{3}} = 11.93\ MVA.$$

The impedance of each conductor is

$$Z_{line} = R_{line} + jX_{line} = 0.274 \times (1 + j\kappa) \times 15 = 4.110 + j4.932\ \Omega.$$

When operating at the rated ampacity, the voltage drop is

$$V_{drop} = I_{rated} \times |Z_{line}| = 313 \times |4.110 + j4.932| = 2.009\ kV$$

and the power loss for the line, accounting for all phases:

$$P_{loss,total} = 3 \times I_{rated}^2 \times R_{rated} = 3 \times 313^2 \times 4.101 = 1.208\ MW.$$

We note that the line-neutral voltage at the sending end is $\frac{22\ kV}{\sqrt{3}} = 12.702 kV$, and the line-to-neutral voltage at the village end of the line is $12.702 - 2.009 = 10.692$ kV. The voltage drop therefore exceeds the typical 5% target. It can be shown that the distribution line can only supply 98.8 A per conductor without violating the voltage drop target, effectively reducing the capability of the line. Verifying this result is left to the reader.

Example 3.3 A cluster of villages is to be supplied by a three-phase distribution line that is 25 km in length. The peak load is predicted to be 3.4 MW with a power factor of 0.85 lagging. This is expected to grow by 2% per year for the next 10 years. Assume the receiving-end line-to-line voltage is 22 kV and the voltage drop limit is 10%. The conductors are separated by 2 m. The choice of conductors are shown in Table 3.1. Select the conductor with the smallest cross-sectional area that satisfies the thermal and voltage limits.

Solution The load after 10 years of 2% of growth per year is

$$P_{total} = 3.4 \times (1 + 0.02)^{10} = 4.145\ MW.$$

Applying (3.3) and (3.5) to find the required per-phase apparent power rating

$$S_{rated} = \frac{P_{total}}{3} \times \frac{1}{0.85} = 1.625\ MVA.$$

(continued)

The current magnitude when supplying the rated power at the nominal line-to-neutral voltage is calculated:

$$I_{rated} = \frac{S_{rated}}{|V_\phi|} = \frac{1.625 \text{ MVA}}{12.702 \text{ kV}} = 127.96 \text{ A}.$$

Consulting Table 3.1, the first two conductors are eliminated as the required current exceeds their ampacity. The next largest conductor (C) is checked to see if it satisfies the voltage drop limit.

$$Z_{line} = (0.869 + j0.484) \times 25 = 21.73 + j12.10 \ \Omega$$

$$V_{drop} = I_{rated} \times |Z| = 3.182 \text{ kV}$$

$$|V_s| = |V_r| + V_{drop} = 15.884 \text{ kV}$$

$$V_{drop,\%} = \frac{V_{drop}}{|V_s|} \times 100 = 20.03\%$$

Although this conductor satisfies the thermal constraint, the maximum allowable voltage drop is exceeded. The next conductor is considered and its voltage drop is calculated. This process repeats until a conductor that satisfies both the voltage and thermal constraints is identified. It can be shown that conductor G satisfies these constraints. This conductor is operating far below its rated ampacity of 590 A.

Table 3.1 Conductor characteristics

	Size (mm^2)	Ampacity (A)	Resistance (Ohm/km)	Reactance (0.5 m spacing) (Ohms/km)	Reactance (1 m spacing) (Ohms/km)	Reactance (2 m spacing) (Ohms/km)
A	13.3	95	2.200	0.419	0.471	0.523
B	21.1	125	1.384	0.401	0.453	0.506
C	33.6	165	0.869	0.379	0.432	0.484
D	107.2	325	0.273	0.340	0.392	0.445
E	135.2	415	0.218	0.323	0.375	0.427
F	201.4	525	0.147	0.318	0.367	0.423
G	241.7	590	0.122	0.308	0.360	0.413

Fig. 3.7 Pole-mounted
three-phase distribution
transformer in Kenya
(courtesy of author)

3.2.6   Transformer Ratings

Once the distribution line reaches the community, the voltage is reduced by using a
transformer. Distribution transformers are often pole-mounted, as shown in Fig. 3.7.

The primary side of the transformer must be compatible with the receiving-end
voltage of the distribution line. The secondary line-to-line voltage is usually 400 V
so that the single-phase voltage supplied to the users is approximately 230 V. There
are other configurations, which will be discussed in Sect. 12.7.4.

The transformer must also be capable of supplying the required apparent power
to the users. The rated power of a transformer is based on the power it can provide
without overheating. Overheating is a concern because it reduces the lifespan of
transformers. Transformers can exceed their ratings for a short period of time.
However, the secondary voltage will be reduced during these periods, and the
lifespan will be somewhat shortened as the transformer's insulation degrades more
rapidly at higher temperatures. In the context of rural electrification, 25 kVA,
50 kVA, and 100 kVA transformers serving multiple users are commonly used.

3.2.7   Low-Voltage Connections

The connection from the low-voltage side of the transformer to the user's home
or business is commonly referred to as the "service connection" or "service drop."
There are two common topologies for the low-voltage connections. The "European"

style uses three-phase distribution transformers. In the "American" style, each phase of the distribution line is connected to another conductor, known as a lateral. Single-phase, center-tapped transformers are connected between the lateral and the distribution line neutral, and users are provided a three-wire service of 120/240 V. This is also known as "split phase." The European style is more common in SSA. We will discuss the advantages and disadvantages of each topology in Chap. 12.

3.3 Infrastructure Cost Model

Grid extension projects are expensive. This is the primary reason that more communities do not have access to electricity. There are three costs associated with building and using an electrical system: construction, operation, and financing. We will consider the first two because they are engineering related. The construction cost can be split into five parts:

1. distribution (medium-voltage) line cost
2. low-voltage line cost
3. transformer cost
4. substation cost
5. user premise equipment (UPE) cost (meters, wiring, outlets, switches, etc.)

Many of these, particularly low-voltage line, transformer and UPE costs are also particularly relevant to off-grid systems. In developing countries in general, the cost of materials tends to dominate the cost of labor, design, and management. The cost for each aspect of grid extension can vary widely across and within countries [2, 5]. Most of the materials must be imported, so the costs are sensitive to foreign exchange rates, import duties, and logistics costs. The standard engineering and construction practices of a particular utility to only utilize components of a certain grade or from a preferred supplier can also influence the costs.

Tables 3.2 and 3.3 show the costs, including material, transportation, and installation, for various components as estimated by the Ghanaian and Zambian rural electrification master plans, respectively. Keep in mind that the actual cost of grid extension can vary substantially from those shown.

3.3.1 Distribution Line Cost

The cost of constructing a distribution line c_{line} can be estimated from its length and capacity:

$$c_{line} = \beta_{line} \times l_{line} \times S_{rated,total} \qquad (3.18)$$

where l_{line} is the length of the distribution line, $S_{total,rated}$ is the rating of the line in megavoltamps, and β_{line} is the cost per MVA/km. The value of the coefficient β_{line}

Table 3.2 Costs from Ghana National Electrification Scheme Master Plan (2010)

Item	Description	Cost
33 kV line	Wood pole, 120 mm^2 conductors	US$26,222/km
11 kV line	Wood pole, 120 mm^2 conductors	US$24,690/km
200 kVA transformer	33/0.4 kV w/accessories	US$16,253
100 kVA transformer	33/0.4 kV w/accessories	US$13,815
50 kVA transformer	33/0.4 kV w/accessories	US$11,851
200 kVA transformer	11/0.4 kV w/accessories	US$13,344
100 kVA transformer	11/0.4 kV w/accessories	US$11,529
50 kVA transformer	11/0.4 kV w/accessories	US$10,243
Low-voltage line	3-phase, 4-wire, wood pole	US$16,597/km
Low-voltage line	1-phase, 3-wire, wood pole	US$14,869/km
Low-voltage line	1-phase, 2-wire, wood pole	US$12,958/km
3-phase user connection	Meter, 25 mm^2 conductor	US$531
1-phase user connection	Meter, 16 mm^2 conductor	US$275

Table 3.3 Costs from Zambia Rural Electrification Authority Master Plan (2005) [5]

Item	Description	Cost
33 kV line	pole, 100 mm^2 conductor w/accessories	US$36,000/km
100 kVA transformer	33/0.4 kV	US$13,700
2.5 MVA substation	–	US$600,000
5 MVA substation	–	US$800,000
10 MVA substation	–	US$1,000,000
15 MVA substation	–	US$1,300,000
33 kV bay	–	US$99,300

varies widely. In general, it ranges between US$1200 and US$6600 per MVA km in developing countries. Some of this variation is due to the design of the distribution line poles, which comprise up to 40% of the total material cost. Pole costs can be reduced by using shorter poles or fewer poles (by increasing the span between the poles). Higher-voltage lines also tend to have lower β_{line} values.

Example 3.4 A 33 kV distribution line is 10 km long. Each conductor has a cross-sectional area of 120 mm^2 with ampacity of 350 A. Compute the rating of the line and the total cost using $\beta_{line} = $ US$1,300/MVAkm.

Solution The rating of the line is found using (3.17):

$$S_{line} = 3 \times 350 \times \frac{33,000}{\sqrt{3}} = 20.005 \text{ MVA}.$$

(continued)

The total cost of the line is found from (3.18):

$$c_{line} = 1300 \times 10 \times 20.005 = US\$260,067.43.$$

This is for the line only. There are still other costs that should also be considered.

If the capacity of the line is not specified, a more general model can be used:

$$c_{line} = \beta_{line,km} \times l \tag{3.19}$$

where $\beta_{line,km}$ is cost per kilometer, which ranges from US\$3,000 to over US\$30,000/km. The lowest cost is found in India. The low cost has been attributed to the use of domestically manufactured components and short, prestressed concrete poles. In SSA, a cost of US\$20,000/km is more typical. Although these costs are high, they are generally less expensive than in the United States, where labor and land costs can be substantially higher than in developing countries.

3.3.2 Low-Voltage Line Cost

Low-voltage lines are needed to connect the transformer secondary to the user. The lines are insulated and can be comprised of two, three, or four wires, depending on the service provided to the user. The cost of the low-voltage line can be modeled as

$$c_{LV} = \beta_{LV} \times l_{LV}. \tag{3.20}$$

The cost coefficient β_{LV} typically ranges from US\$10,000 to US\$18,000/km. The total length of the low-voltage lines depends on the number of users and their proximity to the transformer and each other.

3.3.3 Transformer Cost

The cost of a transformer is largely driven by its power rating. A simple cost model is

$$c_{xmfr} = \beta_{xfmr} \times S_{rated,xmfr} \tag{3.21}$$

where β_{xfmr} is the cost per kilovolt–ampere and $S_{rated,xmfr}$ is the rating of the transformer in kilovolt–amperes. Typical values of β_{xfmr} range from US\$100 to US\$500/kVA. The cost per kilovolt–ampere tends to decrease as the transformer size increases and as the primary voltage decreases.

Example 3.5 A community with 203 households will be served through grid extension. The 33 kV distribution line serving the community will be 12 km long. The peak apparent power of the community is estimated to be 165 kVA. Determine the minimum quantity of 50 kVA transformers needed at the community, assuming the power factor is 0.85. Compute the total transformer cost using the values found in Table 3.2.

Solution The number of 50 kVA transformers required is $165/50 = 3.3$, which we round up to 4. The total transformer cost is

$$c_{\text{xfmr}} = 4 \times 11{,}851 = \text{US\$47{,}404.00}.$$

3.3.4 Substation Cost

The sending end of the distribution line must be connected to the existing grid. In some circumstances, it can be connected directly to an existing distribution line of the same voltage. In many cases, however, a substation must be added or an existing substation modified to accommodate the new line. The substation costs can include switching components and protective equipment and also the cost of land and civil works. The cost of a new substation is related to the rating of the line or lines it serves. However, some of the cost can be considered fixed for a reasonable range of ratings. Therefore, a cost model of the form

$$c_{\text{sub}} = \alpha_{\text{sub}} + \beta_{\text{sub}} S_{\text{sub}} \tag{3.22}$$

is appropriate, where S_{sub} is the power rating of the substation. We will assume the rating of the substation is equal to that of the line. Using the Zambian costs in Table 3.3, the coefficients α_{sub} and β_{sub} are US\$490,000 and US\$53,600/MVA, respectively. If the substation only requires modification, then the cost is considerably reduced. For example, a 33 kV substation bay is US\$90,000 in the Zambian rural electrification master plan.

3.3.5 User Premise Equipment Cost

User premise equipment includes meters, wiring, and other equipment such as circuit breakers and outlets. These costs are also incurred for mini-grid users. A meter is needed to measure energy consumption so that users can be billed properly. Although increasingly rare, some users in developing countries pay a fixed fee—their consumption is not tied to usage. While this saves the cost of installing, maintaining, and reading meters, it tends to encourage waste and excess consumption.

The cost of a meter depends upon its functionality. Premiums are paid for tamper-proofing, prepay functionality, and meters with smart capabilities. The cost of a meter generally does not vary appreciably with the power rating—at least at levels typically used by rural consumers.

UPE costs also include the cost of wiring the premises of new users. Protection devices such as circuit breakers or fuses must be installed, as well as electrical sockets (outlets), lighting receptacles, and the wiring within the house.

The total user premise equipment cost is proportional to the total number of users served N_{user} and is modeled as

$$c_{UPE} = \beta_{UPE} \times N_{user}. \tag{3.23}$$

Again, the cost coefficient can widely vary, but generally ranges from US$90 to US$420 per user. In some situations, the utility provides a panel that includes a meter, a socket, and perhaps a light bulb. The panel is designed so that it can be easily attached to a wall on the interior of the user's premise, reducing or avoiding the need for additional wiring.

3.3.6 Cost Per Connection

The total infrastructure cost of a grid extension project is

$$c_{grid} = c_{line} + c_{LV} + c_{xmfr} + c_{sub} + c_{UPE}. \tag{3.24}$$

We are also interested in the cost per connection. The cost per connection is found by

$$c_{con} = \frac{c_{grid}}{N_{user}}. \tag{3.25}$$

The cost per connection is useful in comparing different grid extension projects. The cost per connection can vary widely but is typically within the bounds of US$500 to US$5000 [3].

Example 3.6 Compute the total and per connection cost of connecting the community in Example 3.5 by grid extension. Assume that 2.5 km of single-phase low-voltage line is needed to connect the households in the community. A new 33 kV substation bay, costing US$90,000, is required. Use Table 3.2 for all other costs.

(continued)

Solution Applying (3.24), the total cost is

$$c_{\text{grid}} = \text{US\$26,222} \times 12 + \text{US\$12,958} \times 2.5 + \text{US\$47,404} + \text{US\$90,000}$$

$$+ \text{US\$275} \times 203$$

$$= \text{US\$540,288.00}.$$

The cost per connection is found using (3.25) to be US\$2661.52.

Fig. 3.8 The share of the upfront grid extension costs are increasingly dominated by the distribution line as its length increases

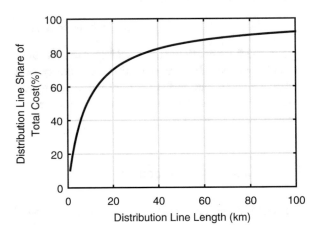

The total cost of a grid extension project and cost per connection are especially sensitive to the length of the distribution line, which can easily be the single greatest cost. Figure 3.8 shows the cost per connection of Example 3.5 as a function of line length. Not reflected in Fig. 3.8 is that as line length increases, larger, more expensive conductor and/or voltage support equipment are required due to the limited voltage reach. This further increases the cost of long distribution lines. The loss along a distribution line also increases with length. The cost associated with losses are not reflected in (3.25).

In terms of infrastructure costs, it is more favorable to connect communities that require short distribution lines. In many countries, grid extension is not considered to be economically viable for communities more than 20 km from the grid.

3.3.7 Lifetime Cost of Grid Extension

The total cost of grid extension must also include maintenance and replacement costs. The total cost over the lifetime of the equipment is

$$c_{\text{grid,life}} = c_{\text{grid}} + t_{\text{grid}} \times m_{\text{grid}} \times c_{\text{grid}} \qquad (3.26)$$

where t_{grid} is the lifespan of the grid extension, in years, and m_{grid} is the annual maintenance cost of the extension, often expressed as a percent of the initial capital. This calculation ignores the time value of money, that is, a dollar today is worth more than a dollar tomorrow or a year from now, due to inflation and other factors.

To account for the time value of money, consider an annual interest rate i (which we will also refer to as the *discount rate*). The value of a sum of money V_0 after 1 year is

$$V_1 = V_0(1 + i) \qquad (3.27)$$

and after Y years

$$V_Y = V_0(1 + i)^Y. \qquad (3.28)$$

We see that $V_Y > V_0$ as long as the interest rate is greater than zero. Equation (3.28) can be rearranged to find the *present value* or *present cost* of a future an income or expense.

As an example, if US\$1000 must be paid for maintenance in 5 years and assuming a discount rate of 4 %, then solving for V_0 in (3.28) yields a present cost of US\$821.93. In other words, US\$821.93 today is worth US\$1000 in 5 years.

For a given stream of fixed payments (an annuity) or expenses of V_{annual} over Y years, the present value is

$$V_0 = V_{\text{annual}} \frac{(1+i)^Y - 1}{i(1+i)^Y}. \qquad (3.29)$$

Example 3.7 The present cost of a grid extension project is US\$100,000. Compute the fixed annual payment equivalent of this cost over a 20-year period at a discount rate of 5%.

Solution The present cost is known and we are asked to find the fixed annual payment. We must rearrange (3.29) to solve for V_{annual}:

$$V_{\text{annual}} = V_0 \frac{i(1+i)^Y}{(1+i)^Y - 1} = 100{,}000 \frac{0.05(1+0.05)^{20}}{(1+0.05)^{20} - 1} = \text{US\$8024.26}$$

We note that the total sum paid over 20 years is $20 \times 8024.26 = $ US\$160,485.20.

Since the maintenance costs of grid extension are modeled as a fixed yearly cost over the life of grid extension, the present cost is

$$c_{grid,0} = c_{grid} + m_{grid} \times c_{grid} \frac{(1+i)^Y - 1}{i(1+i)^Y}. \tag{3.30}$$

From this equation, we see that the maintenance costs, in terms of present dollars, decreases over time.

We can convert the present infrastructure cost of a grid extension project $c_{grid,0}$ into Y fixed annual payments of

$$c_{grid,annual} = \frac{c_{grid,0}}{\frac{(1+i)^Y - 1}{i(1+i)^Y}} \tag{3.31}$$

expressed in present dollars.

Example 3.8 Compute the present cost, fixed annual cost, and fixed annual cost per connection of the grid extension in Example 3.6. Assume the lifespan of the equipment is 30 years, the interest rate is 3%, and the annual cost of maintenance is 1% of the capital cost.

Solution Applying (3.30) yields a present cost of

$$c_{grid,0} = US\$540,288.00 + 0.01 \times US\$540,288.00 \frac{(1+0.03)^{30} - 1}{0.03(1+0.03)^{30}}$$

$$= US\$646,186.83.$$

The fixed annual cost is found using (3.31)

$$c_{grid,annual} = \frac{US\$646,186.83}{\frac{(1+0.03)^{30} - 1}{0.03(1+0.03)^{30}}} = US\$32,967.97$$

which is equivalent to US\$32,967.97/203 = US\$162.40 per connection per year. In other words, US\$162.40 must be collected from each user each year for 30 years to pay for grid extension.

3.3.8 Cost of Energy in Sub-Saharan Africa

There, of course, is a cost associated with producing the energy that is consumed by the users. Unfortunately, the cost in Sub-Saharan Africa tends to be higher than the world average. The production costs in Africa as a whole have been estimated to be US\$0.18/kWh [11]. However, this varies widely, depending on each country's generation resource mix and access to energy sources. In Southern Africa, the costs are estimated to average US\$0.13/kWh and in northern Africa US\$0.24/kWh [7].

There are several reasons for higher-energy costs in Africa. For example, financing costs can be higher due to the perception of greater risk of investments in developing countries due to political and regulatory instability. Improper maintenance and operation can reduce the total production of a power plant, causing the annual cost to be spread across a smaller base of production. The majority of electricity generated in Africa is from thermal fossil-fuel power plants. Thermal-based generators are more efficient at larger capacities. However, power plants in Africa tend to be of a smaller scale than those in other parts of the world, which increases the relative fuel costs. Aging infrastructure increases system losses, which in turn increases the cost to serve each unit of energy. Despite all this, the cost of energy from a large-scale, grid-connected power plant is in almost all cases lower than that of energy supplied by off-grid systems.

The cost of energy is often expressed as the *levelized cost of energy* (LCOE) or *simplified levelized cost of energy* (sLCOE), in dollars per kilowatthour. We will discuss LCOE in detail in Chap. 12; for now, it suffices to note that the LCOE and sLCOE are the price that must be charged for the power plant to financially break even, including the capital costs, fuel costs, maintenance, etc., as well as the time value of money. We will assume that the cost of energy is expressed in LCOE. Note that LCOE can also be calculated for off-grid systems.

3.4 Electrification Cost by Grid Extension

We are now prepared to compute the cost of serving a community through grid extension. The total annual cost consists of two parts: the energy cost, found by multiplying the per kilowatthour energy costs by the annual energy consumption E_{annual}, inclusive of losses, and the annual lifetime cost of the physical grid extension project

$$c_{total,annual} = LCOE \times E_{annual} + c_{grid,annual}. \qquad (3.32)$$

The resulting cost can be used to compare electrification by grid extension with competing electrification options.

Example 3.9 Compute the cost of electrifying the village in Example 3.5, assuming the annual consumption is 365 kWh per user per year and the losses are 10%. Assume the LCOE is US$0.15/kWh.

Solution First compute the annual energy consumption:

$$E_{annual} = 365 \times 203 \times 1.10 = 81,504.50 \text{ kWh}.$$

(continued)

Recall from Example 3.8 that the annual infrastructure cost for this village is US$32,967.97. The annual cost is found using (3.32):

$$c_{total,annual} = US\$0.15/kWh \times 81,504.50\,kWh + US\$32,967.97$$

$$= US\$45,193.65$$

which is equivalent to US$45,193.65/203 = US$222.63 per connection.

3.5 Comparing Electrification Options

There are options to providing access to electricity other than grid extension, for example, off-grid diesel generator sets or solar-powered systems. On the one hand, grid extension is desirable because the LCOE from the grid-connected power plants is usually far lower than for off-grid solutions; however, off-grid solutions do not require the heavy investment in distribution infrastructure. Which is the preferred option? The answer depends on a variety of community- and country-specific factors, some of which can only be estimated. However, we can at least gain insight into this question using the framework we developed.

Two critical factors in determining whether or not a grid extension or off-grid approach is most economical are the total energy consumed and the distance of the community from the grid. As the distance increases, off-grid solutions become cost-effective; as the load increases, grid extension becomes cost-effective.

To illustrate these relationships, consider Fig. 3.9. Here, the annual cost per connection, inclusive of infrastructure and energy costs, is plotted versus the length of the distribution line. The cost associated with three levels of annual consumption

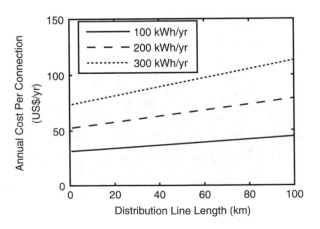

Fig. 3.9 Example annual cost per connection, including energy and infrastructure costs, by grid extension of a hypothetical community, assuming three different levels of consumption

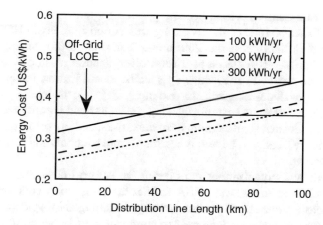

Fig. 3.10 Example total cost per kilowatthour of a hypothetical community, assuming three different levels of consumption. An example LCOE for an off-grid system is shown as the horizontal line

is shown. The grid extension costs increase with the length of the distribution line and annual energy consumption.

The slope of each line is different and increases with consumption, showing that higher-energy consumption requires more expensive, higher-capacity lines and transformers. The vertical axis intercepts are the annual costs for under-the-grid users. It includes the energy costs plus costs that are independent of the distribution line's length, such as metering and wiring.

Another family of curves of interest is the cost per unit energy, as shown in Fig. 3.10. This curve reflects that, although the cost of connecting a user with high consumption is greater than a user with low consumption, the cost per kilowatthour can be less. This curve is used to compare competing electrification approaches. Also shown is the LCOE for a hypothetical off-grid system. The intersection of the horizontal LCOE line with the grid extension cost curve gives an indication of the line length the off-grid system becomes a cost-effective solution. For example, if the consumption is 100 kWh/household/year, then the off-grid system becomes cost-effective if the required distribution line length exceeds 35 km. This is an approximation only as the off-grid solution might itself have localized infrastructure costs that are not included in the LCOE.

3.6 Rural Electrification Programs

Grid extension is often coordinated by the national utility and a Rural Electrification Authority (REA). REAs are government agencies that promote, coordinate, manage funds, train, build capacity, or implement rural electrification.

REAs in Sub-Saharan Africa became popular beginning in the late 1990s to mid-2000s. REAs were established during this period in Kenya (1997), Senegal (1998), Zambia (2003), Nigeria (2005), and Tanzania (2005), among others. In many countries, the REA develops a rural electrification master plan (REMPs). REMPs outline strategies for achieving specific electrification targets that align with broader development goals, for example, 51% electrification by the year 2030. REMPs estimate the cost of achieving the targets and identify technological, regulatory, and financial and economic barriers, risks, and opportunities. The REA identifies priority locations and user types, for example, schools and health facilities, for electrification.

Companies and organizations considering an off-grid electrification program should consult with the relevant REA so that the program is coordinated in the broader electrification efforts in that country. Installing mini-grids or deploying solar home systems and solar lanterns to a community can be wasteful if the grid is extended a short time later.

3.7 Other Considerations

The decision to extend the grid to serve additional communities is largely based on economic factors. However, there are other aspects that should be considered with this mode of electrification.

3.7.1 Prioritizing Electricity Access

With limited budgets, prioritizing which areas to electrify is especially important. Although each country has a different process, higher priority is usually given to communities that:

- have high potential for electricity consumption;
- are near (typically within 20 km) to the existing distribution network;
- are densely populated;
- have industrial, commercial, or tourism potential;
- have medical, educational, or other social institutions;
- have political or cultural significance.

An economic analysis and alignment with other development goals such as access to clean water further guide the prioritization.

3.7.2 Grid Extension to Remote Rural Areas

In the context of developing countries, grid extension to remote rural areas faces the following challenges in particular:

- inadequate supporting infrastructure such as roads makes construction and maintenance more difficult;
- distribution lines must span long distances and rugged terrain, increasing cost and losses;
- low load density and inability/unwillingness to pay for electricity make it difficult to recover infrastructure investment;
- the costs associated with meter reading is high, and so prepaid billing is favored.

The cost of extending the grid to remote communities can be so high that they are never recovered while being affordable to the user. The annual fixed cost in Example 3.9 was US$222.63 per connection. The consumption averaged 365 kWh per connection per year. For this connection to qualify as Tier 5 affordability, the cost cannot exceed 5% of the household income. The average annual household income must exceed US$4452.60. This income is beyond what many rural households can be reasonably expected to earn each year. For this access to satisfy the affordability attribute, the connection must be subsidized.

3.7.3 Cost-Reflective Pricing

In many developing countries, electricity is heavily subsidized by the government. This means that the rate that users pay per unit of energy is lower than it costs the utility to supply it, even in urban areas. The subsidy is a transfer of money from the government to the utility so that the utility is able to operate.

In many countries, subsidies are available to grid-connected users, but not to those served by small-scale off-grid systems. This discourages investment in off-grid solutions. Many mini-grid developers are forced to charge higher rates than offered by the grid in part because of the subsidies available for grid-connected electricity.

The reality in many developing countries is that the utilities or government loses money on many users, particularly those in rural areas. If the benefits of access to electricity are not properly accounted for, for example, the educational, health, and economic benefits associated with electrification, then the utility or government might be reluctant to increase electrification rates.

3.8 Summary

The de facto approach to increasing electricity access is to extend or enhance the existing grid. Connecting consumers in urban and peri-urban areas is more economical and expeditious than those in rural areas far from the existing grid. A grid extension project includes the design of the substation, medium-voltage distribution line, transformer, and user connections through low-voltage lines. The distribution line must be designed to supply the required power at an acceptable voltage while not exceeding the thermal limits of the line.

The economics of grid extension are largely driven by the length of the distribution line, which often exceeds US$20,000/km. The per connection cost is usually at least several hundred US dollars, although the user might pay a lower, subsidized amount. It is important to consider the lifetime cost of a grid extension project, not just the initial construction costs. The energy costs must also be considered. In general, off-grid systems are economically favorable to grid extension for communities far from the existing grid and whose consumption is low. Practical considerations in grid extension include determining which areas to prioritize for access and whether or not to subsidize the cost.

Problems

3.1 A three-phase distribution line whose line-to-line sending-end voltage is 22 kV supplies 2.25 MW at a power factor of 0.85. Compute the magnitude of the current through the a-phase conductor.

3.2 A distribution line has an impedance of $Z = 6.73 + j4.87\ \Omega$. The receiving-end line-to-line voltage magnitude is 30 kV. The load is 1.9 MVA at a power factor of 0.85 lagging. Compute the a-phase conductor current, the magnitude of the sending-end line-to-line voltage, and the voltage drop in percent.

3.3 Verify that 98.8 A is the maximum conductor current that does not exceed the voltage drop limit of 5% in Example 3.2. Compute the current that results in a 10% drop.

3.4 Determine the infrastructure cost c_{grid} and cost per connection of the five villages whose characteristics are shown in Table 3.4. Use the cost coefficients provided in Table 3.5.

3.5 For the five villages described in the previous problem, compute the fixed annual infrastructure cost per connection, assuming a 40-year lifespan, a discount rate of 3%, and an annual maintenance cost of 1%.

3.6 What is the smallest conductor from Table 3.1 that can be used in a three-phase distribution line to supply the village of Dandandu without exceeding a voltage drop of 7.5% or the conductor's ampacity? Assume the line-to-line receiving-end voltage is 33 kV. Assume the conductor spacing is 1 m.

Table 3.4 Village characteristics

	Ababju	Bersoloi	Changola	Dandandu	Ekong
No. of households	500	1,000	5,000	8,000	10,000
Distance to grid (km)	20	8	12	25	10
Distribution line rating (MVA)	1	1	2	3	6
Low-voltage line length (km)	7	8.5	40	60	165
No. of 50 kVA transformers	23	28	45	69	130

Table 3.5 Cost coefficients

Parameter	Value
Distribution line, β_{line} (US\$/MVA)	2000
Low-voltage β_{LV} (US\$/km)	10,500
Transformer β_{xfmr} (US\$/kVA)	180
Substation β_{sub} (US\$/MVA)	40,000
Substation α_{sub} (US\$)	180,000
User β_{UPE} (US\$/User)	120

3.7 The fixed annual grid extension cost for a community of 400 households is US\$20,750. It is estimated that each household will consume 50 kWh of energy per year (primarily for lighting). The LCOE from the grid is US\$0.15/kWh and losses are 11%. Compute the annual total cost per household per year $c_{total,annual}/N_{user}$.

3.8 Consider the community in the previous problem. As an alternative to grid extension, solar home systems (SHS) are considered. Each SHS meets the electricity demands of a single household and costs US\$275. The system requires replacement every 5 years. Assume the cost of the SHS does not change over time. Which solution—grid extension or solar home systems—is more economical?

3.9 A diesel-powered mini-grid is proposed as an alternative to a grid extension project. The mini-grid will serve 100 households and requires 1 km of low-voltage line and UPE for each household. The LCOE is US\$0.50/kWh. Each household will consume 300 kWh per year. The losses are 7%. Compute the annual total (infrastructure plus energy) cost per user using the coefficients in Table 3.5. Assume the discount rate is 4% and the lifespan is 25 years.

3.10 As an alternative to the mini-grid in the previous problem, grid extension is proposed. The LCOE from the grid is US\$0.18/kWh. The losses are 12%. The annual infrastructure cost per user is modeled as

$$c_{total,annual} = 56 + 3.30 \times l$$

where l is the distance of the community to the grid in kilometers. How far from the grid must the community be for the mini-grid to be economically justifiable?

References

1. Berge, A., Vittal, V.: Power Systems Analysis, 2nd edn. Prentice-Hall (2000)
2. Deichmann, U., Meisner, C., Murray, S., Wheeler, D.: The economics of renewable energy expansion in rural Sub-Saharan Africa. Energy Policy **39**(1), 215–227 (2011). doi:https://doi.org/10.1016/j.enpol.2010.09.034. URL http://www.sciencedirect.com/science/article/pii/S0301421510007202
3. Gömez, M.F., Silveira, S.: Rural electrification of the Brazilian Amazon - Achievements and lessons. Energy Policy **38**(10), 6251–6260 (2010). doi:https://doi.org/10.1016/j.enpol.2010.06.013. URL http://www.sciencedirect.com/science/article/pii/S0301421510004763

4. Gönen, T.: Electrical Power Transmission System Engineering, 3rd edn. CRC Press (2014)
5. Government of the Republic of Zambia: Rural electrification master plan for Zambia 2008–2030 (2009)
6. Lee, K., Brewer, E., Christiano, C., Meyo, F., Miguel, E., Podolsky, M., Rosa, J., Wolfram, C.: Electrification for Under Grid households in rural Kenya. Development Engineering 1, 26–35 (2016). doi:http://dx.doi.org/10.1016/j.deveng.2015.12.001
7. Levin, T., Thomas, V.M.: Can developing countries leapfrog the centralized electrification paradigm? Energy Sustain. Dev. 31, 97–107 (2016). doi:https://doi.org/10.1016/j.esd.2015.12.005. URL http://www.sciencedirect.com/science/article/pii/S0973082615301599
8. Mahapatra, S., Dasappa, S.: Rural electrification: Optimising the choice between decentralised renewable energy sources and grid extension. Energy Sustain. Dev. 16(2), 146–154 (2012). doi:https://doi.org/10.1016/j.esd.2012.01.006. URL http://www.sciencedirect.com/science/article/pii/S0973082612000087
9. Sustainable Energy for All: Global tracking framework (2015). URL "http://www.seforall.org/global-tracking-framework"
10. Sustainable Energy for All: Energizing finance. Tech. rep., Sustainable Energy for All (2017). URL https://www.seforall.org/sites/default/files/2017_SEforALL_FR4_PolicyPaper.pdf
11. The African Development Bank Group: The high cost of electricity generation in Africa (2013). URL https://www.afdb.org/en/blogs/afdb-championing-inclusive-growth-across-africa/post/the-high-cost-of-electricity-generation-in-africa-11496/

Chapter 4
Off-Grid System Architectures

4.1 Introduction

As discussed in Chap. 3, the most obvious and direct way to provide electricity to an unserved area is through a connection to the national grid. This may be impractical though, if the community in need of service is too far from the existing grid and/or the load is too little to justify the infrastructure investment. In cases such as these, off-grid systems should be considered. This chapter is concerned with mini-grid systems for off-grid electrification. Solar home systems and solar lanterns are discussed in Chap. 13. By some estimates, as many as 140 million people in Africa alone could be served by mini-grids by 2040 [1]. Scaling to this level will require between 100,000 and 200,000 systems.

The basic function of a mini-grid is to produce and distribute electricity to users in a limited geographic area. The mini-grid is designed to meet the desired electricity access tier. Recall that electricity access tiers are a way of describing the overall quality of the electricity service, as discussed in Sect. 2.7. At a minimum, a mini-grid contains at least one energy source and one load. More complex systems may include multiple sources, energy storage, converters, and controllers and are able to supply AC and DC loads. Other components such as protection devices and monitoring equipment may be present but are not discussed here for the sake of brevity.

Mini-grids can be separated into three systems: energy production, distribution, and end use, as seen in Fig. 4.1. Each system is associated with a separate set of components. Most of the energy production components are located inside a "power house." This may be a dedicated stand-alone structure, or a room inside a building such as school or hospital. Repurposed shipping containers have also been used. The distribution system components connect the energy production system to the end users by way of overhead or underground lines. Larger-capacity mini-grids might

© Springer International Publishing AG, part of Springer Nature 2018
H. Louie, *Off-Grid Electrical Systems in Developing Countries*,
https://doi.org/10.1007/978-3-319-91890-7_4

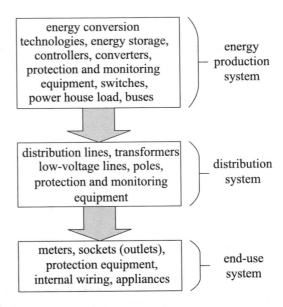

Fig. 4.1 The three systems of a mini-grid

use medium-voltage distribution lines and transformers; however, most mini-grids use low-voltage distribution. Much of this book, including this chapter, is concerned with the energy production system.

The architecture of a mini-grid refers to how its various components are connected together to form a complete system. There are a variety different mini-grid architectures. For the moment, we are not concerned with the technical details of how the individual components function. Rather, we are interested in how they function at a macro-level and how they can be used as building blocks to form a mini-grid.

4.2 Terminology

There is yet to be a universally accepted set of terminology when describing off-grid systems. For example, there is no precise boundary between mini-grids and micro-grids, and any exact definition would be somewhat arbitrary.

However, the use of consistent terminology reduces confusion. The following terms are commonly used in academic literature [2].

- Centralized system (national grid): a large power system that is often a state-owned, vertically integrated and regulated monopoly with centralized control and coordination of generation, transmission, and distribution. Such systems typically serve a large geographic area.
- Decentralized system: composed of autonomous units where generation and distribution have no centrally coordinated interaction with other units.
- Off-grid: an electrical system which is detached from the national grid.

Fig. 4.2 A small-scale, decentralized, hybrid mini-grid using wind and solar power in Muhuru Bay, Kenya (courtesy E. Patten)

- Small-scale: a system whose power production rating does not exceed 5 MW (a mini-grid of this size is actually quite large—it could likely serve several thousand rural households).
- Hybrid: an off-grid system using two or more types of energy conversion technologies to produce electricity.
- Conventional generation: generators that run solely on fossil fuels (usually diesel or gasoline).
- Stand-alone: a system that serves a single user such as a solar home system or solar lantern, typically rated at less than 1 kW.
- Mini-grid: an off-grid system that serves multiple users, typically rated at less than 100 kW and often less than 10 kW.

The mini-grids used in rural electrification tend to be off-grid and small-scale. Many are decentralized. Hybrid mini-grids are common, but not universal. Many of the architectures discussed also apply to stand-alone systems. An example of a small-scale, decentralized, hybrid mini-grid using both wind and solar power is shown in Fig. 4.2.

4.3 Mini-Grid Building Blocks

The basic building blocks of a mini-grid are:

- Energy conversion technology
- Load
- Energy storage
- Converter
- Controller

A basic description of these components is provided in the following sections.

Table 4.1 Common energy conversion technologies for off-grid systems

Energy source	Conversion technology
Biomass	Internal combustion engine, steam turbine
Hydro	Hydroturbine
Natural Gas/LPG	Internal combustion engine
Petroleum/Diesel	Internal combustion engine
Solar	Photovoltaic (PV) module
Wind	Wind turbine

4.3.1 Energy Conversion Technologies

Electric generators of all sorts convert other forms of energy into electricity. Diesel generators, wind turbines, and solar panels all do this. We broadly and generically refer to these types of devices as *energy conversion technologies*. For example, a wind energy conversion system converts the kinetic energy in a mass of moving air to electrical energy, and a diesel generator converts the chemical energy in hydrocarbons to electrical energy. The energy sources and associated conversion technologies typically encountered in mini-grids are listed in Table 4.1. There are other sources and conversion technologies such as geothermal and solar–thermal that might be practical for large-scale electricity generation, but not for small-scale mini-grids. Other technologies such as tidal and wave are not mature or require very particular local conditions to be viable.

Other components such as gear boxes, generators, and controllers are needed to produce electricity. To highlight this, we use the terms "gen sets," "wind energy conversion systems" (WECS), and "micro hydro power" (MHP) instead of "internal combustion engines," "wind turbines," and "hydroturbines." The selection of the energy conversion technology—or technologies in a hybrid system—largely depends on the availability of the energy source, its capital and operation costs, and the reliability and availability requirements of the load.

4.3.2 Loads

The term "load" has multiple meanings in electrical engineering. It commonly refers to the power or energy required by a component or the physical component itself. The context often makes it obvious which meaning is being used. The following description uses the latter meaning.

A load is an end-use device that consumes electrical energy. Lights, mobile phone chargers, and appliances such as refrigerators, computers, televisions, pumps, heaters, and motors are examples of loads. Components whose primary function is to distribute, store, or convert electricity are generally not considered loads. A rectifier, which converts AC to DC, would be an example of such a component. Power absorbed by components such as these is not consumed by the end user, and so it is separately categorized as a loss within the system.

Loads can be classified as being AC or DC. Some mini-grid architectures can only supply AC loads, some only DC, and some can supply both. AC loads must be supplied by an AC source whose voltage magnitude and frequency are within certain ranges to operate properly; similarly, DC loads must be supplied by a DC source of appropriate magnitude and polarity for proper function.

4.3.3 Energy Storage

Mini-grids incorporate energy storage to improve their availability and reliability. Energy storage allows the load to exceed the generation for some period of time. Energy storage is optional in some mini-grid architectures but is especially important in those powered by solar and wind energy. The most common form of energy storage in mini-grid systems is an electrochemical battery. Most often, lead–acid batteries are used, but lithium–ion batteries are now becoming popular.

4.3.4 Controllers

Controllers affect how components of mini-grids operate and interact with each other. Controllers are used to:

- manage the charging and discharging of batteries by regulating their terminal voltage so that the battery lifespan is prolonged;
- regulate the frequency of the system voltage and mechanical revolution speed of the generators;
- regulate the magnitude of the AC voltage;
- maximize the power production from photovoltaic arrays and wind turbines;
- synchronize and manage how power is allocated from generators and inverters connected in parallel;
- coordinate the interaction of different components.

Controllers increase the capital cost and complexity of a mini-grid. These initial the costs are often justified, particularly as system size increases. Controllers increase the lifespan of components and quality of the electricity service. Off-grid system control is further discussed in Chap. 10.

4.3.5 Converters

Most modern controllers use electronic devices to function. Often, they use a converter to achieve the desired control action. Examples of converters include rectifiers, which convert AC to DC; inverters, which convert DC to AC; and DC–

DC converters, which change the magnitude of a DC voltage source. These devices are power electronic-based converters. The converters used in off-grid systems are described in detail in Chap. 9. Transformers are sometimes used in mini-grids serving large geographic regions where voltage drop and losses are a concern.

4.4   System Coupling

The various mini-grid building blocks can be connected in several ways. One common way of categorizing mini-grid architectures is based on how the components are coupled. There are three types of coupling: AC, DC, and AC–DC. The type of coupling depends on how the *energy sources* are connected, independent of whether the load is AC or DC.

4.4.1   AC Coupling

The central feature of an AC-coupled mini-grid is the AC bus, as shown in Fig. 4.3. Note that the schematics shown in this chapter are intended to be illustrative only— it is possible, for example, for an AC-coupled system to have somewhat different components. The schematics show the high-level connection of the components, with the arrows indicating the possible flow of power. For clarity, the distribution and end-use systems are not shown. Do not confuse the schematics with circuit models, even though at times the same symbols are used in both.

In electrical terms, a *bus* is simply a node in the system where various components are connected. National grids can have thousands of buses; mini-grids often have just one. The AC Bus is often just a copper bar inside the circuit breaker box with several cables and switches or circuit breakers attached.

All the components connected to the AC bus are in parallel, and so they operate at the same voltage frequency and magnitude. This means that the voltage output by the generators must be synchronized. Synchronization is discussed in Chap. 10.

Fig. 4.3 Example of an
AC-coupled mini-grid

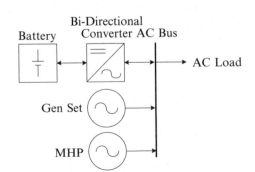

DC components cannot be used unless they are connected to the AC bus through a rectifier or an inverter. The voltage frequency and magnitude at the AC bus should be approximately constant. Certain sensitive loads cannot tolerate deviations beyond a few percent without damage or malfunction. Other loads such as heaters and incandescent lights are more robust and can tolerate variation in the voltage and frequency.

Control of the AC bus voltage frequency and magnitude is an important aspect and consideration of AC-coupled systems. Normally, one component is controlled so that it resembles a voltage source. This component is said to be "forming" the AC bus. The other sources must be able to synchronize to the AC bus voltage and are controlled as current sources to inject power into the bus.

For reasons discussed in the next chapter, only energy conversion technologies capable of adjusting their power output on demand and that have a voltage control system can be used to form the AC bus. This functionality is usually found in conventional- and biomass-fueled internal combustion engines and certain MHP systems. Inverters are also capable of forming the AC bus. One reason why WECSs and PV modules cannot be used to form the AC bus is that they are only capable of producing power when there is sufficient wind speed or sunlight. However, they can be integrated into the system as long as some other source forms the AC bus voltage. AC-coupled systems can be easily expanded by connecting additional load and generation to the AC bus.

4.4.2 DC Coupling

In DC-coupled systems, the energy sources are connected in parallel at a DC bus. In most off-grid systems, there is a single DC bus, as shown in Fig. 4.4. DC-coupled systems almost always include a battery. AC components must include an inverter or rectifier to be integrated into a DC-coupled system. The battery sets the DC bus voltage. Although the problem with forming and maintaining the frequency of the AC bus is eliminated (unless multiple inverters are connected in parallel), the battery must be protected from being over- or undercharged.

For this reason, a system with a DC bus should have charge controllers or diversion loads and diversion load controllers. A charge controller limits the current supplied by a source; a diversion load provides a parallel path for the current, reducing the current into the battery. The DC-coupled architecture is used in smaller-capacity mini-grids and in solar lanterns and solar home systems. Expanding a DC-coupled grid to serve additional AC load is problematic because it might require a larger-capacity inverter.

4.4.3 AC–DC Coupling

An AC–DC-coupled architecture includes at least one AC-coupled source and at least one DC-coupled source, as shown in Fig. 4.5. It offers the most operational

Fig. 4.4 Example of a
DC-coupled mini-grid

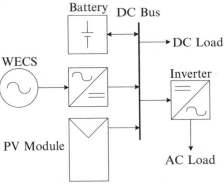

Fig. 4.5 Example of an
AC–DC-coupled mini-grid

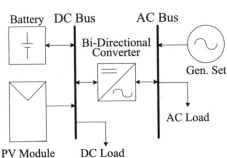

flexibility. In this architecture there are at least two buses: one AC and one DC. The buses are coupled through converters. The flow can be one directional, DC to AC or AC to DC, or bi-directional. The AC–DC-coupled architecture is used in many hybrid mini-grids.

The decision to use a particular architecture is dictated by the choice of energy conversion technology and load. Most often we select the architecture that (1) minimizes AC/DC and DC/AC conversion so as to reduce losses and (2) has the fewest overall components to reduce the cost and complexity. This naturally suggests that systems with AC sources and loads use AC-coupled architecture and those with DC sources and loads use DC-coupled architectures. Hereafter we describe some of the more common AC-, DC-, and AC–DC-coupled architectures.

4.5   Conventional or Biomass Internal Combustion Engine Generator Systems

In this common and simple AC-coupled architecture, a gen set supplies an AC load, as shown in Fig. 4.6. The generator forms the AC bus. The power output is adjusted to regulate the AC bus frequency by controlling the flow of fuel or air to the engine. The power supplied by the gen set flows to the AC bus—which in practice is an

Fig. 4.6 Architecture of an
internal combustion engine
generator system

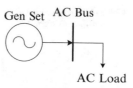

outlet or perhaps a distribution box similar to that found in homes—and then to
the connected loads. There is usually no need for a transformer or other conversion
equipment between the generator and load. The architecture is the same whether the
gen set is fueled by a conventional fuel or a biomass-derived fuel.

Gen sets can be readily obtained in most areas and have low up-front capital cost
but high fuel costs. This architecture is suitable in applications where reliable or
on-demand power is needed, but total energy use is low. For example, an off-grid
system for a community center that only requires electricity for special occasions is
well-served by this architecture. The architecture is more common in areas with low
fuel costs, perhaps through government subsidy.

4.6   Photovoltaic Systems with DC Load

This DC-coupled architecture consists of photovoltaic (PV) modules and a battery.
In its most basic configuration, it only supplies DC loads. The architecture is used in
solar home systems and solar lanterns . It is conceptually simple and its components
are readily available. PV modules naturally output DC current. The battery provides
a stable DC voltage, which many electronic loads require, and allows the system to
provide electricity in the evening and in periods of low irradiance (sunlight). The
battery is charged when the current from the PV module is in excess of the load
current and is discharged when the load current exceeds that of the PV module.

In engineered systems, a charge controller, as shown in Fig. 4.7, is included.
The role of the charge controller is to prevent the battery from being overcharged.
Some charge controllers also prevent the battery from being discharged. In some
PV systems, a component known as a "maximum power point tracker" is connected
between the charge controller and the PV module. This electronic device adjusts the
operating voltage of the PV module to improve its power production.

4.7   Photovoltaic Systems with AC Load

An inverter is required to serve an AC load from a PV system. This is shown in
Fig. 4.8. In this architecture, the inverter forms the AC bus. The inverter converts
DC to AC at a specific frequency and voltage, typically 50 Hz and 230 V, depending
on national standards. The inverter mimics the AC voltage in the national grid

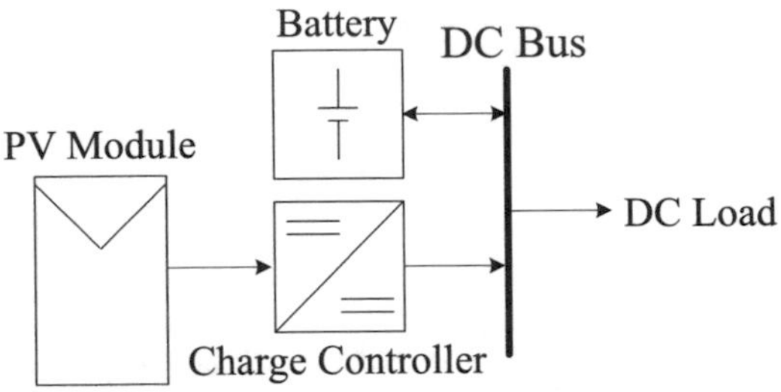

Fig. 4.7 Architecture of a photovoltaic system with DC load

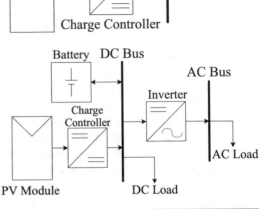

Fig. 4.8 Architecture of a photovoltaic system with AC and DC load

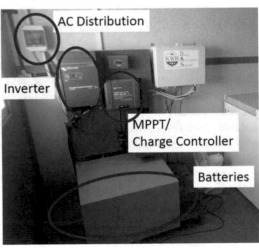

Fig. 4.9 Storage, control and converter components of a 2.25 kW PV system with AC load (courtesy of KiloWatts for Humanity)

so that appliances mass manufactured for the grid-connected market can be used. These are usually less expensive and more widely available than DC appliances. The components of a 2.25 kW mini-grid, excluding the PV array, are shown in Fig. 4.9. This system does not have DC load.

4.8   Wind Energy Conversion Systems

As discussed previously, a WECS cannot form the AC bus. Instead, they are often used in DC-coupled systems, as shown in Fig. 4.10. Most WECSs output AC, and so a rectifier is required to couple them to the DC bus. The battery provides voltage stability and allows for on-demand electricity when the wind speed is insufficient to power the load.

For reasons discussed in Chap. 10, systems with DC-coupled WECSs incorporate a diversion load and a dedicated diversion load controller to prevent overcharging the battery. The controller connects a diversion load—a resistor with a high power rating—in parallel with the battery so that the current from the WECS divides between the battery and diversion load.

4.9   Micro-Hydropower Systems

MHP systems can be AC-coupled and DC-coupled or used in AC–DC-coupled architectures. For systems with capacity less than approximately 5 kW or with irregular water flow, DC-coupled architectures are often used. The architecture mimics that of the wind turbine system in Fig. 4.10, with the hydroturbine replacing the wind turbine. The power house of an AC-coupled MHP system is shown in Fig. 4.11.

In larger-capacity systems with consistent flow, the AC-coupled architecture is common. The AC bus voltage is formed by controlling the excitation system of the MHP system, and the frequency is maintained by using either a mechanical governor, which adjusts the water input to the turbine, or an electronic load controller (ELC) and ballast load. The ELC and ballast load are conceptually similar to a diversion load controller, and diversion load, but are coupled to the AC bus. Unlike a governor, which controls the power output of the MHP system to maintain the AC bus frequency, the ELC controls the power to a load. A system using an ELC and ballast load is shown in Fig. 4.12. If a governor is used, then the system is as in Fig. 4.6, with the MHP system replacing the gen set.

Fig. 4.10 Architecture of a wind turbine system

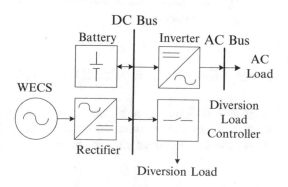

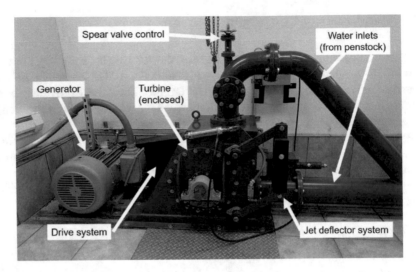

Fig. 4.11 A 25 kW AC-coupled MHP system using a two-jet Pelton turbine (courtesy of author)

Fig. 4.12 Architecture of an
AC-coupled MHP system

4.10   Hybrid Systems

Hybrid systems use two or more different energy conversion technologies for electricity production. Although the term "hybrid" is general, it is most often applied to systems in which one of the sources is a conventional gen set and another is a renewable source. Hybrid systems diversify the energy sources, which can improve reliability and lower operating and capital costs. However, they are more complex to design and operate.

4.10.1   Hybrid Conventional Gen Set/PV Systems

In this AC–DC-coupled architecture, a conventional gen set is coupled with a PV module to serve an AC load. In some systems, the gen set is always or almost always running, in which case the energy produced by the PV array reduces the

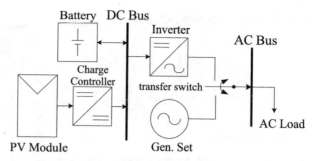

Fig. 4.13 Architecture of a switched hybrid system utilizing a gen set as a backup supply

fuel consumption. In others, the gen set is only operated as a backup, recharging the batteries or to supplying power if the load becomes especially large. There are several possible sub-architectures that can be used, as discussed next.

4.10.1.1   Switched Architecture: Gen Set Backup

Figure 4.13 shows the simplest hybrid gen set/PV architecture. It mimics the architecture of a PV system with AC load. However, there is a transfer switch that connects the gen set to the load while disconnecting the inverter. In other words, the load is served by either the generator or the inverter, not both. There is no need to synchronize the inverter and gen set. The transfer switch can be manually operated or automatic.

4.10.1.2   Switched Architecture: Gen Set Backup with Battery Charging

The switched architecture can be modified as shown in Fig. 4.14 to allow the gen set to charge the battery, even as it is serving the load. This prevents the battery from being undercharged. A rectifier and charge controller, perhaps integrated into the same unit, are needed. Simultaneously supplying the load and recharging the battery can improve the efficiency of the gen set, saving fuel. As with the previous architecture, either the gen set or the inverter supplies the load.

Some systems will use the gen set only for battery charging, eliminating the transfer switch altogether. In this architecture, the inverter exclusively serves the load. Because the AC load is never directly supplied by the gen set, a smaller-capacity gen set can be used. A drawback is that all of the energy provided by the gen set is converted from AC to DC—and potentially stored by the battery—and back to AC. These extra conversion steps are inefficient.

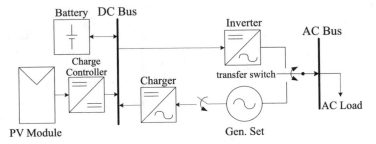

Fig. 4.14 Architecture of a switched hybrid gen set/photovoltaic system where the gen set can be used to charge the battery or supply power directly to the load

Fig. 4.15 Architecture of a hybrid gen set/photovoltaic system where the gen set can simultaneously charge the battery and supply power to the load

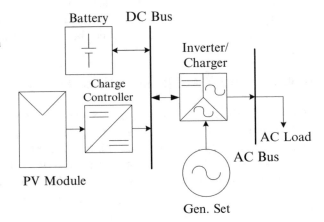

4.10.1.3 Architecture with Battery Charging

The hybrid architectures discussed so far do not allow the AC load to be simultaneously supplied by the gen set and inverter. This functionality is useful when the load occasionally exceeds the power output capability of either the gen set or inverter alone.

Simultaneous supply is possible using the architecture in Fig. 4.15. It requires an inverter that can synchronize its output to that of the gen set. The gen set can charge the battery through a rectifier and charge controller similar to that of Fig. 4.14. However, rather than using a separate rectifier and charge controller, a bi-directional inverter is used, as shown in Fig. 4.15. A bi-directional inverter facilitates power flow to and from the DC bus, allowing the gen set to charge the battery. A charge controller, which is often integrated into the same unit as the bi-directional inverter, is used to manage the charging of the battery from the gen set. Alternatively, the gen set can be directly connected to the AC bus, with the bi-directional inverter facilitating power flow to the DC bus.

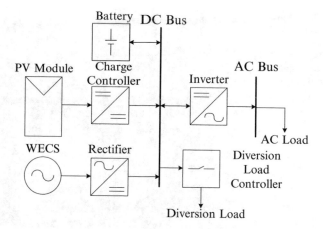

Fig. 4.16 Architecture of a hybrid PV/WECS renewable energy system

4.10.2   Hybrid Renewable Systems

A subset of hybrid systems are those that combine two or more renewable energy
sources, such as PV panels and wind turbines. The sources used dictate the type of
coupling.

Hybrid renewable systems should be selected when the renewable resources have
complementary characteristics, for example, using a WECS with a PV module at
a location where it is windy in the evening when there is no sunlight. If such
a complementarity does not exist, then in general it is preferred to use a single
resource, as it makes the system less complicated. The architecture of a DC-coupled
hybrid PV/WECS is shown in Fig. 4.16. Separate controllers are needed for the PV
module and the diversion load.

4.11   Improvised Systems

Many households in off-grid communities rely on "improvised" or "non-
engineered" systems as shown in Fig. 4.17. These systems often use low-cost
scavenged or repurposed equipment. In some cases, cleverly constructed wind and
hydroturbines have been made, as seen in Figs. 4.18 and 4.19. Improvised systems
generally have low reliability and voltage quality but nonetheless can provide
beneficial electricity access when nothing else is available.

Fig. 4.17 An improvised system consisting of a PV module (not shown), repurposed lead–acid battery, modified sinewave inverter, and AC load (courtesy of author)

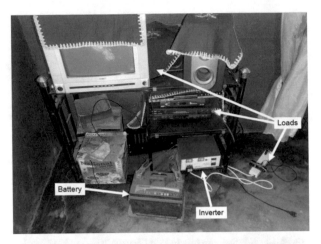

Fig. 4.18 An improvised wind turbine used for battery charging in Zambia (courtesy of author)

4.12 Summary

Mini-grids are electrical systems, isolated from the national grid, that serve multiple users. Mini-grids use energy conversion technologies to convert the energy available in fossil fuels, biomass, moving air and water, sunlight, and other sources to electricity. Various converters, controllers, and energy storage devices are used to facilitate the supply of electricity to the users. A number of architectures are possible, many of which were discussed in this chapter. The architectures vary in their complexity, reliability, flexibility, and relative cost. The architecture is defined by where the energy sources are connected. If all the energy sources are connected to the AC bus, then the system is AC-coupled; if there are only energy sources coupled to the DC bus, then it is DC-coupled; systems with energy sources connected to both AC and DC buses are AC–DC-coupled mini-grids. The voltage magnitude and frequency of the AC bus must be formed by a single source, with the rest synchronizing to it. This adds complexity to the control of the sources. The DC bus avoids this complexity, but instead care must be taken to not overcharge or over discharge the batteries.

Fig. 4.19 An improvised
battery charging scheme
(courtesy of P. Dauenhauer)

Problems

4.1 Describe the basic function of an inverter, rectifier, and charge controller.

4.2 Draw a schematic of a mini-grid consisting of a conventional gen set, PV array, and WECS. The PV panel and WECS are coupled to the DC bus. The gen set forms the AC bus. Include any required controllers.

4.3 Consider a mini-grid with two single-phase, AC-coupled generators. The generators are modeled as a voltage in series with an inductive reactance. The circuit is shown Fig. 4.20. The load is not connected. The generators are controlled so that their terminal voltages V_1 and V_2 are held at:

$$V_1 = 233\angle 0°\,V$$
$$V_2 = 227\angle 10°\,V$$

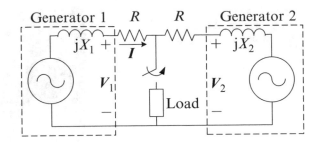

Fig. 4.20 Two generators in parallel

Let the resistance in the conductors connecting the generators each be $R = 0.20\ \Omega$. Compute the resulting current I. What does this result suggest about the importance of synchronizing generators connected to the AC bus.

4.4 An off-grid system will provide overnight lighting to a hospital using LED lights. The system will use PV modules to supply the power. Draw a schematic of the system; assume the LED lights can be connected to a DC bus or AC bus. Justify why you selected the LEDs to be connected to the DC bus or AC bus.

4.5 Draw a schematic of a hybrid off-grid system that is supplied by a PV module, a WECS, and a gen set. Assume there are both AC and DC loads and that the inverter and gen set can be synchronized. Your design should allow for the gen set to charge batteries connected to the DC bus.

References

1. IEA: Africa energy outlook: A focus on energy prospects in Sub-Saharan Africa (2014). URL   https://www.iea.org/publications/freepublications/publication/WEO2014_AfricaEnergy Outlook.pdf
2. Mandelli, S., Barbieri, J., Mereu, R., Colombo, E.: Off-grid systems for rural electrification in developing countries: Definitions, classification and a comprehensive literature review. Renew. Sustain. Energy Rev. **58**, 1621–1646 (2016). doi:https://doi.org/10.1016/j.rser.2015.12.338

Part II
Energy Conversion Technologies

Chapter 5
Off-Grid Generators, Gen Sets, and Biomass Systems

5.1 Introduction

As we discussed in the last chapter, there are several energy conversion technologies that can be used to supply electricity to a mini-grid or other off-grid system. Selecting which particular technology or technologies to use is among the most important design decisions, as it largely dictates the system's architecture, reliability, and economics.

This and the next two chapters are concerned with the principles of operation, technical characteristics, and practical considerations of the energy conversion technologies commonly used in small-scale mini-grids. These technologies include:

- Internal combustion engine generator sets (gen sets)
- Biomass systems
- Wind turbines
- Hydroturbine
- Photovoltaic arrays

Our discussion begins with a description of AC generators, gen sets, and biomass systems. We start with generators because, with the exception of PV arrays, all energy conversion technologies used in off-grid electrification rely on generators to produce electricity.

5.2 Electrical Generators

Generators are electrical machines that convert mechanical energy (typically rotational) into electrical energy. Motors, on the other hand, convert electrical energy into mechanical. In principle, there is no physical distinction between a generator

© Springer International Publishing AG, part of Springer Nature 2018
H. Louie, *Off-Grid Electrical Systems in Developing Countries*,
https://doi.org/10.1007/978-3-319-91890-7_5

and a motor; the only difference is the direction of the energy flow. It is for this reason that scavenged motors are sometimes repurposed as generators in improvised off-grid systems.

The device supplying the input mechanical power to a generator is known as the *prime mover*. Examples of prime movers include wind turbines, hydroturbine, and internal combustion engines. The generator must be carefully designed to match the characteristics of the prime mover, in particular its power, speed, and torque characteristics. In this section, we review the basic principles of generators that are especially relevant to off-grid systems, regardless of the prime mover.

5.2.1 Principle of Operation

All electric generators operate on the same basic principle. When the magnetic flux passing through a coiled wire varies over time, Lenz's Law tells us that a voltage will be induced across the terminals of the coil. There are a variety of ways in which to engineer this interaction. A simple way makes use of the magnetic flux produced by a permanent magnet. The magnet is affixed to a shaft, which allows it to rotate. One or more coils are arranged around the shaft. The coils are stationary so that as the magnet rotates there is a relative motion between each coil and the magnet. The flux through each coil will alternate as the north pole and then south pole pass by. The process repeats with every rotation of the shaft. The alternating flux induces a voltage that can be used to power a load connected to the coils. Of course, some mechanical power is required to rotate the shaft. A prime mover is used for this purpose.

Some generators are designed to output AC voltage, and others DC voltage. We will consider only AC generators as these are the most widely used in off-grid systems. Generators for off-grid applications that output DC are often AC generators with an internal solid-state rectifier, rather than true DC generators. There are two general types of AC generators: synchronous and asynchronous. The most common asynchronous generator is the induction generator. Although induction generators are used in some off-grid applications, in particular small-capacity micro hydro power systems, synchronous AC generators are more common and are the focus of our discussion in this section.

5.2.2 Physical Characteristics

Most generators are constructed using a radial flux arrangement, as shown in Fig. 5.1. There is a cylindrical rotor inside a stationary housing—known as the stator—with a small air gap to allow the rotor to rotate freely. The rotor contains the magnet. The rotor and stator are usually made from highly permeable steel, which provides a path of low reluctance for the flux to flow.

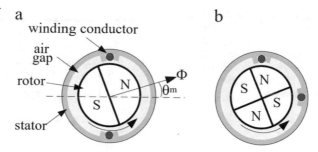

Fig. 5.1 (a) Cross-section of a two-pole synchronous generator with a single stator winding shown. (b) Cross-section of four-pole synchronous generator

The stator houses coils of wire in which the voltage is induced. Each coil has many turns. The coils are inserted into slots around the stator's interior periphery. These coils are referred to as the "stator" windings, also referred to as "armature" windings. Each winding is wrapped around the length of the stator. The two ends of a single-turn stator winding are shown in Fig. 5.1a.

5.2.3 Electrical Characteristics

Electric generators have three important properties that are important to keep in mind:

1. The electrical load on the generator must be equally matched by the mechanical power provided by prime mover (assuming the generator is electrically and mechanically lossless); otherwise the rotating shaft will speed up or slow down. This is simply a consequence of the law of conservation of energy.
2. The frequency and magnitude of the voltage produced by a generator are proportional to the rotational speed of the rotor and the magnitude of the flux linking the coils. This consequence of Lenz's Law, which we discuss in detail later.
3. The voltage that appears at the generator's terminals depends on the size and power factor of the load. This result also depends on the generator's own internal impedance.

These effects are visualized in Figs. 5.2 and 5.3. Any mismatch in electrical and mechanical power causes the generator's speed to change, as shown in Fig. 5.2. This in turn affects the magnitude and the frequency of the generator voltage. Figure 5.3a shows the effect of rotor speed on frequency. Many loads cannot tolerate large fluctuations in voltage magnitude and frequency. Therefore, it is important to have a control system that balances the mechanical and electrical power. With this control system in place, the generator will operate at a constant speed. However, even if the speed is constant, the terminal voltage will vary with the load, as seen in Fig. 5.3b. We therefore need a second control system to regulate the terminal voltage.

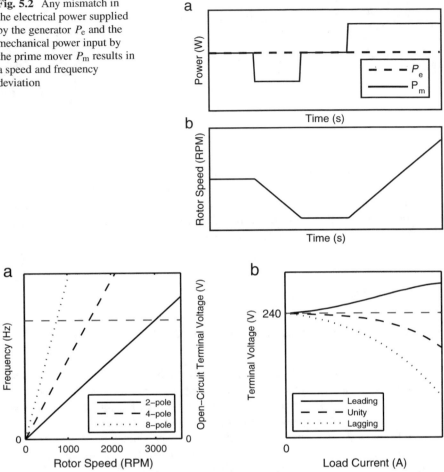

Fig. 5.2 Any mismatch in the electrical power supplied by the generator P_e and the mechanical power input by the prime mover P_m results in a speed and frequency deviation

Fig. 5.3 (a) The frequency and magnitude of the open-circuit voltage of a generator is proportional to the rotational speed of the rotor. (b) The terminal voltage of generator under constant excitation varies with load current and the power factor of the load

The speed and terminal voltage of a generator must be controllable for it to be coupled to the AC bus of a mini-grid. If a generator lacks these capabilities, then it must be coupled to the DC bus through a rectifier. For example, the mechanical power supplied by a wind turbine cannot be increased on demand if it is not windy, and therefore WECS are DC-coupled.[1]

---

[1] An exception to this is if the WECS utilizes an induction generator. AC coupling is common in large-capacity WECS connected to the national grid where other generators form the AC bus voltage.

5.2.3.1 Induced Voltage

Let us now expand on the qualitative description given in Sect. 5.2.1 of how generators are able to produce voltage. For simplicity, we will consider a single stator winding made of n_{turns} turns of wire. Let the flux produced by the rotor be Φ as shown in Fig. 5.1. As the generator's rotor rotates, the magnitude and polarity of the flux passing through the winding varies. For each complete mechanical rotation, the flux passing through the stator's winding alternates from north to south and back to north in a sinusoidal manner. The flux linking the winding ϕ_w at an angular position θ_m of the rotor is

$$\phi_w = n_{turns} \Phi \cos(\theta_m). \tag{5.1}$$

If the rotor spins at an angular velocity of ω_m radians per second, then the angular position at any time t is

$$\theta_m = \omega_m t. \tag{5.2}$$

The subscript "m" reminds us that the quantities are referenced to the mechanical speed and position of the rotor. This time-varying flux induces a voltage e in the winding according to Lenz's Law:

$$e(t) = -\frac{d\phi_w}{dt} = n_{turns} \Phi \frac{d\theta_m}{dt} \sin(\theta_m) = n_{turns} \Phi \omega_m \sin(\theta_m) = n_{turns} \Phi \omega_m \sin(\omega_m t) \tag{5.3}$$

where we have substituted the angular velocity for the derivative of the angular position with respect to time. This description assumes that the generator is open-circuited. When connected to the load, the current through the generator's armature winding also produces flux. This interacts with the flux from the rotor, reducing the total flux linking the coils. This phenomenon is known as "armature reaction" and can be modeled as an inductance as shown later in Sect. 5.2.7.

Sometimes it is advantageous to use several magnets in the rotor. An example of a rotor with two magnets—for a total of four poles, two north and two south—is shown in Fig. 5.1b. With every pair of poles added, the stator windings experience an additional north-to-south and south-to-north flux transition for each mechanical rotation. Therefore, for a given mechanical rotational speed, the flux through the coil varies more rapidly, increasing the magnitude and frequency of the induced voltage. We can rewrite (5.3) to account for the number of poles as:

$$e(t) = n_{turns} \Phi \omega_m \frac{p}{2} \sin\left(\frac{p}{2}\omega_m t\right) \tag{5.4}$$

where p is the number of poles, which is always an even integer. Since the minimum number of poles that a generator can have is two, the fastest speed that a generator can operate at and produce 50 Hz is 3000 revolutions per minute (RPM) and 3600

RPM for 60 Hz output. The conversion from frequency f in cycles per second (hertz) to revolutions per minute N is simply

$$N = f \times 60. \tag{5.5}$$

The speed that the rotor must operate at to provide a desired electrical frequency f_e can be computed from

$$f_e = \frac{f_m p}{2} = \frac{N_m p}{120} \tag{5.6}$$

where f_m and N_m are the speed of the rotor shaft in hertz and RPM, respectively.

Example 5.1 What must the rotational speed, in RPM, of a conventional internal combustion engine coupled to a four-pole synchronous generator be to provide electricity at 50 Hz?

Solution Rearranging (5.6) shows that the engine shaft must rotate at

$$N_m = \frac{120 f_e}{p} = \frac{120 \times 50}{4} = 1500 \text{ RPM}.$$

Example 5.2 A certain hydroturbine is most efficient when rotating at 750 RPM. How many poles should the synchronous generator coupled to the turbine have if the generator is to supply electricity at 50 Hz?

Solution Rearranging (5.6) shows that for optimal turbine efficiency, the generator should have

$$p = \frac{120 f_e}{N_m} = \frac{120 \times 50}{750} = 8 \text{ poles}.$$

It is possible that the number of poles corresponding to the optimal speed of a prime mover is not an even-numbered integer. In this case, the efficiency curve of the prime mover should be consulted to determine if the number of poles should be rounded up or down to the next even-numbered integer.

In practice, generators have multiple stator windings. If the windings are physically displaced by 120°, then the voltage induced in each winding will be out of phase by 120°, resulting in the familiar balanced three-phase waveform. Smaller capacity generators tend to be single-phase.

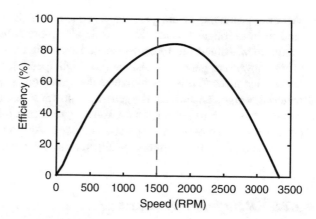

Fig. 5.4 The efficiency of a Pelton hydroturbine depends on its rotational speed

A generator's rotor is usually directly coupled to the shaft of the prime mover—a configuration known as "direct drive." In this case, the rotor and the prime mover's shaft rotate at the same speed. The speed is controlled so that the generator produces the desired frequency. However, the speed required to produce the desired electrical frequency might not overlap with the speed at which the prime mover is most efficient. For example, if a four-pole generator is coupled to a certain Pelton turbine whose efficiency curve is shown in Fig. 5.4, then operating at the 1500 RPM required for 50 Hz voltage output does not maximize the efficiency of the turbine. We could control the turbine so that it operates at its most efficient speed, but the frequency of the voltage would surpass 50 Hz. Gear boxes or pulleys can be used to achieve a desired ratio between rotor shaft and prime mover shaft speeds, but they add complexity and cost and increase maintenance requirements. We should therefore co-design the prime mover and generator so that their shaft speed maximizes the efficiency of the prime mover while producing the required electrical frequency.

5.2.4 Speed, Torque, and Power

As we increase the number of poles of a generator, the speed it must operate at to produce the desired voltage frequency decreases. What are the other implications of this lower-speed operation? Recall from physics that angular speed ω in radians per second, power P in joules, and torque T in newton meters are related by

$$P = T \times \omega. \tag{5.7}$$

This equation tells us that a given amount of power supplied to the rotor shaft can be accomplished at, for example, high torque but low speed, low torque but high speed, or both medium torque and speed. A generator with more poles will experience greater torque to provide the same output power because it rotates slower. A prime mover and generator designed for high-torque, low-speed operation tend to have longer service lives but require more substantial shafts, rotors, and stators to withstand the forces associated with high torque. On the other hand, a prime mover and generator designed for high-speed, low-torque operation require fewer windings and less flux to induce the same voltage per (5.4). This reduces the material, weight, and cost of the generator; however, the higher speed tends to wear components such as bearings more quickly, reducing the service life. In addition, certain losses increase with operational speed. As you can see, design of prime movers and generators requires making several trade-offs.

5.2.5 Rotational Dynamics

We saw in Fig. 5.2 that a mismatch in electrical and mechanical power causes the generator to rotate slower or faster. To understand why, let's briefly review what happens when a net torque T_{net} is applied to a rigid body such as a generator's rotor. The torque causes a change in the rotational speed according to

$$T_{net} = \alpha_m J = \frac{d\omega_m}{dt} J \tag{5.8}$$

where α_m is the acceleration and J is the mass moment of inertia, in kilogram meters squared, of the rotor. When the net torque is positive, the rotor accelerates; when it is negative, the rotor decelerates. Only when the net torque is zero is the speed of the rotor shaft constant.

If we ignore losses, then there are two torques acting on the rotor: the mechanical torque provided by the prime mover and the electromagnetic torque caused by the generator. The torques are in opposite directions so that

$$T_{net} = T_m - T_e. \tag{5.9}$$

The origin of the electromagnetic torque can be explained as follows. Whenever a generator supplies real power to a load, the generator current causes the stator windings to act like electromagnets. Their magnetic fields interact with the magnetic field produced by the rotor. The polarity of each electromagnet is such that it repels a rotor pole rotating toward it and attracts a rotor pole rotating away from it. In other words, the electromagnetic torque is in the opposite direction of the mechanical torque applied to the rotor. The strength of the repulsion and attraction increases with the real power provided by the generator, leading to a similar increase in T_e.

Table 5.1 summarizes how the net torque affects the speed of the rotor. Given (5.7), the same net torque conditions that cause the rotor to accelerate or decelerate also apply to the mechanical and electrical power. With this explanation of rotational dynamics in mind, the reader should return to Fig. 5.2 and verify the rotor speed is as expected.

Table 5.1 Effect of net torque on rotor speed

Net torque	Condition	Result
$T_{net} > 0$	$T_m > T_e$, $P_m > P_e$	Rotor accelerates
$T_{net} = 0$	$T_m = T_e$, $P_m = P_e$	Rotor speed constant
$T_{net} < 0$	$T_m < T_e$, $P_m < P_e$	Rotor decelerates

5.2.6 Speed Control Systems

It is clear that to maintain a constant speed, the mechanical power provided by the prime mover must be adjusted to "follow" the changes in electric load. Different prime movers use different systems for speed control, for example, by using valves and throttles to adjust the flow of water to the turbine or air into a combustion chamber. It is also possible to use a ballast load to keep P_e constant even as users turn appliances on and off. This scheme is commonly used in micro hydro power systems.

Even in systems where the prime mover is not required to operate at constant speed, some precautions are necessary to prevent damaging over-speed conditions. Care should be taken to avoid suddenly disconnecting wind turbines and some hydro turbines from their load without also reducing the input mechanical power. Otherwise the rotor might quickly accelerate to a high speed, potentially causing over voltages. The generator and prime mover might also be damaged due to the associated centripetal forces and vibration.

5.2.7 Circuit Model

We saw in Fig. 5.3b that a generator's terminal voltage depends on the generator current and the power factor of the load. To understand why, consider the circuit model for a single-phase generator shown in Fig. 5.5. A similar model applies for an equivalent single phase of a three-phase generator. The model consists of a voltage source in series with a resistance and inductive reactance. The resistor R_a corresponds to the resistance of the armature windings; the inductive reactance X_s—also known as the synchronous reactance—models the self and mutual inductance of the stator windings. Applying Kirchhoff's Voltage Law using phasor quantities yields:

$$V_a = E_a - I_a (R_a + jX_s) \qquad (5.10)$$

where E_a is the induced voltage from (5.4) in phasor form, V_a is the terminal voltage, and I_a is the armature current. The terminal voltage therefore depends on the current.

Fig. 5.5 Equivalent circuit model of a single-phase synchronous AC generator armature

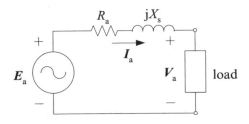

Example 5.3 Consider a mini-grid consisting of a load and single-phase generator coupled to an AC bus. The bus voltage is nominally 230 V. The generator's resistance is 0.75 Ω, and synchronous reactance is 7.5 Ω. Compute the terminal voltage if the induced voltage is $E_a = 230\angle 0°$V and $I_a = 25.0\angle - 70°$A. Compare this to the open-circuit voltage.

Solution The terminal voltage is found by solving for V_a in 5.10:

$$V_a = E_a - I_a (R_a + jX_s)$$
$$= 230\angle 0° - 25.0\angle - 70° (0.75 + j7.5) = 66.40\angle - 44.46° \text{ V}.$$

The terminal voltage has dropped significantly. Most loads designed for a 230 V supply will not function properly or at all at such a low voltage.

The open-circuit voltage is always equal to the induced voltage. In this case it is 230 V.

5.2.8 Excitation Systems

The circuit model described in the previous section offers a hint at how we can maintain a constant terminal voltage magnitude $|V_a|$ even as the load changes. We see that from (5.10), the terminal voltage depends on the induced voltage. If we can adjust the induced voltage, we can control the terminal voltage. Recall from (5.4) that the induced voltage is a function of the rotor speed, the number of poles, and flux. Although the speed can be adjusted, constant speed operation is preferred. We cannot realistically change the number of poles of the generator. We can, however, change the rotor flux if electromagnets are used in the rotor instead of permanent magnets. Most generators use electromagnets. Electromagnets can also produce more flux than a similarly sized permanent magnet. This increases the power density of the generator and lowers its cost.

The electromagnets in the generator's rotor require DC current. The system used to provide the DC current to rotor is known as the "excitation" system [5]. The coils of the electromagnet are known as "field" windings because they produce the rotor's magnetic field. We generically relate the flux produced by the rotor to the field current i_f by a parameter k_f, which itself depends on the physical and magnetic properties of the generator,[2] as

$$\Phi = i_f k_f. \tag{5.11}$$

By controlling the field current, we are able to adjust the flux and thereby control the induced voltage. The flux produced by the electromagnets can magnetize the rotor, so that even with no DC current, there is a small amount of residual magnetism.

There are several excitation systems that are in use today. Generators used in mini-grids tend be "self-excited" in that they use a small portion of their output for excitation.

Example 5.4 Consider the generator in the previous example. Compute the magnitude of the induced voltage required to maintain a terminal voltage of 230 V when the current is $I_a = 25.0\angle - 70.0°$ A. Let the angle of the terminal voltage be -44.46°.

Solution The induced voltage E_a is found by rearranging (5.10)

$$E_a = V_a + I_a \, (R_a + jX_a)$$

$$= 230\angle - 44.46° + 25.0\angle - 70.0° \, (0.75 + j7.5) = 365.21\angle - 18.29° \text{ V}$$

The required magnitude of the induced voltage is 365.21 V. To increase the induced voltage to this level requires additional flux. This can be accomplished by increasing the current to the field (rotor) winding.

5.2.8.1 Static Excitation

A static excitation system uses a separate set of windings known as "excitation" windings mounted in the stator, as shown in Fig. 5.6. Static excitation systems are commonly found in the alternators of most automobiles. During start-up, residual magnetism from the rotor provides a small amount of flux that induces a voltage in the excitation winding. The excitation winding is connected to a rectifier that outputs DC current. The rectifier output is connected to carbon brushes. The brushes

---

[2]The parameter k_f will decrease at high levels of field current as the ferrous material making up the rotor saturates and its permeability drops.

Fig. 5.6 Static excitation
system

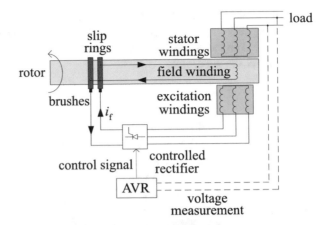

Fig. 5.7 The brushes and
slip rings of a synchronous
generator supply current to
field winding (courtesy of the
author)

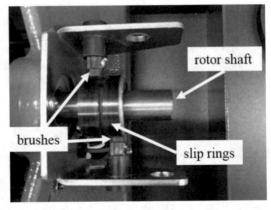

are pressed against rotating slip rings that are connected to the field winding, as
shown in Fig. 5.7. In this way, the stationary rectifier is connected to the rotating
field winding.

The field winding current can be controlled using an Automatic Voltage Regulator (AVR) [5]. The AVR senses the generator's terminal voltage (typically the
line-to-line voltage between any two phases) and adjusts the field current to achieve
the flux needed to keep the terminal voltage constant. The AVR function can be
accomplished with an analog or digital circuit. The excitation scheme shown in
Fig. 5.6 uses a controlled rectifier whose firing angle is set by the AVR to control
i_f. The specifics of controlled rectifiers and AVRs are discussed in more detail in
Chap. 9.

Instead of relying on residual magnetism in the rotor to induce voltage in the
excitation windings during start-up, some generators rely on an external battery
bank that is temporarily connected to the field winding.

A drawback to static excitation systems is that the brushes require maintenance
and replacement, and increase losses.

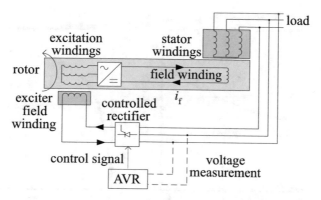

Fig. 5.8 Brushless excitation system

5.2.8.2 Brushless Excitation

To overcome the challenges associated with brushes, brushless excitation systems are sometimes used. In these systems, the excitation winding is on the rotor instead of the stator, as shown in Fig. 5.8. AC voltage is induced in the excitation winding through a stationary electromagnet, which is sometimes called the "exciter field winding." The excitation winding voltage is converted to DC by a rectifier that is attached to, and rotates with, the rotor. The rectifier output is connected to generator's field windings.

The current for the exciter field winding can be supplied by the generator output (using the residual magnetism of rotor during start-up) as shown in Fig. 5.8 or by a pilot exciter. A pilot exciter is a small-capacity permanent magnet generator attached to the end of the main generator shaft. The AC voltage induced in the pilot exciter's stator winding is rectified and supplies DC current to the stationary electromagnet. An AVR adjusts the field current to achieve the desired terminal voltage.

5.2.9 Efficiency

In general, the efficiency of a synchronous generator is high, around 90%, when loaded near rated capacity. The losses that do occur can be categorized as:

- Mechanical: caused by friction of bearings, brushes (if applicable), and windage (aerodynamic drag on the rotor).
- Magnetic: caused by hysteresis and eddy currents associated with the magnetic flux in the rotor and stator.
- Winding (Copper): $I^2 R$ losses associated with the armature current, as well as field and excitation windings (if applicable).

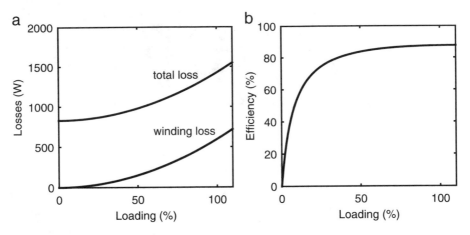

Fig. 5.9 (**a**) Winding losses increase with generator loading. (**b**) The efficiency of a synchronous generator is greatest near rated power

The share of total losses attributed to each category varies with generator loading and speed. Mechanical losses increase with speed but are nearly constant as the load varies. Magnetic losses increase somewhat with speed and load; and copper losses are insensitive to speed but increase rapidly with load as shown in Fig. 5.9a.

The efficiency of a generator is best expressed as a curve rather than a single value. A typical efficiency curve is shown in Fig. 5.9b. The efficiency is low at low loading because the mechanical losses and magnetic losses are high relative to the power produced. The efficiency rapidly rises as the load increases, before plateauing near its rated capacity as the copper losses begin to dominate.

Generators are in fact capable of supplying power in excess of their rating, but this can lead to overheating. High temperatures degrade the generator components, including the winding insulation. An increase in operating temperature over its design value by 10°C can reduce the service life of the generator by as much as 50%. Therefore, generators should only be operated above their rated power for brief episodes—for example, when connected to motors and other devices that require high power during start-up.

5.3   Conventional Internal Combustion Engine Generator Sets

Conventional internal combustion engines (ICE) are commonly used in off-grid systems. In Nigeria alone, there are an estimated 100 million ICE-coupled generators in use. The ICEs used in off-grid systems are most often of the reciprocating type, like those found in automobile engines, rather than turbine type, like those found in large aircraft and power plants. The main components of such a reciprocating ICE generator system are:

Fig. 5.10 Smaller-capacity gen sets are often designed to be portable (courtesy of P. Dauenhauer)

Fig. 5.11 A 660 kW diesel gen set used to power a mini-grid in Haiti (courtesy of Sigora Haiti)

- Fuel tank
- Fuel and air supply system
- Engine
- Cooling and exhaust system
- Generator and excitation system

When these components are packaged together into an integrated unit, it is referred to as a "generator set" or simply a "gen set."

Gen sets range in capacity from a few hundred watts to several hundred kilowatts or even a few megawatts. Gen sets below approximately 10 kW are often constructed to be portable, as shown in Fig. 5.10. Larger gen sets are permanently installed, as shown in Fig. 5.11. Gen sets are usually placed in enclosures that offer protection from the weather and pests, along with acoustic damping.

5.3.1   Principle of Operation

There are two categories of reciprocating ICEs: spark ignition (SI) and compression ignition (CI). Both ignite fuel in a cylindrical combustion chamber, using the resulting expansion to drive a reciprocating piston. The pistons are connected to a crankshaft which rotates and is capable of supplying rotational mechanical power [9].

Spark-ignition engines use a spark plug to ignite a mixture of fuel and air in the combustion chamber. The process consists of:

1. formation of a mixture of air fuel
2. air/fuel mixture intake into the chamber
3. compression (typically by a factor from 8 to 12)
4. spark ignition
5. combustion
6. expansion
7. exhaust

SI engines can be designed to run on gasoline (petrol), natural gas, biogas, syngas, propane, hydrogen (H_2), or alcohols.

Combustion in CI engines occurs by compressing air and a fuel (usually diesel or another oil, including bio-fuels) in the combustion chamber. As the air compresses, its temperature increases, initializing the combustion of the fuel. The process consists of air intake, compression of the air (typically by a factor of 12 to 24), fuel injection, mixture formation, ignition, combustion, expansion, and exhaust.

In SI and CI engines, the crankshaft is directly coupled to the generator's rotor. The designed speed of a gen set is a compromise of several competing factors, including the general speed-versus-torque trade-offs discussed in Sect. 5.2.4 and consideration of the speed-versus-efficiency curve of the engine.

A consequence of these trade-offs is that most small-capacity gen sets—tens of kilowatts or less—use two-pole generators. This means that for 50 Hz systems the engine shaft speed is 3000 RPM. They are often SI engines fueled by petrol, but CI diesel-fueled gen sets in this capacity range are available. Larger capacity gen sets tend to use diesel-fueled CI engines, with four-pole generators operating at 1500 RPM (for 50 Hz systems). The main reasons for this are:

• Higher reliability: CI engines can be operated for 20,000 to 30,000 h before a major overhaul is needed—approximately three times as long as an SI engine. This is due to fewer parts, lower speed operation, and self-lubrication provided by the fuel.
• Decreased fuel consumption: CI engines use higher compression ratios, which allows an improvement in efficiency by about one third over an SI engine; in addition, diesel contains approximately 10% more energy per liter than petrol, so that less fuel is needed.

CI engines are less sensitive to the type and quality of the fuel supplied. This can be relevant in the off-grid context.

5.3.2   Frequency Regulation

Most gen sets are designed to operate at constant speed so that the output frequency is also constant. A speed control device called a "governor" increases the mechanical power to the crankshaft when the crankshaft speed decreases, and decreases the power when the speed increases. The mechanical output power is controlled by throttling (adjusting the amount of the air-gas mixture entering the combustion chamber) in an SI engine or fuel metering (adjusting the amount of fuel injected) in a CI engine. A mechanical governor is typically used in older or lower-quality gen sets; modern and higher-quality gen sets use electronic governors.

5.3.3   Excitation

Larger capacity gen sets use either static or brushless excitation. Voltage regulation can be problematic in lower-quality and less-expensive gen sets. Some portable gen sets do not include an AVR; these gen sets should only be used if there are no sensitive electronic loads and if no other generators or inverters are connected to the AC bus.

Small-capacity gen sets typically rely on the residual magnetism in the rotor at start-up. Generators that have been idle for extended periods will need their residual magnetism to be restored in a process known as "field flashing." Field flashing is the temporary application of DC current from an external source such as a battery to the field winding.

5.3.4   Fuel Consumption and Efficiency

Gen sets are expensive to operate. However, their upfront costs are low, and they are readily available in developing countries, making them popular for off-grid electrification. The bulk of the operating costs are for fuel. Some off-grid systems are so remote that the cost to transport the fuel is more expensive than the fuel itself. As such we are concerned with operating them as efficiently as possible. The main losses associated with the ICE portion of a gen set are:

- Friction: from the pistons, bearings, valve train, pumps, and other moving parts
- Heat transfer: heat from combustion that is rejected to the cooling system
- Exhaust: heat in the gases that are emitted from the engine

A general rule of thumb is that a third of the input energy to an ICE is converted to useful mechanical output, a third is rejected to the coolant, and a third is contained in the heat of the exhaust, yielding a mechanical efficiency of 33% [9]. While an easy

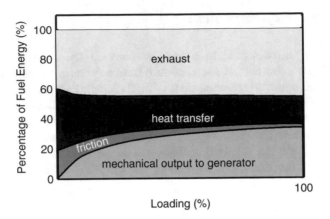

Fig. 5.12 Energy allocation in an ICE

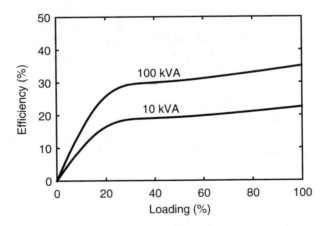

Fig. 5.13 Efficiency curve of gen sets of different rated capacities

rule to remember, it is overly simplistic, as the efficiency of a gen set varies with the load it supplies. Figure 5.12 shows the relative portion of input fuel associated with these losses as load increases [3].

The overall efficiency of a gen set is the product of the ICE and generator efficiencies, both of which depend on the load:

$$\eta_{\text{genset}}(P_e) = \eta_{\text{ICE}}(P_e) \times \eta_{\text{gen}}(P_e) \tag{5.12}$$

with a resulting efficiency curve shown in Fig. 5.13. The maximum efficiency occurs near the rated power and ranges from 20 to 40%.

Table 5.2 shows how diesel fuel consumption, in liters per hour, varies with gen set loading and capacity. The fuel supply and storage aspects should not be overlooked. The fuel requirements of a 50 kW gen set operating at 50% loading continuously for 1 week is nearly 1500 liters.

Table 5.2 Typical gen set fuel consumption

Capacity (kW)	Loading			
	25% (l/h)	50% (l/h)	75% (l/h)	100% (l/h)
10	1.3	2.5	3.5	4.3
50	4.9	8.7	12.5	16.4
100	8.3	15.9	22.3	27.6
500	39.7	73.8	89.7	118.1

Table 5.3 Typical gen set efficiency

Capacity (kW)	25%	50%	75%	100%
10	19	20	22	23
50	26	29	30	31
100	30	32	34	36

Table 5.3 shows the typical efficiency of gen sets of various capacities. The efficiency increases with loading and with generator capacity. Gen sets that are designed to be portable typically have lower efficiencies, as low weight, not efficiency, is prioritized.

Due to the low efficiency at low loading, a rule of thumb is to avoid operating a gen set below 50% of its rated power. If possible, loads should be scheduled simultaneously to concentrate the gen set's operation into fewer hours of high-load operation, instead of low-load operation spread over more hours. Another reason to avoid low loading is "wet stacking." When diesel gen sets operate at low loading, some of the fuel is not combusted. This can cause an oily material to accumulate in the exhaust system, which needs to be removed. In hybrid systems with batteries, a viable operational strategy is to only operate the gen set to charge the batteries when they are at a low state-of-charge. This control strategy and others are described in Chap. 10.

Example 5.5 Consider a 10 kW gen set supplying power to four 2.5 kW pumps. Each pump must operate for 1 h each day. Compute the fuel savings if the pumps are operated simultaneously for 1 h, compared to if the pumps are operated individually over a 4-h period. The gen set fuel consumption is provided in Table 5.2.

Solution When the gen set supplies the combined load of 10 kW, it is fully loaded (100%). From Table 5.2, it consumes 4.3 liters for the 1 h it is on. When the generator supplies each pump individually, it is loaded at 25% and consumes 1.3 liters per hour for a total of 4 h. A total of $4 \times 1.3 = 5.2$ liters is consumed. By operating the pumps simultaneously, the fuel consumption is reduced by nearly 20%.

Example 5.6 Consider an 800 W petrol generator supplying a mini-grid whose load is a continuous 200 W. Compute the monthly fuel cost assuming: petrol costs US$1.0 per liter, the efficiency at 200 W is 13%, and the energy density of the fuel is 31.5 MJ/l.

Solution The monthly energy consumption is

$$Monthly\ Energy\ Consumption = 200\ W \times 24\ h \times 30\ days/month$$

$$= 144\ kWh/month.$$

The input energy is found by dividing the output energy by the efficiency at the loading level:

$$E_{in} = \frac{144}{0.13} = 1107.7\ kWh = 1107.7\ kWh \times 3.6\ MJ/kWh = 3987.7\ MJ.$$

The liters of fuel required each month is:

$$Liters\ per\ Month = \frac{3987.7\ MJ}{31.5\ MJ/liter} = 126.59\ liter.$$

The fuel cost of the generator is therefore US$126.59 per month. Note that an 800 W generator can be purchased for approximately US$200. In less than two months of operation, the fuel costs will have exceeded the capital costs.

5.3.5 *Practical Considerations*

There are several features that make gen sets a common choice for off-grid systems. Gen sets can be purchased in a range of capacities, from a few hundred watts to hundreds of kilowatts. They are widely available and have lower capital costs than most renewable-based energy conversion technologies, at about US$100 to US$300 per kilowatt. Their mechanical nature offers a possibility of repair for minor failures. Gen sets can be operated on-demand with negligible start-up time. When operating as a backup supply for a hybrid renewable system, they can reduce the required battery size. This is done by operating the gen set during occasions with high-demand or low-energy resource availability (e.g., cloudy days for a PV-based system). Portable gen sets are easy to install as they do not require civil works. Most gen sets contain integrated control systems that govern the frequency and magnitude of the voltage they output, making their use convenient.

Gen sets, however, have several disadvantages. Their operating costs are high, primarily due to the fuel expense. They also require a fuel supply chain that must be managed. In remote locations it might be challenging and costly to refuel the generator. Large holding tanks add costs and the potential for leakage and environmental spoilage. The fuel can also only be stored for a certain period of time, perhaps 3 months for petrol to 1 year for diesel.

Gen sets emit air pollution—in the form of carbon dioxide but also particulates and nitric oxides NO_x and sulfur oxide SO_x—as well as noise pollution.

Gen sets have many moving parts which require maintenance, repair, and occasional replacement. Regular oil changes, replacement of air and fuel filters, valve or spark gap adjustment, and other maintenance are regularly required. Trained technicians and replacement parts must be available. The service life of a gen set should be carefully monitored. A typical large-capacity gen set will have a service life of about 3 years of continuous operation, if properly maintained, before a major refurbishment is needed. This is considerably less than a PV array, which might last 15 to 20 years or longer.

5.4 Biomass Systems

Although most gen sets use conventional fossil fuels, they can also be powered from biomass-derived fuel. Biomass systems convert the energy accumulated in organic matter to generate heat and/or electricity. Biomass is distinguished from fossil fuels in that the organic matter in biomass was recently alive. As we discussed in Chap. 2, biomass in the form of fuel wood, charcoal, and crop and animal residue already constitutes a large portion of the energy used by rural off-grid households. This traditional use of biomass is not the subject of this section. Instead, we focus on the biomass systems capable of producing electricity.

Biomass must be processed before it can be efficiently or conveniently used to generate electricity. Depending on the feedstock and process used, different types of biofuels result. Electricity is generated when the biomass-derived fuel is used in an engine to drive a generator as is done in a gen set. The following are biomass-derived fuels used in off-grid electrification:

- Biogas: biomass undergoes a biological process known as anaerobic digestion to produce a gas that is primarily a mixture of methane and carbon dioxide.
- Synthesis gas: biomass is processed through a thermal-chemical reaction to produce a gas consisting of carbon monoxide and hydrogen.
- Solid biomass: biomass is burned as a solid, after being dried and cut.

The input biomass, whatever it may be, is known as the *feedstock*.

5.4.1   Principle of Operation

The energy found in biomass originated as solar radiation (sunlight), which was converted to plant matter[3] through photosynthesis. The energy can be passed up the food chain, albeit inefficiently, from plant to animal and from prey to predator.

In a photosynthetic process, the energy in sunlight combines carbon dioxide and water to form plant matter—sugars, cellulose, and starches, among others. The chemical reaction for glucose, for example, is

$$6CO_2 + 6H_2O + \text{light energy} \rightarrow C_6H_{12}O_6 + 6O_2. \qquad (5.13)$$

When plant matter decays, energy is released in the form of heat. The energy can also be released through combustion. In combustion, oxygen reacts with the plant matter to produce carbon dioxide, water, and energy. In the case of glucose:

$$C_6H_{12}O_6 + 6O_2 \rightarrow 6CO_2 + 6H_2O + \text{energy}. \qquad (5.14)$$

We note that this is simply (5.13) run in the opposite direction. Biomass systems ultimately rely on combustion to generate heat and electricity. Although carbon dioxide is released, many consider the combustion of biomass to be carbon neutral, arguing that the carbon released was recently removed from the atmosphere by the feedstock.

5.4.2   Biomass Resource

The amount of energy stored in the earth's biosphere is immense, approximately 1.5×10^{22} J. There are numerous sources of biomass that can be available at the local level, as shown in Table 5.4 [1]. Both plant matter and animal waste can be used in biomass systems.

Table 5.4 Biomass sources

Gasification	Bagasse, bamboo, cashew nut shells, coconut shells, cotton stalks, forest pruning, grass/bushes, maize cobs and stalks, rice husk, saw dust, wheat and millet straw, wood pulp
Biogas	Bread, cattle and pig manure, chicken litter, eggs, human excreta, slaughterhouse waste, stalks, straw, vegetables and grain leaves

---

[3] We use the term "plant matter" in a broader sense so that it includes all algae, even those species not classified as plants.

Most biomass calculations involve relating the required quantity of input feedstock to produce a certain amount of electrical energy. A critical factor affecting these calculations is the assumed moisture content of the feedstock. The moisture content of fresh cut trees by mass is about 50%. Moisture in manure is higher. Moisture makes up about 75% of the mass of chicken manure and 90% of pig manure. Moisture adds weight and volume, but does not contribute to the energy content of biomass. If the biomass is to be directly burned, then the moisture must be removed beforehand; otherwise the combustion will be inefficient and smoky. Moisture then not only does not contribute to the energy content of biomass, it requires energy to be removed. In applications requiring low-moisture biomass such as gasification and direct combustion, sun drying is an economical option. Sun drying can reduce the moisture of wood to about 10 to 15%. Doing so takes up space and requires oversight.

The energy content of biomass is usually based on the heat produced through combustion. This ranges from about 8 MJ/kg for freshly cut wood to 15 MJ/kg for dried wood [7]. Certain dried crops such as sugarcane and maize may release 18 MJ/kg when combusted. This is low compared to methane, whose specific energy is 56 MJ/kg. The heat released from combustion of dried manure is about 10 MJ/kg, but this varies across different animal species. Daily wet manure production varies from about 0.2 kg for a hen, 3.0 kg for a pig, and about 40 kg for a cow. The biomass yields from crops range from about 5 to 80 dry tonnes per hectare (10,000 m^2) per year, assuming one crop per year. There are many variables that affect the energy content and production rates of biomass, so the values presented in this section should be used for rough estimations only.

The relatively low specific energy of biomass makes its collection, transportation, and preparation burdensome. It is difficult to make a biomass system economically viable unless the biomass has already been concentrated [7]. Biomass is concentrated at agricultural and forest product processing facilities and animal enclosures. To limit transportation costs, the biomass facility should be located close to the feedstock source.

Example 5.7 The electricity requirements of a certain household are 401.5 kWh per year. A biogas system yields 0.3 m^3 of biogas for every 8 kg of input fresh (wet) pig manure. The energy density of the resulting biogas is 23 MJ/m^3. The biogas is used to power a gen set. The efficiency of the gen set is 20%. How many pigs are required if each pig produces 2.3 kg of fresh manure per day?

Solution The average daily electrical energy consumption of the household is

$$E_{\text{daily,elec}} = \frac{401.5 \text{ kWh/yr}}{365 \text{ days/yr}} = 1.1 \text{ kWh/day}.$$

(continued)

Each day the manure from a single pig produces biogas with volume:

$$V_{\text{gas,pig}} = 2.3 \text{ kg} \times \frac{0.3 \text{ m}^3}{8 \text{ kg}} = 0.0863 \text{m}^3.$$

The biogas produced from each pig each day is converted to electrical energy according to

$$E_{\text{gas,pig}} = 0.0863 \text{m}^3 \times 23 \text{ MJ/m}^3 \times 0.20$$

$$= 0.397 \text{ MJ} = \frac{0.397 \text{ MJ}}{3.6 \text{ MJ/kWh}} = 0.110 \text{ kWh/pig/day}$$

The total number of pigs required is therefore

$$\frac{1.1 \text{ kWh/day}}{0.11 \text{ kWh/day/pig}} = 10 \text{ pigs.}$$

The family must own at least ten pigs for this plan to work.

5.4.3 Biogas

Solid biomass is an inconvenient fuel source. Converting it into gas can increase its energy density and allow it to be efficiently used in gas turbines and gas engines. Biomass is converted to biogas—a mixture composed primarily of methane (CH_4) and carbon dioxide—through anaerobic digestion. The biogas can then be combusted in a gen set to produce electricity. It can also be used in a stove for cooking and for heating and lighting. Anaerobic digestion relies on microorganisms to process the biomass under oxygen-free conditions. Digestion occurs over a period time, usually tens of days. Little energy is consumed, and so the efficiency of converting the biomass to biogas is relatively high. Around 85% of the energy from the input biomass is available in the produced biogas. However, in most small-scale settings, and with shorter digestion times, the efficiency is closer to 60%[7].

Anaerobic digestion is a biological process, relying on various types of bacteria to occur. The general process can be broken down into three parts: (1) in *hydrolysis,* the complex organic polymers are broken down to form sugars, fatty acids, and amino acids; (2) during *acidiogenesis,* and *acetogenesis,* the sugars, fatty acids, and amino acids are converted to acetates, single-carbon compounds, hydrogen, and carbon dioxide; and (3) in *methanogenesis,* bacteria produce biogas from the hydrogen, carbon and acetate [7, 8].

The basic reaction inputs are biomass and water, and the result is a mixture of carbon dioxide and methane. The reaction for a general organic polymer with n

Fig. 5.14 Biogas digester vessel

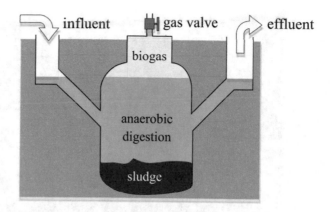

carbon, a hydrogen and b oxygen atoms is

$$C_nH_aO_b+(n-a/4-b/2)H_2O \rightarrow (n/2-a/8+b/4)CO_2+(n/2+a/8-b/4)CH_4$$
$$(5.15)$$

The resulting gas (biogas) is typically 50–80% methane by volume.

Anaerobic digestion is promoted by wet, dark, and warm—but not hot—conditions. The requirement of water in the reaction means that the input biomass is typically a slurry. The feedstock is often manure or crop waste. Woody biomass is not suitable for digestion due to its high content of lignin, which cannot be broken down by anaerobic bacteria. The high moisture content of manure makes it a convenient feedstock for creating biogas—if it were used to create synthesis gas or directly combusted, it would have to be dried.

Anaerobic digestion occurs in a reactor vessel called a "digester." The basic scheme is shown in Fig. 5.14. Digesters can be simply constructed. They are not exposed to high temperatures or require mixing, although some mixing can improve the yield of biogas. Digesters are commonly made from brick, concrete, and/or steel on or below ground, as seen in Fig. 5.15. Digesters should be gas tight, with a piping system to remove the biogas and a hatch to introduce additional biomass. For digestion to commence, the anaerobic bacteria must be introduced into the digester. This is called inoculation. Conveniently, the bacteria are naturally found in some manure.

The rate at which biogas is produced depends on the temperature in the digester. Increasing temperature promotes methane production—every 10°C increase in temperature approximately doubles the process speed. This reduces the physical size requirements of the digester. However, at around 60°C, the microbial activity quickly drops. Most digesters operate at temperatures between 30 and 40°C. A typical fermentation (digestion) period ranges from 10 to 30 days, depending on the feedstock and digester conditions. Each dry kilogram of biomass input will yield about 0.2 to 0.4 m^3 of biogas. Biogas has an energy content of approximately 20 to 23 MJ/m^3 at atmospheric pressure. This is lower than pure methane (28 MJ/m^3) due to the relatively high carbon dioxide content. A few cubic meters of biogas per

Fig. 5.15 A below-ground biogas digester being constructed in Malawi (courtesy P. Dauenhauer)

day is enough to meet the electricity requirements of a household. The minimum required volume of the digester is found from

$$Digester\ Volume = input\ flow\ (m^3/day) \times fermentation\ period\ (days). \quad (5.16)$$

Biogas can be produced in batches or continuously—with the removal and addition of perhaps 5% of the material per day. Continuous production tends to be more efficient. Biogas often contains of hydrogen sulfide, which should be removed before use in an internal combustion engine. The carbon dioxide can be removed by spraying the biogas with water. The residue from digesters using manure feedstock is a valuable fertilizer. In fact, this is a main motivator for using anaerobic digestion.

Because of their simplicity in construction and management, tens of millions of digesters have been deployed in rural settings. The digesters tend to be small, less than 10 m^3 in volume and cost just several hundred dollars to build. Most use the biogas for cooking and heating, not electricity generation.

5.4.4 Biomass Gasification

Biomass gasification differs from biogas production in that thermal-chemical reactions rather than a biological process are used [2, 4, 6, 7]. The resulting product is a gaseous fuel, known as "synthesis gas" or simply "syngas." The term "bio syngas" is also used to distinguish it from syngas derived from other processes. The exact nature of the syngas depends on the specific process used. The gasification process is complex but fundamentally involves the use of heat, steam, and oxygen

Fig. 5.16 Downdraft
biomass gasification process

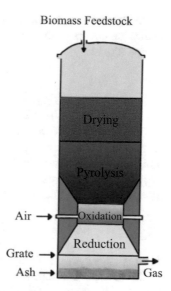

Biomass Feedstock

to convert the biomass to syngas. Syngas is composed primarily of hydrogen and carbon monoxide, with some carbon dioxide, and a small amount of methane.

It is important that the feedstock has low moisture content. Crop residue and woody biomass are often used. The basic gasification reaction using glucose as an example is

$$C_6H_{12}O_6 + O_2 + H_2O \rightarrow xCO + yCO_2 + zH_2 + \text{other species}. \tag{5.17}$$

Note that the relative quantities of carbon monoxide, carbon dioxide, and hydrogen vary.

Biomass gasification occurs in a reactor (also known as a "gasifier"), which itself resembles a large metal vessel. Most reactors in rural electrification applications are of the fixed-bed type, of which there are several varieties including updraft, downdraft, and cross-draft. We shall consider the downdraft variety shown in Fig. 5.16. Downdraft gasifiers are best-suited for the small-scale systems used in off-grid electrification. These are typically between 10 and 100 kW of capacity.

Prior to gasification, the biomass must be prepared. This typically involves harvesting, cutting to size, and sun drying. It is then loaded into the top of the reactor. The top of the reactor is at an elevated temperature—100 to 160°C—due to the reactions occurring below. The elevated temperature dries the biomass further. As we proceed down into the reactor, the temperature increases and there is less air (due to the biomass above and below it). As the biomass temperature reaches several hundred degrees Celsius, pyrolysis occurs. Recall from Sect. 2.4.2 that pyrolysis is the thermal-chemical process used to manufacture charcoal. Volatiles and tar are produced, including methane and hydrogen. The remaining biomass is known as "char," which is primarily carbon.

As we continue down the reactor, we reach a section where air is introduced. Oxidation occurs as the air reacts with the carbon in the char, producing heat and carbon dioxide

$$C + O_2 \rightarrow CO_2. \qquad (5.18)$$

The temperature in this section is the highest. Only a portion of char is converted to carbon dioxide. The heat produced by the oxidation in (5.18) supplies energy for the remaining char to react with steam. The steam comes from various sources: residual moisture in the biomass, in the introduced air, and the combination of hydrogen and oxygen gases. The reaction between char and steam forms hydrogen and carbon monoxide in the following reduction reaction

$$C + H_2O \rightarrow H_2 + CO. \qquad (5.19)$$

The resulting syngas is at high temperature ($>700°C$). It is removed near the bottom of the reactor. The leftover ash collects at the bottom of the reactor where it is periodically removed. The resulting gas must be filtered to remove ash and tar. The syngas must also be cooled before it can be used in a combustion engine. These processes create waste which must be appropriately managed and also require a reliable source of water.

The energy content of syngas is typically 3 to 6 MJ/m^3; the low end of this range is appropriate for syngas made in smaller-scale settings. This is much lower than natural gas (approx. 30 MJ/m^3) and biogas. The energy content can be increased to 10 to 19 MJ/m^3 by injecting steam or oxygen instead of air during the oxidation stage. But this adds to the cost and complexity and is not typically done in small-scale systems. For each kilogram of dried input biomass, approximately 1.5 m^3 of syngas is produced.

The efficiency of gasification is determined by dividing the energy content of the syngas produced by the energy in the biomass consumed. This is known as the "cold gas efficiency," as it ignores the thermal energy of the syngas. The cold gas efficiency is typically between 20 and 70%. Again, small-scale gasifiers tend to have efficiencies on the low end of the range. This does not include the efficiency losses in the engine and generator.

Biomass gasification units are commercially available. They typically range from tens of kilowatts to a few hundred kilowatts in capacity. Electricity generation from biomass gasification has proven technically and economically successful in certain areas of South Asia. India is reported to have installed over 150 MW of gasification-based power plants, most of which are tens of kilowatts in capacity or less.

Syngas can be directly combusted in a gas turbine or gas engine to drive a generator. Engines designed to run on petrol or diesel can be modified to run on syngas, although their rated power is usually decreased. It is also common to use a mixture of syngas and diesel in an engine. In this "dual fuel" mode, diesel serves as the pilot fuel, and the syngas is mixed with the air prior to combustion. This relatively easy to implement modification can reduce diesel consumption by 80%.

Example 5.8 Consider a biomass gasification system serving 400 houses. Each house consumes 300 Wh of electricity per day. The gasification system is 60% efficient, and the combined efficiency of the engine and generator is 33.3%. The energy content of the dry husk feedstock is 12.6 MJ/kg, which costs US$25 per metric ton. Compute the mass of husk required each day and cost of fuel per kilowatthour of electricity consumption.

Solution The solution process requires several conversions. The daily electrical demand, expressed in megajoules, is

$$(0.300 \times 400) \, \frac{3.6 \, \text{MJ}}{1 \, \text{kWh}} = 432 \, \text{MJ}.$$

The required energy content of the input husks is found by accounting for the efficiency of the gasification and the engine/generator:

$$432 \times \frac{1}{0.6 \times 0.333} = 2162.2 \, \text{MJ}.$$

The total mass of husk required each day is therefore

$$2162.2 \times \frac{1 \text{kg}}{12.6 \text{MJ}} = 171.60 \, \text{kg}.$$

The cost of a single kilogram of husk is US$0.025. Therefore, the cost per day is

$$171.60 \, \text{kg} \times \text{US}\$0.025/\text{kg} = \text{US}\$4.29/\text{day}.$$

The cost per kilowatthour is

$$\frac{\text{US}\$4.29}{0.30 \, \text{kWh/house/day} \times 400 \, \text{houses}} = \text{US}\$0.0358/\text{kWh}$$

At a glance, the fuel cost per kilowatthour provided by this system is favorable to grid-provided electricity. However, the fuel cost does not include capital or other operational costs.

5.4.5 Solid Biomass Direct Combustion

Biomass can also simply be combusted to produce heat. The energy released is used to create steam, which drives a steam turbine. The steam turbine is coupled to a generator. This scheme is only practical for larger-capacity systems, typically hundreds of kilowatts to about 20 MW. Solid biomass direct combustion systems are commonly colocated in agricultural and forest product processing facilities. The crop and forest waste serves as a feedstock, eliminating the need for their removal, and a portion of the heat and steam produced can be used in the processing facility. This arrangement is known as "co-generation" or "combined heat and power" (CHP). CHP should be considered whenever possible—it makes use of heat that would otherwise be wasted. Some of the electricity is used to power the facility; any excess can be used to supply a mini-grid.

The biomass used in direct combustion applications must be processed. This typically includes chopping and drying. Like gasification facilities, direct combustion facilities required trained technicians to manage the process.

5.4.6 Practical Considerations

The use of biomass for off-grid electricity generator offers several advantages, depending on the process:

- plant-based biomass can be considered a renewable resource if the rate of use does not exceed the rate of plant growth;
- some biomass processes produce a number of usable by-products, including fertilizer;
- biomass systems can make use of waste products that otherwise would have to be removed from a facility;
- biomass fuels can be stored, allowing generators to produce power on demand;
- in direct combustion systems operating as CHP facilities, the waste heat and steam can be used locally for agricultural or manufacturing processes;
- the collection, transportation and processing of biomass creates employment at the local level and can improve cash flow in rural areas.

The disadvantages include:

- maintaining a consistent supply chain of biomass is problematic; many crop-based agricultural inputs are seasonal and a biomass system might be unable to operate at full capacity during certain times of the year;
- managing digesters, gasifiers, and direct combustion systems requires some training and, in some cases, full-time operators;
- certain biomass processes require waste to be removed;
- certain biomass processes require a reliable source of water;

- the price of the feedstock can be volatile—it should not be assumed that crop residue and other wastes can be obtained for free;
- biomass systems coupled with an agricultural or forest processing facility will likely not continue to be economically viable if that facility shutters.

From a broader perspective, biomass has many uses including as food, to prevent erosion, and as raw material. Replacing food crops or converting forest land to grow biomass crops should generally be avoided.

5.5 Summary

This chapter provided the basic technical principles and characteristics of generators and internal combustion engine-coupled generator sets using conventional and biomass fuels. A generator outputs AC voltage whose magnitude and frequency depend upon the rotational speed of the rotor. Matching the required generator speed with the speed-power or efficiency curve of the prime mover is an important design consideration.

Voltage is induced in a generator's armature windings using a rotating electro-magnet. An exciter is used to supply current to the electromagnet. An Automatic Voltage Regulator can be used to control the terminal voltage of the generator. If the load increases without a similar increase in the mechanical power supplied to the generator, then the rotational speed will decrease; similarly, if the load decreases, the speed will increase. The voltage frequency and voltage magnitude are proportional to the generator's rotational speed. It is important for AC-coupled generators to be operated at constant speed, for example, using a governor, and have voltage control. Generators are most efficient when operating at or near their rated power.

Gen sets use internal combustion engines as their prime mover. The engine can either be spark ignition or compression ignition. Gen sets are relatively inexpensive compared to other energy conversion technologies but are more expensive to operate due to high fuel costs. Their fuel consumption can be decreased by operating them at or near full load for shorter periods of time rather than lightly loaded for longer periods of time.

Biomass can be converted to fuels suitable for use in internal combustion engines using anaerobic digestion to produce biogas or gasification to produce syngas. Biogas is primarily methane and carbon dioxide; syngas is primarily hydrogen and carbon monoxide. Biomass can also be directly combusted and used in steam turbines. Biomass systems critically rely on a consistent supply of its feedstock, which can be forest products, crops and crop residue, or animal excrement.

Problems

5.1 Compute the rotational speed in RPM and radians per second, and the torque of a four-pole generator whose developed electrical power is 10 kW at 60 Hz.

5.2 Consider a three-phase 20 kVA gen set whose line-to-neutral induced voltage is $E_a = 223\angle 0°$ V. The single-phase current supplied is $I_a = 35\angle - 60°$ A. The armature impedance is $0.6 + j3.6$ Ω. Compute the terminal line-to-neutral voltage and the total (three-phase) apparent power provided to the load.

5.3 Repeat Problem 5.2 but with $I_a = 35\angle 60°$ A.

5.4 A large mini-grid is expected to have a daily load of 980 kWh. Assume the energy is consumed at a constant rate throughout the day. Let the fuel cost be US$1.25/liter. Three options are being considered to serve the mini-grid:

- five equally loaded 10 kW gen sets;
- one 50 kW gen set;
- one 100 kW gen set.

Compute the annual cost associated with each design. Refer to Table 5.2 for the fuel consumption for each gen set. Assume the fuel consumption varies linearly between the loading percentages given in Table 5.2.

5.5 Describe the difference in the processes that convert solid biomass to biogas and to syngas.

5.6 A biogas system is being planned for a community. The system should be sized to provide 12 kWh/day of electricity. The biogas is expected to have a density of 20 MJ/m^3. Determine the volume of biogas that must be supplied each day to a gen set whose efficiency is 19%.

5.7 The feedstock to a gasification system is bagasse (sugarcane crop residue). Each day 1000 kg of the feedstock is available. After drying, the dried sugarcane's specific energy is 16.5 MJ/kg. Compute the energy available for input to the gasification system if the moisture content of bagasse is 40%.

5.8 Consider the gasification system in Problem 5.7. Assume the efficiency of the gasification process is 50% and the gen set efficiency is 24%. Compute the daily electrical energy provided by the system.

5.9 A mini-grid is to be implemented for the community of Ababju. The community has 500 households. Each household is expected to consume 600 Wh per day. Assume the power consumption is constant throughout the day. The mini-grid will be powered using two 10 kW gen sets. Let the price of diesel be US$1.35/liter. Perform the following analyses:

- Compute the annual fuel cost of the gen sets. Assume the gen sets are equally loaded and the fuel consumption is in Table 5.3 (use linear interpolation between the data points). Repeat this calculation if one gen set is fully loaded and other is loaded such that the required power is supplied. Comment on which loading strategy is more fuel efficient.

- The crop resource around Ababju has a specific energy of 15.4 MJ/kg (dry) and costs US$0.02/kg (dry). A gasification system with an efficiency of 57% can be used to convert the biomass to syngas. The syngas will be used in a dual-fuel gen set. The energy supplied to the gen set is 60% from syngas and 40% from diesel. Compute the annual fuel cost (biomass plus diesel) of using this system to power the mini-grid if the gen sets are equally loaded and if one gen set is fully loaded and other is loaded such that the required power is supplied.

References

1. Bhattacharyya, S.C. (ed.): Rural Electrification Through Decentralised Off-Grid Systems in Developing Countries. Springer (2013)
2. Boyle, G. (ed.): Renewable Energy, 2nd edn. Oxford University Press (2004)
3. Caton, J.: An Introduction to Thermodynamic Cycle Simulations for Internal Combustion Engines. Wiley (2016)
4. Goswami, Y. (ed.): Alternative Energy in Agriculture, vol. 2. CRC Press (1983)
5. McLean, G.: Chapter 5 - generators. In: Warne, D. (ed.) Newnes Electrical Power Engineer's Handbook, 2nd edn., pp. 105–133. Newnes, Oxford (2005). doi:https://doi.org/10.1016/B978-075066268-0/50005-6. URL http://www.sciencedirect.com/science/article/pii/B97807506 62680500056
6. Sørensen, B.: Renewable Energy, 3rd edn. Elsevier Academic Press (2004)
7. Twidell, J., Weir, A.: Renewable Energy Resources, 2nd edn. Taylor & Francis (2006)
8. Wilkie, A.C.: Biomethane from biomass, biowaste, and biofuels. In: Bioenergy, pp. 195–205. American Society of Microbiology (2008). URL http://www.asmscience.org/content/book/10.1128/9781555815547.ch16
9. Wong, E., Whitall, H., Dailey, P.: Distributed Generation: The Power Paradigm for the New Millennium, chap. 2. CRC Press (2001)

Chapter 6
Off-Grid Wind and Hydro Power Systems

6.1 Introduction

This chapter continues our discussion on energy conversion technologies. The on-going cost and difficulty in maintaining a reliable supply of fossil or biomass fuel make gen sets impractical for some off-grid applications. Wind turbines, hydroturbines, and photovoltaic arrays have no fuel costs. Instead, they make use of locally available and replenishable energy sources. This chapter describes the principle of operation, modeling, analysis, and practical considerations of wind energy conversion systems and micro hydro power systems.

6.2 Wind Energy Conversion Systems

There are an estimated 800,000 small-scale stand-alone wind turbines used for off-grid electricity access worldwide [6]. A wind energy conversion system (WECS) uses a wind turbine as a prime mover. The wind turbine converts energy in moving air into rotational energy that is used by a generator to produce electricity. The term "wind turbine" commonly refers to the complete energy conversion system, including the generator. However, because we will discuss the generator and turbine separately, we use "wind turbine" to strictly refer to the prime mover portion of the WECS. The WECS used in off-grid applications are much smaller in physical size and electrical capacity than those in utility-scale applications. WECS used in off-grid systems typically range from a few hundred watts to perhaps 30 kW in capacity.

© Springer International Publishing AG, part of Springer Nature 2018
H. Louie, *Off-Grid Electrical Systems in Developing Countries*,
https://doi.org/10.1007/978-3-319-91890-7_6

Fig. 6.1 A 400 W
three-blade horizontal axis
wind turbine with blade
length of 0.5 m. An
anemometer is used to
measure the wind speed

6.2.1 Principle of Operation

The basic components of a WECS are the blades, hub, rotor, and nacelle, as shown
in Fig. 6.1. The nacelle is an enclosure that houses the generator, controls, and other
related components. The wind turbine typically has three blades, which rotate about
a horizontal axis. The WECS is mounted atop a tower to expose it to faster and more
consistent wind.

The blades are connected to the hub. The hub is connected to the turbine's shaft
inside the nacelle. The blades are aerodynamically designed to generate lift as air
passes over them. The lift exerts a mechanical torque on the blades, which causes
the blades, hub, and shaft to rotate. The shaft is either coupled to the rotor of the
generator directly or through a gearbox. Only large-capacity WECS, typically those
over 10 kW, use gear boxes. Most WECS have a tail vane that points the hub into
the wind to maximize power production.

6.2.2 Power in the Air

The electrical power a WECS can produce is dependent on the power of the air
passing through the area swept by the blades. Consider a wind turbine whose blades
have length r_{WT} when measured from their tip to center of the hub, as shown in
Fig. 6.2. As the hub rotates, the blades sweep out a circular disk with area A_{WT}. The
swept area is found by simple geometry $A_{WT} = \pi r_{WT}^2$.

Let the air passing through the swept area have constant and uniform velocity
v_{air}. After t seconds, some length l of air will have passed through the swept area.

Fig. 6.2 Swept area of a wind turbine

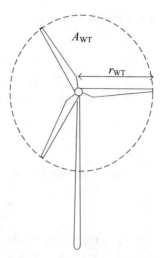

Fig. 6.3 The volume of air passing through the area swept by the blades of a wind turbine

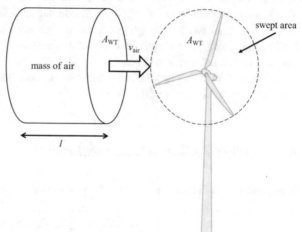

The length is $l = v_{air} \times t$. The shape of the air passing through the swept area is cylindrical as shown in Fig. 6.3. Its mass is found by multiplying its volume by its density:

$$m_{air} = A_{WT} \times l \times \rho_{air} = A_{WT} \times v_{air} \times t \times \rho_{air} \qquad (6.1)$$

where ρ_{air} is the density of the air, typically 1.23 kg/m^3. This varies somewhat with pressure and temperature. The air passing through the swept area has kinetic energy:

$$E_{air} = \frac{1}{2} m_{air} v_{air}^2 = \frac{1}{2} A_{WT} l \rho_{air} v_{air}^2 = \frac{1}{2} A_{WT} v_{air} t \rho_{air} v_{air}^2 = \frac{1}{2} A_{WT} t \rho_{air} v_{air}^3.$$
$$(6.2)$$

Fig. 6.4 Power flow and
losses in a WECS

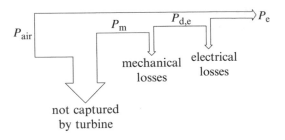

To determine the power of the air, (6.2) is differentiated with respect to time:

$$P_{\text{air}} = \frac{dE_{\text{air}}}{dt} = \frac{1}{2} A_{\text{WT}} \rho_{\text{air}} v_{\text{air}}^3. \qquad (6.3)$$

We make the following observations. First, P_{air} is the power available to the wind
turbine, not the mechanical power produced by the wind turbine or the generator.
The relationship between the power in the air and the output power is shown
in Fig. 6.4. Second, P_{air} is proportional to the area swept by the wind turbine
blades. Increasing the length of the blades increases the available power. Third,
and importantly, there is a cubic relationship between the speed of the air and the
power available to the turbine. In other words, doubling the wind speed increases
the power by a factor of eight (2^3).

6.2.3   Wind Turbine Mechanical Power

The wind turbine is only able convert a portion of P_{air} to usable mechanical power:

$$P_{\text{m}} = C_{\text{p}} P_{\text{air}} = \frac{1}{2} C_{\text{p}} A_{\text{WT}} \rho_{\text{air}} v_{\text{air}}^3 \qquad (6.4)$$

where P_{m} is the mechanical power input to the turbine shaft and C_{p} is the power
coefficient. The power coefficient can be thought of as the efficiency of the wind
turbine in extracting power from the air. Although the derivation is beyond the scope
of this book, C_{p} cannot exceed the Betz Limit[1] of 0.593. In practice, the power
coefficient is normally between to 0.2 and 0.4.

There are several factors that influence the power coefficient. The most significant are:

- the pitch angle of the blades β;
- the tip speed ratio (TSR) λ;
- the number of blades.

---

[1]The Betz Limit states that a wind turbine, regardless of design, cannot extract more than 16/27
(0.593) of the kinetic energy of wind.

Fig. 6.5 Pitch angle of the
blade of a wind turbine

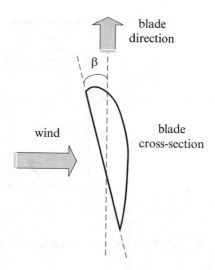

6.2.3.1   Pitch Angle

The pitch angle β is the angle between the plane of rotation and the chord of the blade, as shown in Fig. 6.5. The pitch affects the aerodynamics of the turbine, and consequentially C_p. Small-capacity wind turbines use fixed-pitch blades; in larger-capacity turbines, the pitch angle can be controlled, but this adds additional cost and complexity.

6.2.3.2   Tip Speed Ratio (TSR)

The TSR λ is a unitless quantity defined as the ratio of the tangential velocity of the tip of the rotor blade U to the velocity of the incident air v_{air}:

$$\lambda = \frac{U}{v_{air}}. \tag{6.5}$$

The tip speed depends on the blade length and rotational speed:

$$U = \omega_m r_{WT} = 2\pi N_m r_{WT} \times \frac{1}{60} \tag{6.6}$$

where ω_m and N_m are the rotational speeds of the blade in radians per second and RPM, respectively. To achieve the same TSR at a given wind speed, turbines with shorter blades must rotate more quickly (greater revolutions per minute) than those with longer blades.

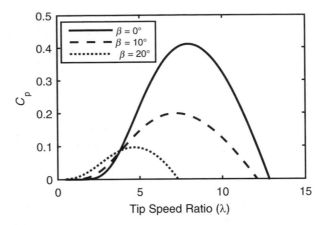

Fig. 6.6 Power coefficient versus tip speed ratio for various pitch angles

6.2.3.3 Number of Blades

The TSR that maximizes the power coefficient depends on the number of blades the wind turbine has. As the number of blades increases, the optimal TSR decreases. So, for a given blade length and wind speed, a wind turbine with more blades will rotate slower, but with greater torque, than a wind turbine with fewer blades. We will assume that the wind turbine has three blades hereafter. This is the most common design.

6.2.3.4 C_p-Curve

A plot of the power coefficient versus the TSR is known as the C_p-curve. An example C_p-curve for a three-blade wind turbine is shown in Fig. 6.6 for various pitch angles. The optimal efficiency typically occurs when the TSR is between six and eight, meaning that the velocity of the tip of blade is six to eight times that of the incoming air. Rotating faster or slower than this reduces the power coefficient and therefore reduces the power available for electricity generation.

6.2.3.5 Speed–Mechanical Power Curve

The C_p-curve is a function of the TSR and the pitch angle, and so it is sometimes written as $C_p(\lambda, \beta)$. We are interested in expressing the power coefficient in terms of rotation speed $C_p(N_m, v_{air}, \beta)$ where N_m is the speed of the rotor shaft, in RPM. This can be done for a given wind speed by substitution using (6.5) and (6.6)

$$N_m = \lambda \times v_{air} \times 60 \times \frac{1}{2\pi\, r_{WT}}. \tag{6.7}$$

Fig. 6.7 Example speed–mechanical power curve for a three-bladed wind turbine with a pitch angle of 0° at different wind speeds

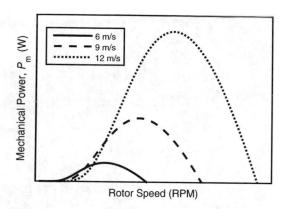

The mechanical power in (6.4) can then be re-written as

$$P_m(N_m, v_{air}, \beta) = \frac{1}{2} C_p(N_m, v_{air}, \beta) A_{WT} \rho_{air} v_{air}^3. \tag{6.8}$$

The plot of $P_m(N_m, v_{air}, \beta)$ versus rotational speed is known as the speed–mechanical power curve of the wind turbine. There is a different curve for each wind speed, as shown in Fig. 6.7. The optimal rotor speed increases with the wind speed. Note that Fig. 6.7 is for a specific wind turbine and with a pitch of $\beta = 0°$. A different pitch will result in different curves. Each wind turbine has a different speed–mechanical power curve.

Example 6.1 An energy kiosk uses a wind turbine to recharge lead-acid batteries. The blade length when measured from the center of the hub to the tip of a blade is 3.2 m. The turbine has three blades, and C_p is maximized when the TSR is 6.84. Compute the required rotor speed of the generator, in RPM, for the turbine to operate at optimum efficiency when the wind speed is 12 m/s. Compute the frequency of the produced AC voltage, assuming the turbine is directly coupled to an eight-pole synchronous generator.

Solution The speed that the tip of the rotor blade must travel to maximize the performance coefficient is found from (6.5):

$$U = v_{air}\lambda = 12 \times 6.84 = 82.08 \text{ m/s.}$$

The length of the blade is 3.2 m so that the turbine must be rotating at:

$$\omega_m = \frac{82.08}{3.2} = 25.65 \text{ rad/s.}$$

(continued)

Converting to RPM:

$$N_m = 25.65 \times \frac{60}{2\pi} = 244.9 \text{ RPM}$$

The frequency of the voltage is found using (5.4):

$$f_e = 244.9 \times \frac{8}{2 \times 60} = 16.33 \text{ Hz.}$$

This frequency is too low for direct coupling to most loads.

The speed–mechanical power curve of the turbine shows that for a given wind speed, the turbine might rotate at a range of speeds. What will the rotation speed actually be? There is nothing about the wind turbine that forces it to rotate at the speed that maximizes P_m (or equivalently C_p). Rather, the wind turbine will operate at the speed that balances the mechanical power input (assuming no mechanical losses) with the developed electrical power. The wind turbine will accelerate or decelerate until this condition is matched. This follows from the law of conservation of energy as discussed in Sect. 5.2.5. The electrical power developed depends on the characteristics of the generator and the load it supplies, as discussed next.

6.2.4 Generator

WECS in off-grid systems often use a synchronous generator with permanent magnet excitation (PMSG). The generator is usually three-phase. Induction generators can be used, but they require reactive power to function. They are only used in hybrid mini-grid architectures in which another generator, for example, a gen set, supplies reactive power or if auxiliary excitation capacitors are included in the WECS. However, most commercially available WECS designed for off-grid applications use PMSG. We therefore limit our following discussion to PMSG.

The magnets used in PMSG are often made from neodymium or other rare earth metals. These magnets are substantially stronger than the ceramic magnets found in household items. The magnets can be mounted on the rotor so that their flux is parallel (axial flux) or perpendicular (radial flux) to the axis of rotation. In the axial flux configuration, the rotor and stators are on parallel planes and are shaped like flat disks with a small air gap between them. Axial flux generators are used in "homemade" WECS that have been deployed with varying success in off-grid systems [1, 8].

6.2.4.1 Speed–Developed Power Curve: Resistive Load

The speed that the generator in a WECS operates increases and decreases with the wind speed. The resulting fluctuating voltage magnitude and frequency prohibits the direct coupling of a WECS to most loads. Rather, the output is typically rectified to DC and used to charge batteries.

As the rotational speed of the generator changes, so does the power it outputs. To understand why, consider the equivalent circuit of the PMSG, which is the same as in Fig. 5.5. From (5.4), the voltage magnitude and frequency increase in proportion to the rotational speed. We can therefore express the induced voltage in a phase of a wye-connected generator as the phasor:

$$E_a = k_g \omega_m \angle 0° \tag{6.9}$$

where k_g is a constant whose value depends on the flux linking each winding and the physical characteristics of the generator, including how the windings are wound within the stator and how they are connected.

For simplicity, assume that the generator is connected to a resistive load R_L such as a water heater as shown in Fig. 6.8. The current produced by a single phase of the generator is:

$$E_a = I_a \left(R_a + j\omega_e L + R_L \right) \tag{6.10}$$

$$k_g \omega_m = I_a \left(R_a + j\frac{p}{2}\omega_m L + R_L \right) \tag{6.11}$$

$$I_a = \frac{k_g \omega_m}{R_a + j\frac{p}{2}\omega_m L + R_L} \tag{6.12}$$

where p is the number of poles, R_a is the stator winding resistance and L is the inductance corresponding to the synchronous reactance X_s.

The real electrical power output by a three-phase generator is

$$P_e = 3|I_a|^2 R_L = \frac{k_g^2 \omega_m^2}{|\left(R_a + j\frac{p}{2}\omega_m L + R_L \right)|^2} R_L = 3\frac{k_g^2 \omega_m^2}{(R_a + R_L)^2 + \frac{p^2}{4}\omega_m^2 L^2} R_L. \tag{6.13}$$

Fig. 6.8 Equivalent single-phase circuit of a WECS connected to a resistive load

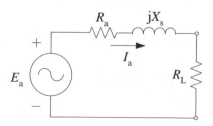

Fig. 6.9 Electrical power
developed (output power plus
electrical losses) by a WECS
connected to resistive loads

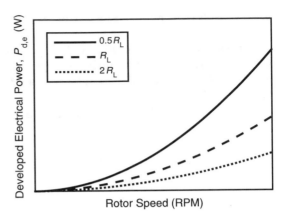

Fig. 6.9 Electrical power developed (output power plus electrical losses) by a WECS connected to resistive loads

As shown in Fig. 6.4, if we ignore mechanical losses, then the mechanical power output by the turbine must equal $P_{d,e}$, the power *developed* by the generator. The power developed by the generator is the total electrical power associated with the generator: its power output P_e and the electrical losses $P_{e,loss}$ associated with the winding resistance, as shown in Fig. 6.4. The developed power is found from:

$$P_{d,e} = P_e + P_{e,loss} = 3|I_a|^2 (R_L + R_a) \tag{6.14}$$

$$P_{d,e} = 3 \frac{k_g^2 \omega_m^2}{(R_a + R_L)^2 + \frac{p^2}{4} \omega_m^2 L^2} (R_L + R_a). \tag{6.15}$$

A plot of $P_{d,e}$ versus ω_m is known as the speed–developed power curve of the generator. An example is shown in Fig. 6.9. Keep in mind that this is the developed power, not the output power. The presence of ω_m^2 in the numerator and denominator shows the opposing influence of rotational speed on power output. On the one hand, increased speed increases the induced voltage, which tends to increase power. On the other hand, the inductive reactance also increases with speed. This acts to reduce the current and therefore the power.

Figure 6.9 shows the effect of different resistive loads on the speed–power curve. As the resistance decreases, the WECS develops more power at a given RPM. It will be shown later how this can be used to control the operational speed of the WECS. However, decreasing the load resistance also decreases the efficiency of the generator. The efficiency of the generator portion of the WECS can more generally be written as

$$\eta_{gen} = 100 \times \frac{P_e}{P_{d,e}}. \tag{6.16}$$

For a resistive load, it can be shown that the efficiency is

$$\eta_{gen} = 100 \times \left(1 - \frac{R_a}{R_a + R_L}\right). \tag{6.17}$$

Example 6.2 A three-phase WECS is connected to a 5 Ω resistive load. The generator has a stator resistance of 0.3 Ω and inductance of 3 mH. The wind turbine has 8 poles and k_g is 2 V/rad/s. Compute the power developed, power supplied to the load, the electromagnetic torque, and the generator efficiency when the wind turbine rotates at 200 RPM and at 300 RPM.

Solution The mechanical frequency corresponding to 200 RPM is:

$$\omega_m = 2\pi \frac{200}{60} = 20.94 \text{ rad/s}.$$

The power to the load is found using (6.13).

$$P_e = \frac{3k_g^2 \omega_m^2 R_L}{(R_a + R_L)^2 + \frac{p^2}{4}\omega_m^2 L^2} = \frac{3 \times 2^2 \times 20.94^2 \times 5}{(0.3 + 5)^2 + \frac{8^2}{4} \times 20.94^2 \times 0.003^2} = 934.85 \text{ W}.$$

The power developed includes the power losses associated with the stator winding. Using (6.14)

$$P_{d,e} = \frac{3k_g^2 \omega_m^2 (R_L + R_a)}{(R_a + R_L)^2 + \frac{p^2}{4}\omega_m^2 L^2} = \frac{3 \times 2^2 \times 20.94^2 \times (5 + 0.3)}{(0.3 + 5)^2 + \frac{8^2}{4} 20.94^2 \times 0.003^2} = 990.94 \text{ W}.$$

The torque is found by dividing the power developed by the mechanical speed:

$$T = \frac{992.61}{20.94} = 47.31 \text{ N m}.$$

The efficiency can be calculated from (6.16) or (6.17) to be

$$\eta_{gen} = \frac{936.43}{992.61} = 94.34\%.$$

Repeating the calculations for 300 RPM shows that the power output increases to 2097.5 W. The power developed becomes 2223.4 W, with corresponding torque of 70.78 N m, and efficiency of the generator is the same: 94.4%.

6.2.4.2 Speed–Developed Power Curve: Battery Charging

In most off-grid applications, a WECS charges a battery through a rectifier as shown in Fig. 6.10. The analysis is more complicated as it involves a mixing of AC and DC circuits. For now, we will only present the speed–developed power curve, saving a detailed analysis for Chap. 9.

Fig. 6.10 Wind energy conversion system charging a battery through a rectifier

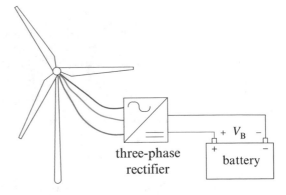

Fig. 6.11 Electrical power developed by a WECS connected to batteries of different voltages

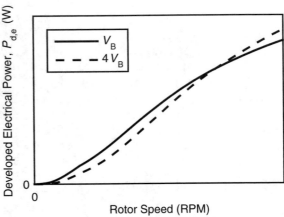

Figure 6.11 shows the power developed by the generator $P_{d,e}$ as a function of rotor speed for two different battery voltages, V_B and $4V_B$. For example, this could be a 12 V battery and a 48 V battery. No power is developed by the generator until its output voltage is high enough to charge the battery. This happens slightly above the battery's open-circuit voltage. The power developed increases with speed but is nonlinear. At very high speeds, the synchronous reactance dominates and the increase in developed power tappers off.

The shape of the curve will vary somewhat as the battery's state-of-charge increases. We also see that the selection of an appropriate battery voltage is important. The rotational speed that the generator begins developing power is four times greater for a 48 V battery than a 12 V battery; however, at high speeds more power is developed with the 48 V battery.

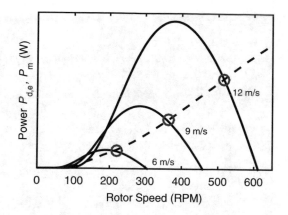

Fig. 6.12 The intersection of the speed–mechanical power and speed–developed electrical power curves determines the rotational speed of a wind turbine at a given wind speed

6.2.5 Operating Speed

The rotational speed of a WECS depends on a somewhat complicated interaction between the speed-mechanical power of the wind turbine and the speed-developed electrical power of the generator. We will consider a direct drive WECS; the extension to WECS with gearboxes follows a nearly identical process.

In a direct drive system, the turbine and generator rotate at the same speed. In addition, and assuming steady-state operation, the mechanical torque must equal the electrical torque, and the mechanical power must equal the developed power.

If we ignore the mechanical losses, then the operating point of the WECS that satisfies these conditions is at the point that the speed–mechanical power curve of the wind turbine and speed–developed power curve of the generator intersect. Figure 6.12 shows an example for three different wind speeds. The speed–developed power curve for the $4V_B$ battery in Fig. 6.11 is shown as the dashed line.

For this scenario, if the wind speed is 6 m/s, then the steady-state rotor speed will be approximately 220 RPM. If the wind speed increases to 9 m/s, then the rotor speed will increase to 380 RPM, and if the wind speed is 12 m/s, then the rotor speed will be 520 RPM. It is obvious from Fig. 6.12 that these operating points do not maximize the mechanical power available at each wind speed. In each case, the rotor is rotating faster (to the right) of the peak speed–mechanical power curve. In other words, its TSR is too high, and the turbine is said to be overspeeding. The overspeed condition occurs when the wind turbine is too powerful for the generator. This may or may not be a concern. It does indicate, however, that the turbine can be coupled with a larger generator or that the blades are longer than needed.

Example 6.3 The wind turbine whose speed–mechanical power curve is shown in Fig. 6.12 has a blade length of r_{WT} = 2.4 m. Compute the TSR when the wind speed is 12 m/s when used to charge a battery whose speed-mechanical power and speed-developed electrical power curves are in Fig. 6.12.

Solution By inspection, we see that when the wind speed is 12 m/s, the rotor rotates at 520 RPM. Rearranging (6.7) shows that the TSR is:

$$\lambda = \frac{N_m \times 2\pi \times r_{WT}}{v_{air} \times 60} = \frac{520 \times 2\pi \times 2.4}{12 \times 60} = 10.89.$$

This is much larger than the optimal TSR, which is usually between six and eight.

6.2.5.1   Open-Circuit Operation

Disconnecting the load from a WECS under high wind conditions can be dangerous and damaging. Under open-circuit conditions, the electrical power developed by the generator is zero. The imbalance between the mechanical power and the developed power causes the wind turbine to accelerate. Acceleration continues until the mechanical power matches the friction and windage losses, which we have previously ignored. At this speed, the centripetal force can be large enough to destroy the WECS. In addition, because the induced voltage is proportional to the rotor speed, an overvoltage can occur. Controllers that use series-type switches, like those used for PV arrays, should not be used with WECS as they can cause an open-circuit condition.

6.2.5.2   Short-Circuit Operation

When a WECS is short-circuited, the generator will develop a large amount of power at a low RPM. The developed power is entirely absorbed by the stator windings. The speed-developed power curve of a short-circuited generator is steep, and the intersection point between it and the turbine's speed-mechanical power curve is at or near zero RPM. This effectively stops the wind turbine, often in a manner of seconds.

A short circuit acts as an electromagnetic brake for the WECS. This is useful for maintenance and when especially high wind speeds are expected. A multipole switch can be used to simultaneously short all three phases together. Care should be taken to ensure the switch is installed on the WECS side of a rectifier, not the battery side. Otherwise, serious damage to the battery can occur.

6.2.6   Control

The objectives of WECS control are to:

- improve power production;
- avoid damage caused by over-speed operation;
- avoid damage caused by power production in excess of the rated power.

There are three general ways of controlling a WECS: adjusting the yaw, changing the pitch angle, or varying the generator's speed–developed power curve.

6.2.6.1   Yaw Control

Yaw and pitch are forms of mechanical control. Yawing is when the WECS rotates about the vertical axis. Yawing can turn the hub directly into the wind to increase the power available to the turbine or away from the wind to decrease it. The latter action is called furling. Small WECS use passive yawing systems through a specially designed tail vane. Some WECS automatically tilt the hub upward on a hinge during episodes of dangerously high wind. Larger-capacity turbines, approximately 20 kW and above, are more likely to use active control employing various wind sensors, electronic controllers, and servos.

6.2.6.2   Pitch Control

Larger-capacity turbines often include a form of pitch control. As the pitch of the blade changes, the aerodynamic characteristics of the turbine change, either increasing or decreasing the generated lift. This affects the C_p curve as shown in Fig. 6.6. The pitch can be adjusted to increase or decrease C_p for a given wind speed and generator RPM.

6.2.6.3   Generator Control

Generator control is accomplished by changing the generator's speed–developed power curve. This causes the intersection point of the generator's speed–developed power curve and the wind turbine's speed–mechanical power curve to change. The generator's speed–developed power curve can be changed by adjusting R_L or the voltage on the DC side of the rectifier if the WECS is used for battery charging. Devices known as maximum power point trackers can be used to adjust the voltage of the DC side of the rectifier. These devices are specially controlled DC/DC converters and are discussed in Chap. 10.

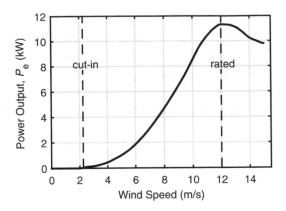

Fig. 6.13 Power curve of a small wind turbine with cut-in wind speed of 2.5 m/s and rated wind speed of 12 m/s

6.2.7   Power Curve

The relationship between wind speed and power output by the generator is shown by the WECS's power curve. A typical power curve is shown in Fig. 6.13. The power curve can be constructed by identifying the intersection point of the speed-mechanical power and speed-developed electrical power curves at various wind speeds and multiplying the power by the corresponding generator efficiency. In practice, the power curves are empirically determined by the manufacturer. The power curve can be divided into four regions.

Below Cut-In Wind Speed
When the wind speed is below the cut-in wind speed, no power is produced. This is because there is insufficient wind for the blades to overcome friction and rotate. If the wind turbine is used to charge batteries, then no power will be produced until the voltage exceeds that of the battery. The wind turbine might rotate, but no power is produced.

Between Cut-In and Rated Wind Speeds
The WECS begins producing power in this region. A well-designed WECS will operate at or near the maximum C_p. The power produced is approximately cubically related to the wind speed, depending on the efficiency curve of the generator.

Between Rated and Cut-Out Wind Speeds
The rated wind speed for many WECS is between 12 and 16 m/s. This is very windy, corresponding from 43.2 to 57.6 km/h. Although the power in the air continues to increase with the cube of the wind speed, we should limit the power developed by the generator. If we did not, the internal losses of the generator would continue to increase and the generator might overheat and fail. In addition, the rotational speed would also continue to increase, possibly leading to the destruction of the WECS or tower. The rated power can be increased by using a higher-capacity generator. Very high wind speeds, while having a lot of power, occur infrequently. The occasional

increase in energy production might not warrant the added cost of a larger generator and the more robust tower it requires.

Instead, the power from a smaller generator is intentionally limited using one or more of the discussed control methods (yaw, pitch, or generator control). Informally, this is known as "spilling wind" as the WECS is not using as much of the wind resource as it could. WECS with sophisticated control systems are able to more tightly regulate the power around the rated value than shown in Fig 6.13.

We must carefully interpret the rated power advertised by WECS manufacturers. A WECS rated at 10 kW will only produce 10 kW at the rated wind speed and load conditions. There is no guarantee that the turbine will ever encounter the rated wind speed. Some organizations such as the International Electrotechnical Commission (IEC) and the American Wind Energy Association (AWEA) have protocols aimed at ensuring that WECS are rated consistently, but not every manufacturer follows these protocols.

Above Cutout Wind Speed
Some WECS will stop producing power at high wind speeds to avoid damage. Pitching, yawing, and electromagnetic breaking are methods to accomplish this.

Power Curve Model
The power curve can be reasonably represented by a simple continuous piecewise model:

$$
P_e(v) = \begin{cases}
0 & : v \leq v_{\text{cut-in}} \\
\eta_{\text{gen}} \frac{1}{2} C_p(\lambda, \beta) A_{WT} \rho_{\text{air}} v^3 & : v_{\text{cut-in}} < v \leq v_{\text{rated}} \\
P_{\text{rated}} & : v_{\text{rated}} < v \leq v_{\text{cut-out}} \\
0 & : v > v_{\text{cut-out}}
\end{cases}
\tag{6.18}
$$

where $v_{\text{cut-in}}$, v_{rated}, and $v_{\text{cut-out}}$ are the cut-in, rated, and cutout wind speeds. The model assumes the WECS is able to tightly regulate the power at the rated value.

6.2.8   Energy Production

The power produced by WECS is usually inconsistent and difficult to forecast more than a day in the future. On most days, the power will fluctuate for example as shown in Fig. 6.14.

6.2.9   Wind Resource

The suitability of a WECS for an off-grid application depends on the characteristics of the local wind resource. The characteristics of interest include the distribution of wind speeds, the wind direction, consistency, and short- and long-term patterns. The

Fig. 6.14 Power output from a 1 kW WECS in Kenya over a 24-h period

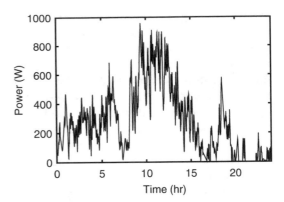

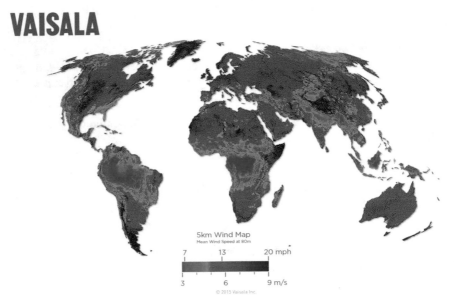

Fig. 6.15 Wind resource map (courtesy of Viasala, Copyright (c) 2017 Vaisala)

wind resource varies geographically and temporally, and is influenced by regional as well as local conditions. This variation is apparent in a wind resource map, as shown in Fig. 6.15, depicting the average wind speed at 80 meters above ground.[2]

The generally low wind speeds near the equator have implications of the use of wind energy in developing countries. Large swaths of equatorial South America, Africa, and South Asia have wind resources unsuitable for wind generation. There

---

[2]This is a typical tower height for utility-scale wind turbines; wind turbine towers used in off-grid applications are usually do not exceed 30 m.

Table 6.1 Roughness length

Terrain	Roughness length (m)
Snow surface	0.003
Rough pasture	0.010
Fallow field	0.03
Crops	0.05
Few trees	0.10
Forest and woodlands	0.50

are exceptions. Areas near the Horn of Africa—Ethiopia, Eritrea, Tanzania, and Kenya—are notably windy, as are areas near the coast in South Africa and west Saharan Africa.

The local wind resource should be assessed before committing to using a WECS. The local wind resource is influenced by topography, vegetation, and land/water interfaces. These conditions are not usually captured in computer-generated wind resource maps, whose finest resolution is usually about 1 km by 1 km.

6.2.9.1 Effect of Height

Wind speed varies with height. This is known as wind shear. The wind speed is slower close to the ground due to turbulence caused by the interaction of the air with obstacles and the ground. The increase of wind speed with height depends upon the surrounding terrain and is estimated by

$$v_1 \ln\left(\frac{z_2}{z_0}\right) = v_2 \ln\left(\frac{z_1}{z_0}\right). \tag{6.19}$$

The equation shows how the velocity of wind v_1 at a height of z_1 meters above ground is related to the velocity v_2 at z_2 meters above ground based on the roughness length z_0. Table 6.1 shows the roughness length for different terrain types. Depending on the roughness length, the wind speed above the surface can be considerably greater than at ground level.

Example 6.4 The measured wind speed at 3 m above ground is 6 m/s. Compute the estimated speed at 15 m above the ground. The surrounding terrain has few trees. Compute and compare the resulting power using the power curve in Fig. 6.13 if the hub height of the is WECS 3 m and 15 m.

(continued)

Solution Re-arranging (6.19) to solve for v_2:

$$v_2 = v_1 \ln\left(\frac{z_2}{z_0}\right) \times \frac{1}{\ln\left(\frac{z_1}{z_0}\right)} = 6 \ln\left(\frac{15}{0.1}\right) \times \frac{1}{\ln\left(\frac{3}{0.1}\right)} = 8.84 \text{ m/s}.$$

Using Fig. 6.13, the power output by the WECS at a height of 3 m (wind speed of 6 m/s) is approximately 1.9 kW. If the WECS is at 15 m (wind speed of 8.84 m/s), the power is approximately 4.7 kW. The power increase by installing the WECS at a greater height is notable.

6.2.10 Tower

The tower is an important element of a WECS. The cubic relationship between wind speed and power suggests that the WECS be placed in a stream of consistent high-speed wind. Towers are typically 10 m to 20 m tall. The tower can cost as much as the WECS and should be budgeted for.

Large-capacity WECS are heavy—a WECS rated at 10 kW typically weighs over 450 kg. The static and dynamic forces on the tower can be large. The tower should be properly engineered, installed, and maintained. Tower failure will almost surely destroy the turbine and may cause damage to nearby property or injury to people who might be in the way. A local site survey is needed before the tower can be designed. Tower design is beyond the scope of our discussion, but the general tower types are described in the following.

6.2.10.1 Freestanding

Freestanding towers are typically either made from a metal pole or lattice structure. An example of a freestanding tower for a large wind turbine is shown in Fig. 6.16. They require the least amount of land space of all the tower types. They also require substantial material and are the most expensive tower option. The foundation must be set deep in the ground. And, because wind turbines require at least annual maintenance, provisions must be made for climbing the tower.

Fig. 6.16 A 30 m
freestanding tower for a
30 kW WECS (courtesy of
author)

6.2.10.2   Guyed Lattice

Guyed lattice towers use high tensile strength wires to prevent the tower from
moving. The wires are separated by 120° and extend out to an equal radius from
the center of the tower. The guys must be firmly anchored in the ground, typically at
distance of 50 to 80% the height of the tower. The tower can be less substantial due
to this added support. These towers must also be climbed to perform maintenance
on the wind turbine.

6.2.10.3   Tilt-Up

Tilt-up towers are typically made from steel tube. Several sections of tube are
sometimes used. The tube is mounted on a hinge at the base of the tower. Guy
wires are used to stabilize the tower. The tower can be raised and lowered using a
winch or a gin pole for leverage, as seen in Fig. 6.17. Tilt-up towers require flat land.
Importantly, maintenance can be done safely at ground level and no crane is needed
for installation.

Fig. 6.17   A 10 m tilt-up tower for a 1 kW WECS being raised by a gin pole (courtesy E. Patten)

6.2.11   Siting

The variability and dependency of a wind resource on local conditions are more extreme and less predictable than with solar or hydro resources. It is therefore important to assess the wind resource through direct meteorological measurement over a prolonged period of time. More information about wind resource assessment is provided in Chap. 11.

It is important to locate wind turbines away from buildings, trees, and other obstacles. A general rule-of-thumb is that when the blades are at the lowest position, they should be 10 m higher than anything within a 150 m radius. In systems with multiple WECS, care must be taken to avoid placing one WECS downwind from another.

6.2.12   Practical Considerations

The use of WECS in off-grid electrification appears to be waning as prices of PV arrays have sharply decreased in recent years. Nonetheless, there are several reasons to consider WECS for off-grid electrification:

- there are no fuel costs;
- in some locations, the wind resource coincides with the load and so the system might require less battery backup;

- it is possible to construct and repair WECS locally, provided skilled workers, equipment, and supplies are available;
- the size and conspicuousness of a WECS makes their theft unlikely.

Balancing these advantages are several disadvantages:

- the upfront costs of a WECS are high;
- the variability and uncertainty of the wind resource make selecting and an appropriately size WECS challenging;
- it is difficult, expensive, and time-consuming to assess the wind resource;
- wind turbines and their towers require periodic maintenance and inspection;
- there can be a high risk of lightning strikes that must be protected against by surge arrestors and enhanced grounding;
- the tower might require a substantial footprint for guy wires;
- WECS are visible from far away. This can attract unwanted attention and perhaps incite jealously;
- birds and bats can strike the WECS raising wildlife concerns and damaging the turbine;
- climbing freestanding and guyed lattice towers is a safety risk.

Some have proposed to use vertical axis wind turbines in off-grid systems. These have not proven to be successful except for in niche applications.

Typical issues requiring maintenance include replacing cracked hubs and blades, lubricating and replacing bearings, and periodically inspecting and retightening tower connections. WECS must also be carefully managed in the event of high winds and storms. This might require lowering the tower and applying the electromagnetic brake.

6.3   Micro Hydro Power Systems

In favorable conditions, small-scale hydro power systems can be the most economic and reliable of all the energy conversion technologies. The high density of water— more than 800 times that of air—means that hydroturbines can be surprisingly compact (see Fig. 6.18). Hydro power systems with capacities from 5 kW to 100 kW are often referred to as "micro hydro power" (MHP). Capacities less than 5 kW are sometimes referred to as "pico hydro." We will broaden the use of MHP to include these smaller systems. The term "MHP system" refers to the turbine, controls, and generator collectively.

Hydro power systems can be designed to make use of run-of-river or reservoir water supplies. Reservoir water supplies require an impoundment structure such as a dam. Run-of-river systems rely on the natural flow of the water. Because run-of-river systems do not store water, they require less expensive civil works. Most run-of-river systems use a conveyance system to provide a separate pathway for the water to flow. The water flows through a pipe known as a "penstock" to the turbine. Most MHP systems in off-grid applications use run-of-river water supplies, which are the focus of this section.

Fig. 6.18 This small (22 cm diameter) runner for a Pelton turbine can be used to supply over 10 kW of power

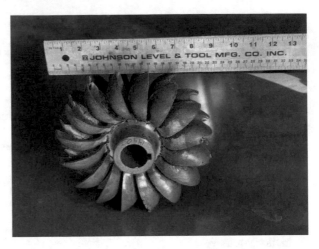

6.3.1 Hydro Resource

Hydro power systems harness energy in the water cycle. The energy in the hydro resource is provided by the sun. Sunlight causes water molecules at the surface of oceans, rivers, and lakes to separate from each other, transforming from liquid water to water vapor.

The water vapor, being less dense than the surrounding air, rises into the atmosphere. Under the right conditions, the water vapor condenses back into liquid form. The liquid water now has a greater amount of potential energy than when it evaporated owing to its higher elevation. The water falls to the earth as precipitation.

A portion of the precipitation falls on land. Some of the water flows downhill into a river on its way to a lake or ocean. The water flowing in a river has both kinetic energy and, because it is above sea level, potential energy. This energy can harnessed in a hydro power scheme.

Although more consistent than wind or solar resources, river flow is prone to seasonal and long-term variation. MHP systems are typically sized to only make use of a fraction of the available flow, so that the penstock is always full. This allows the MHP system to have access to a constant source of input energy. Nonetheless, care must be taken to understand seasonal variability before implementing any MHP scheme. Locations with conditions conducive to MHP are those where the elevation drops sharply, with high and consistent annual precipitation, and large catchment areas.

6.3.2 Principle of Operation

The prime mover in a MHP system is a hydro turbine. Hydroturbines convert the energy in a volume of water into rotational energy of the turbine shaft. The shaft is coupled to the rotor of an electric generator either directly or through a gearbox or a belt and pulley system.

6.3.2.1 Bernoulli's Equation

Consider a volume of water upstream from a hydroturbine. The total energy in a volume of water with density ρ_{wa} arises from three sources:

- kinetic energy: increases with the velocity of water v_{wa};
- potential energy: increases with the elevation z of the water above a reference plane (typically taken as the location of the turbine);
- internal energy: increases with the pressure p_{wa} of the water.

If we ignore losses, then the total energy in the water is unchanged as it flows down a river or a penstock. In other words, the total energy is constant. This is a simple result of the conservation of energy.

Bernoulli's equation formally describes this relationship.

$$\frac{1}{2}\rho_{wa}v_{wa}^2 + \rho_{wa}gz + p_{wa} = K \tag{6.20}$$

where g is the acceleration due to earth's gravity, 9.81 m/s^2, ρ_{wa} is the density of the water, and p_{wa} is in pascals (Pa). Atmospheric pressure is 101.325 kPa. The mass of 1 liter of water is 1 kilogram, so that ρ_{wa} is 1000 kg/m^3. The units of each term in (6.20) are equivalent to joules per cubic meter. In other words, each term on the left hand side of (6.20) is a contribution to the water's total energy density. The constant K is not a universal constant. Its value is determined from the velocity, elevation, and pressure of the water. Bernoulli's equation states that, if we ignore losses, then as water in open air flows downhill, its potential energy decreases, but the kinetic energy increases by an equal amount. In other words, its velocity v_{wa} increases as z decreases. The pressure component remains nearly constant as the water is exposed to the atmosphere both upstream and downstream.

6.3.2.2 Total Head

Bernoulli's equation is often re-written by dividing both sides by ρ_{wa} and g:

$$\frac{1}{2g}v_{wa}^2 + z + \frac{p_{wa}}{\rho_{wa}g} = H_t. \tag{6.21}$$

The unit of each term is now in meters. The first term is referred to as the "velocity head," the second as the "elevation head," and the third term as the "pressure head." The term H_t is referred to as the "total head." Equation (6.21) is convenient because it allows the energy density of a water resource to be expressed by a single quantity, the total head. For a given total head, the total energy in a volume of water V_{wa} is equal to

$$E_{wa,total} = \rho_{wa} \times V_{wa} \times g \times H_t. \tag{6.22}$$

This is the same as the potential energy of water with mass $\rho_{wa} \times V_{wa}$ at a height H_t. We can thereby conceptually replace water with a certain velocity, elevation, and pressure with an equivalent mass (or volume) of water with no velocity, no pressure, and at an elevation of H_t. Hydro power engineers often use the total head to succinctly describe the energy density of a water resource.

Example 6.5 Compute the total head of two cubic meters of water whose velocity is 1 m/s and is located 38 meters above the reference plane. Assume the water is exposed to atmospheric pressure (101.325 kPa). Show that the total energy using (6.22) and calculated from Bernoulli's equation are equal.

Solution Applying (6.21) to find the total head:

$$H_t = \frac{1}{2g}v_{wa}^2 + z + \frac{p}{\rho g} = \frac{1}{2 \times 9.81}1^2 + 38 + \frac{101,325}{1000 \times 9.81} = 0.05 + 38 + 10.33 = 48.38 \text{ m}.$$

Note that the volume of water does not affect the total head. The greatest contributor to the water's total head is associated with its elevation.

The energy is found using (6.22)

$$E_{wa,total} = \rho_{wa} \times V_{wa} \times g \times H_t = 1000 \times 2 \times 9.81 \times 48.38 = 0.949 \text{ MJ}.$$

Bernoulli's equation (6.20) has units of energy density, so it must be multiplied by the volume of water to convert to energy:

$$E_{wa,total} = V_{wa}K = 2\left(\frac{1}{2}\rho_{wa}v_{wa}^2 + \rho_{wa}gz + p_{wa}\right)$$

$$= 2\left(\frac{9.8}{2}1^2 + 1000 \times 9.81 \times 38 + 101,325\right) = 0.949 \text{ MJ}.$$

We see that $E_{wa,total}$ is the same from both calculations, just as it should be.

6.3.2.3 Effective Head

The *effective head* is the head available to a hydroturbine for energy conversion. The effective head is always less than the total head. The difference arises from both practical and theoretical reasons. First, losses such as friction in the conveyance system reduce the energy available to the turbine, which can be expressed as reduction in the total head by a value H_f. Second, the energy associated with the velocity head is assumed not to be available to the turbine. The reason is that its contribution to the total head is generally minimal (see previous example) and for convenience is assumed to be zero. Third, the turbine is not able to extract all of the energy of the water it interacts with. The water at the source and at the outlet of the turbine are both at atmospheric pressure. The water exiting the turbine therefore must at least have internal energy corresponding to the pressure head. Therefore, a portion of the total head equal to $p_{wa}/(\rho_{wa}g)$ is unavailable for energy conversion.

The result is that the effective head H is

$$H = z - H_f = z\eta_{convey} \tag{6.23}$$

where η_{convey} is the efficiency of the conveyance system. In a lossless conveyance system, the effective head is equal to the vertical geometric difference between the turbine and the upstream water source.

6.3.2.4 Flow Rate

In addition to the head, we are also interested in the flow rate of the water resource. If the flow rate is Q cubic meters per second, then the effective power of the water is

$$P_{wa} = \frac{dE_{wa}}{dt} = \rho_{wa} \times g \times H \times \frac{dV}{dt} = \rho_{wa} \times g \times H \times Q. \tag{6.24}$$

Flow rates for MHP systems are often expressed in liters per second, which when multiplied by 0.001 is converted to cubic meters per second.

6.3.2.5 Power Extracted by the Turbine

The mechanical power extracted (developed) by the turbine is equal to the difference between the power of the water at the turbine's inlet and outlet.

$$P_{d,turbine} = P_{wa} - P_{outlet} = \eta_{turbine} P_{wa} \tag{6.25}$$

where $\eta_{turbine}$ is the efficiency of the turbine.

Example 6.6 The water source described in Example 6.5 is used in a MHP system. The flow rate into the turbine is 0.005 m³/s. Assume the conveyance system efficiency is 90%, and the turbine efficiency is 85%. Compute the power extracted by the turbine.

Solution The effective head is found from (6.23):

$$H = z\eta_{convey} = 38 \text{ m} \times 0.90\% = 34.2 \text{ m}.$$

(continued)

The corresponding power at the inlet to the turbine is

$$P_{wa} = \rho \times g \times H \times Q = 1000 \times 9.81 \times 34.2 \times 0.005 = 1677.5 \text{ W}.$$

The power extracted at the turbine is found from (6.25)

$$P_{d,turbine} = \eta_{turbine} P_{wa} = 0.85 \times 1677.5 = 1425.9 \text{ W}.$$

Table 6.2 Effective head and turbine type

High head	Medium head	Low head
Pelton	Crossflow	Crossflow
Turgo	Turgo	Propeller
Multi-Jet Pelton	Multi-Jet Pelton Francis	Kaplan

6.3.3   Hydroturbines

Hydroturbines extract energy from the water at their inlet and use it to rotate a shaft. Several different turbine types have been devised that can be matched with a particular water resource's effective head, as shown in Table 6.2. One way of visualizing the suitability of a particular turbine for given flow and effective head conditions is the *turbine application chart*, as shown in Fig. 6.19. The diagram shows the general regions of the flow–head plane that a turbine type can operate efficiently in. For reference, constant power lines are usually shown. The lines show the combination of head and flow that result in a given electrical power output of the generator, assuming some efficiency of the turbine and generator. From (6.24), the lines appear as downward sloping diagonal lines, showing that a given power can be achieved with a low flow and high head, or high flow and low head. Application diagrams often use logarithmic scales.

6.3.3.1   Turbine Types

Hydroturbines are classified as either being "impulse" turbines or "reaction" turbines. Impulse turbines use the kinetic energy of the water to rotate a shaft; reaction turbines use both the kinetic and pressure energy of the water to rotate a shaft. Most MHP schemes use impulse turbines. Reaction turbines are more common in large-capacity hydro schemes. There are several types of turbines within

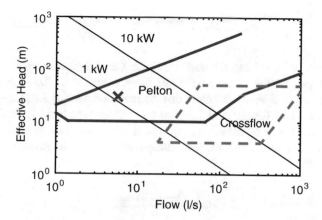

Fig. 6.19 Turbine application chart of Pelton and crossflow turbines for MHP applications; the constant power lines assume a 75% efficiency of the turbine and generator [9]; the operating point of the turbine in Example 6.7 is indicated as the "x"

each class. Pelton, Turgo, and crossflow turbines are impulse turbines. Francis and Kaplan turbines are reaction turbines.

The selection of which turbine type to use is an important early design decision. The decision is primarily based on maximizing the efficiency of the turbine while satisfying the power and rotational speed requirements of the generator. A common way of selecting the turbine type is by calculating the specific speed, as described next.

6.3.3.2 Specific Speed

The *specific speed* is an index that is useful for selecting the type of turbine that is appropriate for a particular water resource and required turbine rotational speed. The basic approach is to compute the specific speed based on (1) the required mechanical power developed by the turbine $P_{d,turbine}$, (2) the required rotational speed, and (3) the effective head. The specific speed therefore encapsulates the input conditions (effective head) and output requirements (developed power and rotational speed) of the turbine.

Each turbine type has a range of specific speeds that it is suitable for. Within this range, the turbine will operate at or near its maximum efficiency. Keep in mind that the specific speed itself is not a speed. It is perhaps less confusing to consider it an index. Once the specific speed has been calculated, the turbine type is selected by matching the specific speed to the range of specific speeds each turbine is suitable for.

Confusingly, there are at least three formulations of the specific speed. The *dimensionless* specific speed $\mathscr{S}$ is:

$$\mathscr{S} = \frac{\omega_m \sqrt{P_{d,\text{turbine}}/\rho_{wa}}}{(gH)^{5/4}} = \frac{\omega_m \sqrt{Q}}{(gH)^{3/4}}. \tag{6.26}$$

Thus, the specific speed can be calculated using the mechanical power developed or the flow rate in the numerator. Note the difference in the exponent in the denominator. As the name implies, this formulation results in a specific speed that is unitless. This formulation contains two values which are typically taken as constants: the density of water ρ_{wa} and the acceleration due to gravity g. These constants are removed from the calculation to simplify it, resulting in another specific speed:

$$N_s = \frac{N_m \sqrt{P_{d,\text{turbine}}}}{H^{5/4}}. \tag{6.27}$$

This formulation also expresses the rotational speed of the turbine N_m in RPM rather than radians per second.

Eliminating the constants ρ_{wa} and g from (6.26) means that (6.27) is not dimensionless. To confuse the matter further, the specific speed is commonly calculated using different units. The so-called "SI" version uses units of RPM, meters, and kilowatts; in the United States it is customary to use RPM, horsepower[3], and feet. These formulations are related to the dimensionless specific speed as:

$$N_{s,\text{SI}} = \frac{\text{RPM}\sqrt{\text{kW}}}{\text{m}^{5/4}} = 165.7 \times \mathscr{S} \tag{6.28}$$

$$N_{s,\text{US}} = \frac{\text{RPM}\sqrt{\text{hp}}}{\text{ft}^{5/4}} = 43.4 \times \mathscr{S}. \tag{6.29}$$

Even though these formulations result in a value for the specific speed that is not dimensionless, the units are typically omitted. These different formulations of specific speeds adds confusion. Always be diligent in using consistent formulations of specific speed when selecting turbines.

Example 6.7 Compute the dimensionless specific speed and the specific speed using RPM, kilowatts, and meters, for a water resource with an effective head of 30 m. Assume the turbine will rotate at 1500 RPM with a developed mechanical power of 1.25 kW.

Solution The dimensionless specific speed is computed using (6.26):

(continued)

---

[3]One kilowatt is equal to 1.34 hp (horsepower).

$$\mathscr{S} = \frac{\omega_\mathrm{m}\sqrt{P/\rho_\mathrm{wa}}}{(gH)^{5/4}} = \frac{\frac{2\pi}{60} \times 1500\sqrt{1250/1000}}{(9.81 \times 30)^{5/4}} = 0.144.$$

From Fig. 6.19, we see that a Pelton turbine is suitable for this location and that the corresponding flow will be approximately 6 m/s. Using the formulation (6.28):

$$N_\mathrm{s,SI} = \frac{1500\sqrt{1.25}}{(30)^{5/4}} = 23.89.$$

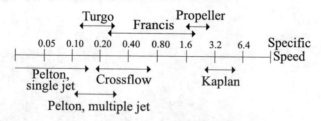

Fig. 6.20 Approximate suitability ranges of specific speed $\mathscr{S}$ for various turbines [2]

Each turbine type is suitable for a range of specific speeds, as shown in Fig. 6.20. The suitability generally refers to the ability of the turbine to operate efficiently under the given rotational speed, effective head, and power (or flow). Note that the scale is not linear. Inspecting (6.26) shows that as the head increases, the specific speed decreases; and as the power output decreases, so does the specific speed. For MHP, the water resource generally has a high head, and the power output is relatively low. Hence, the specific speed is also low for a given rotational speed. In order to match a turbine to the resource, the turbine's specific speed should then also be low. Therefore, from Fig. 6.20, Pelton, Turgo, or crossflow turbines should be considered.

6.3.4 Pelton Turbine

Pelton turbines are a common choice for MHP as they are suitable for high-head, low-flow water resources. Lester Pelton, an American inventor, patented the Pelton turbine in 1890. His original design has subsequently been improved upon.

Fig. 6.21 A large horizontal axis Pelton turbine with its casing and manifold removed. This turbine has been decommissioned and its buckets painted. It serves as a display model. It is larger than would typically be found in MHP systems (courtesy of the author)

Fig. 6.22 A 25 kW AC-coupled MHP system using a two-jet Pelton turbine (courtesy of the author)

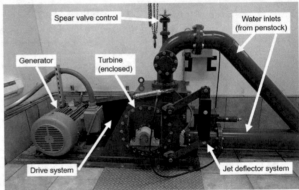

6.3.4.1   Physical Description

A Pelton turbine is shown in Figs. 6.21 and 6.22. Pelton turbines resemble a water wheel with polished metal buckets located around the periphery. Each bucket contains two symmetric cups with a sharp ridge known as a "splitter" between them. Pelton turbines can be designed to either rotate around the vertical or horizontal axis. Horizontal rotation makes maintenance easier. The rotating portion of a hydro turbine is known as the "runner." Pelton runners are characterized by their Pitch Circle Diameter (PCD), as shown in Fig. 6.23. For MHP, the PCD is generally less than 0.5 m. This is much smaller than WECS or PV arrays of the same capacity. A water-tight housing known as a "casing" surrounds the runner. There is water and air inside the casing, both of which are at atmospheric pressure.

The manifold of a Pelton turbine consists of the pipes and valves that connect the outlet of the penstock to one or more nozzles. Each nozzle creates a high-speed jet of water. Each jet is aimed at the splitter of the bucket in front of it, as shown in Fig. 6.23. The splitter separates the water into two streams. Each stream flows into one of the cups of the same bucket, as shown in Fig. 6.24. The force of the water into the bucket and cups causes the runner to rotate. The buckets have notches at their

Fig. 6.23 A horizontal axis
Pelton turbine with one
nozzle

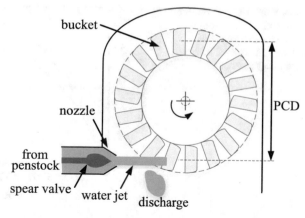

Fig. 6.24 Relative velocity
of the water jet impacting a
bucket of a Pelton turbine

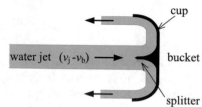

periphery to minimize the interference of adjacent buckets with the water jet as the runner rotates. The water exits the turbine through the tailrace where it rejoins the river.

6.3.4.2   Principle of Operation

The water at the inlet to the Pelton turbine has lost its potential energy but increased its pressure due to the mass of the water in the penstock above it. As the water exits the nozzle, the pressure decreases to atmospheric pressure but its velocity increases. The pressure energy is converted to kinetic energy. This is consistent with Bernoulli's equation. The transfer of kinetic energy from the water to the turbine is highly efficient.

6.3.4.3   Efficiency Curve

The buckets are shaped to maximize the amount of kinetic energy transferred from the water to the turbine. Recall that the water jet is aimed at the splitter of a bucket, separating it into two streams which flow into the two cups. The streams are redirected along the smooth curvature of the cups, effectively making a U-turn.

The force from the water jet onto the bucket is found from the change in momentum M of the water:

Fig. 6.25 Typical efficiency
curve of a Pelton turbine
designed to operate at
1500 RPM (solid), theoretical
efficiency (dashed)

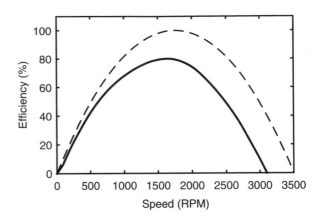

$$F = \frac{\mathrm{d}M}{\mathrm{d}t} = 2\rho_{\mathrm{wa}} Q \left(v_{\mathrm{j}} - v_{\mathrm{b}} \right) \tag{6.30}$$

where the momentum is the product of the mass and velocity of the water, v_{j} is the
velocity of the jet, and v_{b} is the tangential velocity of the bucket. The difference
$v_{\mathrm{j}} - v_{\mathrm{b}}$ is the relative velocity between the water jet and the bucket, which is moving
away from it. The multiplier of two in (6.30) is a result of the water exiting the
bucket in the same direction that it came.

Power is force multiplied by velocity so that

$$P_{\mathrm{d,turbine}} = 2\rho_{\mathrm{wa}} Q \left(v_{\mathrm{j}} - v_{\mathrm{b}} \right) v_{\mathrm{b}}. \tag{6.31}$$

It can be shown that (6.31) is maximized when

$$\frac{v_{\mathrm{b}}}{v_{\mathrm{j}}} = 0.50 = y \tag{6.32}$$

with corresponding maximum power $P_{\mathrm{d,turbine}}^{*}$

$$P_{\mathrm{d,turbine}}^{*} = \frac{1}{2}\rho_{\mathrm{wa}} Q v_{\mathrm{j}}^{2}. \tag{6.33}$$

The ratio of tangential velocity of the bucket to that of the water jet is known as
the "speed ratio" y. Figure 6.25 shows how the theoretical efficiency of the turbine
varies with the speed ratio. The efficiency of the turbine is:

$$\eta_{\mathrm{turbine}} = \frac{P_{\mathrm{d,turbine}}}{P_{\mathrm{wa}}}. \tag{6.34}$$

In theory, a Pelton turbine is 100% efficient when the speed ratio is $y = 0.50$. In
reality the optimal speed ratio is typically $y = 0.46$ due to friction, turbulence, and
mechanical misalignment. For small Pelton turbines the peak efficiency is 50% to

85%. The efficiency is somewhat insensitive to variations in speed near the optimal speed ratio, for example, by $\pm15\%$. Larger deviations lead to increasingly inefficient operation. If the power supplied by the water to the turbine is constant, then the speed–mechanical power curve of the Pelton turbine resembles Fig. 6.25 but scaled by P_{wa}.

6.3.5 Pelton Turbine Design

Consider a Pelton turbine with one input nozzle. The velocity of the water leaving the nozzle depends on the effective head. If we assume that the nozzle is lossless, then the energy in the water on either side of it must be the same. The velocity of the water is found by setting the energy associated with the effective head equal to the kinetic energy in the jet of water:

$$mgH = \frac{1}{2}mv_j^2 \qquad (6.35)$$

$$v_j = \sqrt{2gH}. \qquad (6.36)$$

The velocity of the jet of water only depends on the effective head of the water resource if the acceleration due to gravity is taken as a constant.

6.3.5.1 Speed Equation

The speed that the turbine rotates at can be determined from basic geometry. Let d_{PCD} be the length of the PCD of the turbine. The speed of the bucket is the tangential speed of the runner at a distance $d_{PCD}/2$ from its rotational axis so that:

$$v_b = \frac{d_{PCD}}{2}\omega_m \qquad (6.37)$$

$$\omega_m = \frac{2v_b}{d_{PCD}} \qquad (6.38)$$

$$N_m = \omega_m \frac{60}{2\pi} = \frac{2v_b \times 60}{2\pi \times d_{PCD}} = \frac{y \times v_j \times 60}{\pi d_{PCD}}. \qquad (6.39)$$

The result (6.39) is the *speed equation* of a Pelton turbine. From (6.36), we can express the speed of the turbine in terms of the head of the water resource:

$$N_m = \frac{y \times \sqrt{2gH} \times 60}{\pi d_{PCD}}. \qquad (6.40)$$

Example 6.8 A Pelton turbine is connected to a generator that is designed to operate at 1800 RPM. What should the PCD be for the turbine to operate at its maximum efficiency if the effective head is 50 m? Assume the turbine is directly coupled to the generator rotor and let the speed ratio be 0.50. Repeat the calculation using an effective head of 100 m.

Solution The optimal generator speed N_m^* is 1800 RPM. Re-arranging and solving (6.40) for a head of 50 m yields:

$$d_{\text{PCD}} = \frac{y \times \sqrt{2gH} \times 60}{\pi N_m} = \frac{0.50 \times \sqrt{2 \times 9.81 \times 50} \times 60}{\pi \, 1800} = 0.166 \text{ m}.$$

For a head of 100 m, the same approach returns a PCD of 0.235 m, showing that the Pelton turbine is physically larger as the head increases. This is because with a higher head, the velocity of the jet of water is faster. From (6.37), to keep the turbine operating at the optimal RPM, the PCD must increase.

6.3.5.2   Turbine Power

The power available to the turbine is found from (6.24). The only variable that can be controlled by the designer is Q, the flow rate, assuming there is sufficient flow of the river. The flow rate can be increased by making the cross-sectional area of the nozzle larger. Increasing the flow rate increases the power output of the turbine. However, the flow cannot be made arbitrarily large. For a Pelton turbine, the flow rate is limited by size of the water jet relative to the PCD. For most bucket designs, the diameter of the water jet d_{jet} cannot be greater than 10 or 11% of the PCD

$$d_{\text{PCD}} \geq \frac{d_{\text{jet}}}{0.11}. \tag{6.41}$$

This leads to a trade-off. To increase the power of the turbine, the flow rate must increase. However, this requires a larger PCD. Increasing the size of the PCD affects the rotational speed for a given jet velocity. If the rotational speed is held constant by a speed controller, then the turbine will no longer be operating at the optimal speed ratio y, and the efficiency will decrease. There is thus a "sweet spot" in which the power and effective head will be such that they match the optimal speed of the turbine. The matching is exactly what is done when the turbine is selected based on its specific speed as previously discussed.

The flow from the nozzle, and hence the penstock intake, depends on the circular area of the nozzle A_{noz}

$$Q = A_{\text{noz}} v_{\text{jet}} = A_{\text{noz}} \sqrt{2gH}. \tag{6.42}$$

From basic geometry, and making the simplifying assumption that the area of the jet is equal to the area of the nozzle, the flow can be expressed in terms of the jet diameter

$$Q = \frac{\pi d_{jet}^2}{4} \sqrt{2gH} \tag{6.43}$$

and solving for d_{jet} yields

$$d_{jet} = \left(\frac{4Q}{\pi \sqrt{2gH}} \right)^{0.5}. \tag{6.44}$$

Example 6.9 Determine the PCD, diameter of the jet, and flow rate for a Pelton turbine used in a mini-grid. The power output by the MHP generator is 20 kW and rotates at 1500 RPM; the water resource has an effective head of 80 m. The turbine should operate at its maximum efficiency, which corresponds to a speed ratio $y = 0.46$. The efficiency at this speed ratio is 80%. The efficiency of the generator is 90%.

Solution Begin by determining the required power available to the turbine, accounting for the efficiencies:

$$P_{wa} = 20 \times \frac{1}{0.80 \times 0.90} = 27.78 \text{ kW}.$$

From (6.24), the flow rate must be

$$Q = \frac{P_{wa}}{\rho_{wa} \times g \times H} = 0.0354 \text{ m}^3/\text{s}.$$

The PCD is found by re-arranging (6.40):

$$d_{PCD} = \frac{y \times \sqrt{2gH} \times 60}{\pi N_m} = \frac{0.46 \times \sqrt{2 \times 9.81 \times 80} \times 60}{\pi \, 1500} = 0.2320 \text{ m}$$

We must make sure that the diameter of the jet does not exceed 11% of the PCD. The diameter of the jet is found from (6.44)

$$d_{jet} = \left(\frac{4Q}{\pi \sqrt{2gH}} \right)^{0.5} = \left(\frac{4 \times 0.0354}{\pi \sqrt{2 \times 9.81 \times 80}} \right)^{0.5} = 0.0337 \text{ m}$$

(continued)

This value exceeds 11% of the PCD, and so the design must be reconsidered. One option is to design the turbine based on a higher speed ratio. This will increase the corresponding PCD, but the efficiency will be reduced. A somewhat larger flow rate will be required. A gearbox or pulley system could also be used to allow the turbine to rotate slower than the generator shaft, allowing a turbine with a larger PCD to be used.

6.3.5.3 Multiple Jets

It is possible to design a Pelton turbine to use more than one jet. Multi-jet systems offer higher rotational speeds, flow control (by shutting off some of the nozzles), and a reduced chance of total blockage. Multi-jet systems, however, are more complex and expensive. Multi-jet systems with more than two nozzles are usually not found in horizontal axis Pelton turbines due to the complexity of the manifold.

In multi-jet systems, the volume of water entering the penstock is split among the jets. The turbine can accommodate greater flow while not exceeding the 11% of PCD threshold. The specific speed increases with the number of jets.

6.3.5.4 Turgo and Crossflow Turbines

Some water resources are not suitable for Pelton turbines, for example, water resources where the head is lower but the required flow is higher. In these cases, multi-jet Pelton, Turgo, or crossflow impulse turbines should be considered as their specific speed is higher than a single-jet Pelton turbine (see Fig. 6.20).

The Turgo turbine resembles a Pelton turbine, but the runner consists of single cups. The buckets are also shallower. Unlike a Pelton turbine, the water jet does not strike the bucket head on; rather, it is aimed at the plane of the runner at angle, typically 20°. The water flows in one side of the bucket and out the other. The flow in Turgo turbines can be larger than for similarly sized Pelton turbines, but the efficiency is somewhat lower. The specific speed is about twice that of a single-jet Pelton turbine. Turgo turbines are generally more compact and operate at higher RPM than Pelton turbines. They are also easier to manufacture.

The runner of a crossflow turbine is shaped like a cylinder that rotates along a horizontal axis, as shown in Fig. 6.26. Crossflow turbines are used in off-grid electrification because of their simple construction and because they are suitable for a wide range specific speeds. They tend to rotate faster than Pelton and Turgo turbines, which reduces the need for gearboxes, belts, or pulleys to couple the turbine shaft to the generator's rotor. Crossflow turbines are also suitable for lower-head applications. An example of an improvised turbine is shown in Fig. 6.27.

Fig. 6.26 The runner of a crossflow turbine resembles a squirrel cage (courtesy of Canyon Industries)

Fig. 6.27 An improvised hydroturbine in Malawi. Notice the use of plastic beverage bottles for nozzles (courtesy P. Dauenhauer)

6.3.6   Generator

MHP systems often use single- or three-phase synchronous generators. The excitation can be any of those discussed in Sect. 5.2.8. The circuit model is of the same form as any other synchronous generator, as shown in Fig. 5.5. Induction generators can also be used, provided there is a sufficient source of reactive power.

6.3.7   Control

The electrical architecture of an MHP dictates the control scheme that must be used. As described in Chap. 4, there are two common architectures for MHP systems: one in which the power is used for charging batteries (DC-coupled) and the other in which AC loads are directly connected (AC-coupled).

DC-coupled MHP systems behave similarly to DC-coupled WECS. They are often used to charge batteries. Just like a WECS, the operating speed of the MHP will be such that the mechanical power produced by the turbine matches the developed electrical power. Unlike a WECS, the mechanical power to the MHP turbine is relatively constant. It is the changes in load that cause the rotational speed to change and perhaps deviate from its optimal speed.

For AC-coupled systems, the frequency must also be maintained within specific limits as required by the load. There are two general approaches to hydroturbine speed control: supply side and load side. A supply-side control strategy adjusts the input mechanical power to the turbine in response to change in rotational speed; a load-side strategy adjusts the load so that the electric power supplied by the generator is constant.

6.3.7.1 Spear Valves

One supply-side approach is to use a spear valve instead of a nozzle to create a jet of water. A spear valve allows for continuous adjustment of the flow rate into the turbine. When the electrical load decreases, the spear reduces the flow of water to the turbine. Consequentially, P_{wa} also decreases. The velocity of the water jet does not change, however, so the same rotational speed can be maintained. Spear valves can be manually controlled if tight speed regulation is not required, but more often are controlled automatically by a governor. Being a mechanical system, spear valves require periodic replacement and maintenance. They are also somewhat less efficient than a nozzle.

6.3.7.2 Deflector

A deflector is a supply-side method of speed control. As the name implies, the water jet is deflected from striking the runner. A movable metal plate inside the casing is often used. Deflectors are more common in manually controlled systems. They are also useful for quickly stopping the turbine in case of an emergency.

6.3.7.3 Electronic Load Controller

Rather than controlling the input mechanical power, the electrical power output can be controlled to regulate the speed of the turbine. The basic principle is for the electronic load controller (ELC) to control the power to a resistive ballast load P_{BL} so that the power from the generator P_e remains constant as the load P_{BL} changes:

$$P_e = P_L + P_{BL} \tag{6.45}$$

The details will be covered in Chap. 10.

Fig. 6.28 Typical hydro system used in a small-scale mini-grid

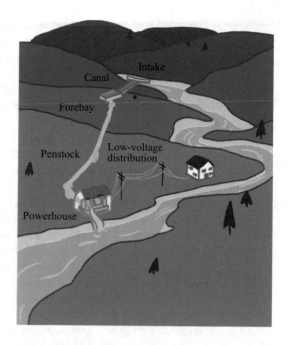

6.3.8 Conveyance Systems

In a run-of-river MHP scheme, a conveyance system is used to transport the water from upstream to the downstream turbine. A representative system is shown in Fig. 6.28. In some cases, the river itself can be used as the conveyance system, but this is generally avoided as there is less control of the flow. The components are also exposed to the environment, including river debris.

Instead, a weir is used upstream to divert a portion of the river flow away from the river. A settling basin can be used to remove particles such as sand from the water. A forebay tank can be used to provide a modest amount of water storage. The elevation change between the weir and the forebay tank should be minimum so that the potential energy of the water is preserved.

6.3.8.1 Penstock

Connected to the forebay tank is the penstock. Penstocks can be hundreds of meters long in some cases. A screen or sluice box is used to prevent debris from entering the penstock. Water flows down through the penstock, losing potential energy but gaining kinetic and pressure energy.

Plastic ABS (acrylonitrile butadiene styrene) or PVC (polyvinyl chloride) pipe is often suitable for MHP systems. The pipes are lightweight and easily obtainable and can be assembled using glue. It is recommended that the penstock be buried to protect it from disturbances and degradation from sunlight. In areas with low temperatures, burying will protect the penstock from freezing. Steel or aluminum pipes should not be buried due to worries of rusting or corrosion if the soil is acidic. To save on material costs, a common practice is to use pipe with thinner wall thickness at the top of the penstock (where the pressure is lower) and only use thicker-walled pipe toward the bottom of the penstock.

It is recommended to include one or more pressure gauges along the penstock for troubleshooting purposes, and a drain valve near the outlet at the bottom of the penstock. Should the penstock inlet become clogged, the water leaving the penstock will not be replenished by water entering it. A vacuum is formed inside the penstock which could damage it. A simple solution is to install a vacuum valve which opens when the penstock pressure drops.

If the outlet of the penstock becomes clogged, or the valve to the turbine is closed too quickly, a high-pressure surge occurs. The high pressure is caused by the sudden reduction of kinetic energy of the water. Per Bernoulli's equation, this energy must be converted to either pressure energy or potential energy. With no physical path to increase its potential energy, the water pressure suddenly increases.[4]

This phenomena, also known as a "water hammer," can increase the pressure in the penstock by 150%. Using slow-closing valves such as a gate or slide valve can be used to reduce the risk of damage.

As the water flows through the penstock, it experiences losses due to the friction of the water along the penstock's inner walls. The severity of the loss increases with the length and surface roughness of the penstock, and decreases with the diameter. The penstock diameter, however, can be increased to reduce losses. For the same flow rate, the water through a penstock with a larger diameter will travel slower. Since friction losses are speed-dependent, the losses decrease. Increasing the penstock diameter adds material and installation costs to the system, and so the benefit must be carefully considered. The length of the penstock is dictated by the terrain, but should be minimized. Therefore, sites where the terrain drops steeply are favorable. Straight penstocks are preferred. If bends are necessary, they should be gradual, rather than sharp 90-degree "elbows."

6.3.9 Efficiency

The losses associated with an MHP system can be grouped into intake, penstock, manifold, nozzle, runner, turbine, drive, and generator. The efficiencies are multiplied to determine the total efficiency of the MHP η_{MHP} from the water intake to the terminals of the generator.

$$\eta_{MHP} = \eta_{intake} \times \eta_{penstock} \times \eta_{manifold} \times \eta_{nozzle} \times \eta_{runner} \times \eta_{gen} \qquad (6.46)$$

---

[4]In some larger systems, a surge tank is used that allows the kinetic energy to transform to potential energy, thus reducing the pressure increase.

Note that the effective head already considers the intake and penstock losses. The efficiency of the turbine includes the nozzle, manifold and runner losses. In well-designed systems, as capacity increases, efficiency improves. In practice, the efficiencies are not exactly known, and so estimates are used. A common rule-of-thumb is to assume the total efficiency is 50%. This is greater than the efficiencies of other energy conversion technologies.

6.3.10 *Practical Considerations*

MHP systems have a long history in rural electrification; it is estimated that several hundred thousand units have been installed. MHP systems have the following favorable characteristics:

- MHP systems are inexpensive, simple to operate, and have no fuel costs;
- the operational life is long, due to the low temperature and consistent operating conditions; systems using induction generators and electronic load controllers are particularly long-lasting;
- power production can be consistent, and constant speed operation is possible, eliminating the need for inverters, batteries, and charge controllers;
- the water resource is renewable, and electricity is generated without emissions.

    The disadvantages include:

- a relatively small percentage of communities without electricity access have a water resource suitable for MHP;
- the up-front costs are high, as the conveyance system must be designed and installed;
- the water resource characteristics—flow rate, head, and effects of seasonality— must be determined beforehand, requiring time and resources;
- each system must be custom-designed, requiring expertise, time, and resources;
- commercially available turbines come in a limited number of discrete sizes, which might not match the site characteristics [7];
- because many stakeholders are affected by micro hydro projects, they are often subject to greater regulatory scrutiny than other renewable resources.

Readers considering installing MHP systems should review books dedicated to the topic, especially those that provide detailed descriptions of systems already in operation [3–5, 9].

6.4 Summary

Wind energy conversion systems (WECS) and micro hydro power (MHP) can be used to supply electricity in off-grid systems. They offer the primary advantage of having zero fuel costs. However, they require very specific conditions to be viable.

Further, in the case of WECS in particular, the power production can vary rapidly as the weather conditions change. This makes WECS unsuitable for AC-coupled systems.

The power available to a WECS is proportional to the area swept by the rotor blades and varies with the cube of the wind speed. The latter effect makes it extremely important to install WECS in areas with favorable wind resources and on top of high towers. The speed that a wind turbine rotates at in relation to the wind speed affects its efficiency. For most three-bladed wind turbines, the maximum power is extracted from the air when the tip speed ratio is between 6 and 8. However, the actual rotational speed also depends on the characteristics of the WECS generator and the load.

The power available to a MHP system depends on the head and flow rate of the water resource. There are several hydroturbine types that can be used. For most off-grid systems, impulse turbines—Pelton, Turgo, or crossflow—are used. MHP requires a conveyance system to supply water from the uphill water resource to the turbine. This adds complexity and cost. However, MHP systems are extremely durable and can be designed to require little intervention. Unlike WECS, the power into a hydroturbine is nearly constant, allowing it to be AC-coupled producing constant voltage and frequency, as long as a governor or electronic load controller is used.

Problems

6.1 Compute the required hub-to-tip blade length of a wind turbine whose developed electrical power is 6 kW when C_p is 0.28 and the wind velocity is 9 m/s. Ignore mechanical losses within the WECS.

6.2 The hub-to-tip length of a certain wind turbine blade is 1.2 m. What must the rotational speed be to achieve a TSR of 7.0 when the wind speed is 8 m/s?

6.3 A three-phase WECS is connected to a 10 Ω resistive load. The generator has a stator resistance of 0.7 Ω and inductance of 5 mH. The wind turbine has 12 poles and k_g is 3 V/rad/s. Compute the power developed, power supplied to the load, the electromagnetic torque, and the generator efficiency when the wind turbine rotates at 100 RPM and at 280 RPM.

6.4 A WECS is used to charge a battery. The power to the battery is measured, along with the wind speed and the rotational speed. The computed TSR is 12.1, much greater than the optimal value. Provide a credible reason why the TSR is greater than the optimal value.

6.5 Compute the flow rate required to provide 1, 5, and 10 kW of power to a hydroturbine assuming a conveyance efficiency of 85% and an elevation head z of 45 m. Repeat the calculation with an elevation head of 65 m.

Table 6.3 Water resource and conveyance characteristics

Parameter	Location A	Location B	Location C	Location D
Elevation head (m)	5	25	40	65
η_{convey} (%)	90	86	81	76

6.6 Consider the water resources for the four sites in Table 6.3. Compute the specific speed if the turbine is directly coupled to a four-pole generator and supplies 8 kW of power at 50 Hz. Assume the turbine and generator efficiencies combined are 80%. Suggest a turbine type for each location based on the calculated specific speed.

6.7 A Pelton MHP system supplies 30 kWh per day at 50 Hz. The elevation head is 60 m. The efficiency of the conveyance system is 85%; the efficiency of turbine is 86%; and the generator efficiency is 94%. Determine the PCD of the turbine. The diameter of the jet cannot exceed 11% of the PCD. Assume the turbine rotates at 1500 RPM and the optimal speed ratio is 0.46.

6.8 Repeat Problem 6.7 but assume the MHP must supply electricity at 60 Hz.

6.9 A Pelton turbine has a PCD of 0.20 m. The water resource has an elevation head of 60 m, and the conveyance system has an efficiency of 85%. The efficiency of the turbine is described by the equation $\eta_{turbine} = 0.85 \left(1 - \frac{1}{0.46^2}(y - 0.46)^2\right)$ for speed ratios y between 0 and 0.92. The water flow is constant at 8 l/s. The generator has four poles and an efficiency of 94%. Compute the speed of the turbine, in RPM, if the electric power supplied to the load is 3194 W. Compute the frequency of the output voltage.

6.10 Consider again the scenario in Problem 6.9. Now assume the load reduces to 2000 W. Compute the new speed and voltage frequency. Comment on what this implies regarding the importance of governing or electronic load control for MHP systems.

6.11 A mini-grid is to be implemented for the community of Ababju. The community has 500 households. Each household is expected to consume 600 Wh per day. Assume the power consumption is constant throughout the day. The mini-grid will be powered either using wind or hydro power. Perform the following analyses:

WECS: the wind resource at Ababju was measured at a height of 3 m. The terrain near where the measurement was made is a fallow field. The measured wind speed was consistently 4.5 m/s. The WECS considered for the mini-grid has a hub-to-tip blade length of 1.75 m. Assume the turbine can be controlled such that C_p is 0.31 for all wind speeds. Determine the number of WECS required to supply power to Ababju if the turbines are mounted at a height of 10 m. Repeat the calculation if they are mounted at a height of 15 m.

MHP: the water resource in Ababju has an elevation head of 90 m. Assume the conveyance efficiency is 85%; the turbine efficiency is 85%; and the generator efficiency is 90%. The generator used in the MHP will have four poles. The Pelton turbine's optimal speed ratio is 0.46. Determine the required flow rate, the specific speed of the MHP, and compute the Pelton turbine's PCD and diameter of the water jet.

References

1. Bartmann, D., Fink, D.: Homebrew Wind Power. Buckville Publications (2009)
2. Csanady, G.T.: Theory of Turbomachines. McGraw-Hill, New York (1964)
3. Davis, S.: Microhydro. New Society Publishers (2003)
4. Davis, S. (ed.): Microhydro. New Society Publishers (2010)
5. Harvey, A., Brown, A., Hettiarachi, P., Inversin, A.: Micro-Hydro Design Manual. Practical Action Publishing (1993)
6. International Renewable Energy Agency: Off-grid renewable energy systems: Status and methodological issues (2015).   URL http://www.irena.org/publications/2015/Feb/Off-grid-renewable-energy-systems-Status-and-methodological-issues
7. Jawahar, C., Michael, P.A.: A review on turbines for micro hydro power plant. Renew. Sustain. Energy Rev. **72**, 882–887 (2017). doi:https://doi.org/10.1016/j.rser.2017.01.133. URL http://www.sciencedirect.com/science/article/pii/S1364032117301454
8. Latoufis, K.C., Pazios, T.V., Hatziargyriou, N.D.: Locally manufactured small wind turbines: Empowering communities for sustainable rural electrification. IEEE Electrification Mag. **3**(1), 68–78 (2015). doi:10.1109/MELE.2014.2380073
9. Thake, J.: The Micro-Hydro Pelton Turbine Manual. Practical Action Publishing (2000)

Chapter 7
Photovoltaic Arrays

7.1 Introduction

Photovoltaic (PV) arrays are commonly used in off-grid systems (see Fig. 7.1) and are becoming the default choice of energy conversion technology in such applications. This is primarily driven by falling costs, and the above average sunlight in Sub-Saharan Africa and South Asia, where electrification rates are the lowest. Whereas relatively few energy-impoverished communities are in locations whose climate and terrain can support wind-, hydro-, or biomass-based generation, most have enough sunlight for PV-based generation to be practical.

7.2 Solar Resource

Each year, 3.8×10^{24} joules of energy in the form of electromagnetic radiation from the sun passes through Earth's atmosphere. This is about 10,000 times our present annual energy consumption.

The amount of solar energy a site receives depends on the location, time of day, and weather conditions. It is important to understand a site's solar resource as this ultimately dictates the required capacity of the PV array. The power provided by sunlight is known as "irradiance", whose units are watts per square meter. Figure 7.2 is a solar resource map that shows the average irradiance around the world. The single measurement that most concisely, but not entirely, describes the solar resource of a location is the *insolation*. This should not be confused with "insulation." Insolation is the energy provided by sunlight per unit area over a period of time. It is commonly expressed as kilowatthours per meter squared per day or kilowatthours per square meter per year.

© Springer International Publishing AG, part of Springer Nature 2018
H. Louie, *Off-Grid Electrical Systems in Developing Countries*,
https://doi.org/10.1007/978-3-319-91890-7_7

Fig. 7.1 A 3 kW PV array split into two strings in Kenya (courtesy E. Patten)

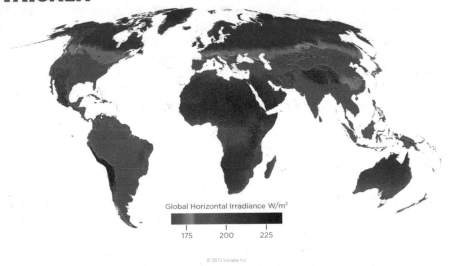

Fig. 7.2 World solar resource map showing average irradiance (courtesy of Viasala, Copyright (c) 2017 Vaisala)

Insolation varies greatly, typically ranging from 3.5 to 6.0 kWh/m²/day when averaged across a year. The monthly variation in insolation can be substantial. This is especially noticeable in regions with pronounced rainy seasons or are far from the equator.

Fig. 7.3 A 350 W
monocrystalline PV module
with 72 cells in series
(courtesy of Itek Energy)

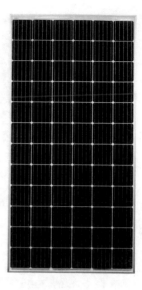

7.3 Physical Description

Solar power has become quite common in recent years. Nearly everyone has seen or is at least familiar with solar panels. One is shown in Fig. 7.3. On the front side are a number of dark rounded square areas. These are photovoltaic cells, the basic building blocks of a PV system. PV cells are typically made from doped silicon crystals that have been cut into thin, flat wafers. PV cells are produced in several standard sizes, for example, 0.125 m × 0.125 m or 0.156 m × 0.156 m.

The current and power output of a PV cell increases in proportion to its area. The voltage, however, is generally unaffected by size. A typical 0.156 m × 0.156 m cell can output approximately 2 to 4 W. To increase the power output, PV cells are connected together to form *modules*. The cells are often connected in series to increase the total voltage. For example, a typically sized 350 W module might contain 72 series-connected cells.

PV modules can be easily obtained all over the world from a variety of manufacturers. When multiple PV modules are connected together they form a *PV array*. The modules can be connected in series, parallel, or combination thereof.

Mini-grid systems tend to use modules that are larger—both in the physical sense and in terms of their power output capability—than used in solar home systems or solar lanterns. The dimensions of a typical 300 W module used in a mini-grid are approximately 1.95 m × 1.00 m × 0.046 m and weigh about 24 kg. PV modules are also called PV panels because they physically resemble flat rectangles.

Figure 7.4 shows the physical layout of a PV module. The cells are all connected in series and are arranged in four columns. The cells in each column are connected to each other using a conductive tabbing ribbon. The PV module is encapsulated within a metal frame to prevent the intrusion of water and pests. The face of the

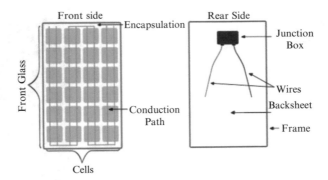

Fig. 7.4 Components of a PV module

module is covered by protective glass with an antireflective coating. On the rear side of the module are a backsheet and a junction box. The positive and negative wires from the string of PV cells extend from the junction box. These wires are used to connect the PV modules together to form an array or to connect the PV module to other components. The junction box often contains by-pass diodes, as discussed later in Sect. 7.12.1.

7.4 Principle of Operation

A PV cell is a solid-state device that converts a portion of the incident sunlight into DC power. The intensity of the sunlight is described by its *irradiance G*. It follows that irradiance is a measure of power density. For context, on a clear sunny summer day in the mid-continental United States, the irradiance will be about 1000 W/m^2. A PV module whose area is 2 m^2 would receive 2000 W of solar power under these irradiance levels. However, only a portion of this power is converted to electricity. The efficiency of commercially available cells is typically between 12 and 18%. This may seem low, but it is improving. Modules with higher efficiency are commercially available but come at a higher cost. In off-grid electrification, cost is often a primary consideration.

There are several types of PV cells. The most common are made from either monocrystalline or polycrystalline silicon (Si) material. Monocrystalline cells are made from silicon that has a single, continuous lattice structure. Polycrystalline cells on the other hand contain several smaller silicon crystal structures. The coloration and shape of the cells allow one to immediately tell if a module uses poly- or monocrystalline cells. The surface of a monocrystalline cell has a consistent coloring, but a polycrystalline cell appears patchy. Polycrystalline cells are typically rectangular; monocrystalline cells are rectangular but with rounded edges as a result of the manufacturing process. Monocrystalline cells are somewhat more efficient than polycrystalline cells but are more expensive to manufacture and hence cost more. Other common PV cell materials include gallium arsenide (GaAs), cadmium telluride (CdTe), and amorphous silicon, which has the advantage of being flexible. We will focus on mono- and polycrystalline silicon cells because they are the most common.

The basic mechanism that allows a PV cell to generate electricity is the photovoltaic effect.[1] The photovoltaic effect is the generation of voltage when light is shined onto a material.

In order to operate as intended, PV cells require light. You may remember from basic physics that light has a wave/particle duality. Most often we think of light as an electromagnetic wave that emanates from a source and varies sinusoidally in time and space. In the visible spectrum oscillations at high frequencies correspond to violet or blue light, lower frequencies correspond to red light.

In the early twentieth century, quantum physicists developed a particle theory of light. The photon theory of light relates well to our understanding of how PV cells operate. In this understanding, light is made up of photons. Photons are characterized by energy. High-energy photons are associated with high-frequency light. Lower-energy photons are associated with lower-frequency light.

Interesting things happen when light shines on a PV cell. If a photon of the right energy comes close to a valence electron of silicon, then there is a high probability that the electron will be excited into the conduction band. Valence electrons are virtually immobile; when an electron is in the conduction band, however, they can move quite freely. In this way light significantly changes the electrical properties of a PV cell.

In order to power an external circuit, PV cells are made from doped silicon—usually with boron and phosphorous—so that they have a *p-n junction*. A p–n junction is a fundamental building block of semiconductor electronic devices, including diodes, transistors, and integrated circuits. A PV cell is conceptually the same as a diode. We will not get into the details of p–n junctions as this is covered in most textbooks on electronics, but it suffices to note that the p–n junction results in a built-in electric field inside the PV cell. The built-in electric field within the PV pushes electrons that are in the conduction band through an external circuit, providing DC current. Energy is transferred from the light to the now-mobile electrons, effectively converting light into electricity.

7.4.1   Unilluminated PV Cell

A PV cell in the absence of light behaves like a diode, which itself is just a p–n junction. The current–voltage (I–V) characteristic of an ideal diode is

$$I_D = I_0 \left(e^{V_D/V_T} - 1 \right) \tag{7.1}$$

---

[1]Not be confused with the similar, yet different, photoelectric in which the electrons are freed into space.

where I_D is the current through the diode, I_0 is the reverse saturation current, V_D is the voltage across the diode, and V_T is the "thermal voltage." The magnitude of the reverse saturation current is small, around 10^{-9} A, and depends on the physical characteristics of the diode. When a diode is reversed biased, the current through it is near zero. This can be readily observed from (7.1). As V_D becomes negative, the exponential term approaches zero, and the current I_D approaches $-I_0$.

The thermal voltage is:

$$V_T = \frac{q V_D}{nkT} \tag{7.2}$$

where q is the charge of an electron, 1.602×10^{-19} C, T is the temperature, in Kelvin, k is Boltzmann's constant, 1.38×10^{-23} J/K, and n is the ideality factor. The ideality factor is 1 in an ideal diode. For a silicon diode at approximately room temperature (26.85°C), $V_T = 25.8$ mV.

As stated before, an unilluminated PV cell is simply a p–n junction and so can be modeled as a diode. We will refer to (7.1) as we construct a model for the illuminated PV cell.

7.4.2 Illuminated PV Cell

Now assume light is shined on the PV cell. Some of the photons will excite electrons into the conduction band. The built-in electric field will push some of the excited electrons to the n-side of the p–n junction. This concentration of charge on either side of the junction results in a measurable voltage across the cell. We therefore expect an illuminated PV cell to exhibit a non-zero open-circuit voltage.

If an external conduction path is provided, then the electrons on the n-side of the junction will travel through it. The magnitude of the resulting current depends on the resistance of the external circuit. The current under short-circuit conditions is known as the *illumination current* I_G. The illumination current is always positive, and its magnitude is proportional to the irradiance G and the area and efficiency of the PV cell. The illumination current under clear sunny conditions will range from about 0.5 A for small cells to about 9 A for 0.156 m x 0.156 m cells.

We must modify the ideal diode equation (7.1) to account for the presence of the photogenerated current:

$$I_{cell} = I_G - I_0 \left(e^{V_{cell}/V_T} - 1 \right) \tag{7.3}$$

$$I_{cell} = I_G - I_D \tag{7.4}$$

where I_{cell} is the current output by the PV cell. We have also replaced the diode voltage with the PV cell voltage V_{cell}. The two voltages are equal if we ignore losses, as we have so far.

Equation (7.3) is known as the *characteristic equation* of the PV cell. Note that when the illumination current is zero (7.3) is equivalent to the ideal diode equation (7.1). We can rearrange (7.3) to solve for the cell voltage:

$$\frac{I_G - I_{cell}}{I_0} + 1 = e^{V_{cell}/V_T} \tag{7.5}$$

$$V_T \ln\left(\frac{I_G - I_{cell}}{I_0} + 1\right) = V_{cell}. \tag{7.6}$$

7.4.3 Open-Circuit Voltage

We can manipulate (7.3) to determine the open-circuit voltage $V_{cell,OC}$ when the PV cell is illuminated by setting the cell current to zero:

$$V_{cell,OC} = V_T \ln\left(\frac{I_G}{I_0} + 1\right). \tag{7.7}$$

From (7.2) to (7.7), we see that the open-circuit voltage depends on the temperature, the ideality factor, the illumination current, and the reverse saturation current. The open-circuit voltage of an illuminated cell is typically 0.6 to 0.7 V. The natural logarithm in (7.7) shows that the open-circuit voltage is relatively insensitive to changes in the illumination current—and therefore it is also relatively insensitive to changes in irradiance. Also note that from (7.2) and (7.7), as temperature decreases, the open-circuit voltage increases.

7.4.4 Short-Circuit Current

Under short-circuit conditions, the PV cell voltage is zero ($V_{cell} = 0$), and from (7.3) we see the short-circuit current I_{SC} is:

$$I_{SC} = I_G - I_0\left(e^{0/V_T} - 1\right) \tag{7.8}$$

$$I_{SC} = I_G - I_0\left(1 - 1\right) \tag{7.9}$$

$$I_{SC} = I_G \tag{7.10}$$

In other words, the short-circuit current is equal to the illumination current.

Example 7.1 Compute the open-circuit voltage and short-circuit current of a PV cell whose reverse saturation current is 9^{-9} A, illumination current is 8.46 A and whose thermal voltage is 28 mV.

Solution From (7.10) the short-circuit current of a PV cell is equal to the illumination current so that $I_{SC} = 8.46$ A. The open-circuit voltage is found by solving (7.7):

$$V_{cell,OC} = V_T \ln\left(\frac{I_G}{I_0} + 1\right) = 0.028 \times \ln\left(\frac{8.46}{9^{-9}} + 1\right) = 0.613 \text{ V}$$

Of course the purpose of a PV panel is to deliver power. Under either short-circuit or open-circuit conditions, the output power is zero. To understand the power delivered under practical conditions, we must first consider the cell's current–voltage (I–V) characteristic.

7.5 I–V Curve

A plot of the characteristic equation of a PV cell or module is known as the I–V *curve*. The I–V curve provides information on the quality of the PV cell or module, and its open-circuit voltage and short-circuit current, which are needed to properly design a PV system.

The I–V curve for an ideal PV cell is shown in Fig. 7.5. The I–V curve is unlike that of a voltage source or a current source. It resembles a hybrid of the two. The current is nearly constant for a wide range of terminal voltages, similar to a current source. However, as the knee of the curve is passed, the I–V curve resembles a voltage source, with little change in voltage for a wide range of current. An important characteristic of PV cells is that they are self-limiting. PV cells and modules can be short-circuited or open-circuited without damaging themselves.

7.5.1   *Circuit Model*

We are now prepared to develop a circuit model of a PV cell. We start with a simplified ideal model that captures the general characteristics. The model consists of an ideal current source in parallel with a diode, as shown in Fig. 7.6. The value of the current source is equal to the illumination current. This is simply the circuit equivalent of (7.3).

Fig. 7.5 *I–V* curve of a
single ideal PV cell

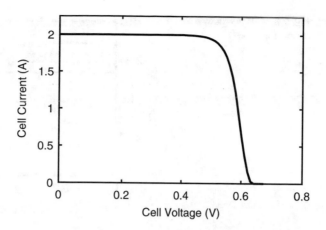

Fig. 7.6 Simplified lossless
PV cell circuit model

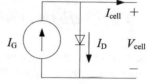

Fig. 7.7 PV cell circuit
model including losses

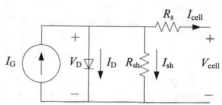

The simplified model ignores losses internal to the cell. These losses can be
included in the model as lumped series and shunt resistances shown in Fig. 7.7.

The corresponding current and voltage relationship is found by application of
Kirchhoff's Current Law (KCL) at the top node, followed by substitution

$$I_{\text{cell}} = I_G - I_D - I_{\text{sh}} \tag{7.11}$$

$$I_{\text{cell}} = I_G - I_0 \left(e^{(V_{\text{cell}} + I_{\text{cell}} R_s)/V_T} - 1 \right) - \frac{V_{\text{cell}} + I_{\text{cell}} R_s}{R_{\text{sh}}}. \tag{7.12}$$

The resulting expression is an implicit equation, with the PV cell current I_{cell}
appearing on both the left- and right-hand sides. The equation is therefore solved
by using numerical means rather than analytically. The power loss associated
with shunt resistance is quite small, about 0.1% of the power output; the series
resistance loss is larger, perhaps 0.3% [3]. These losses are typically considered to
be negligible, and the simplified model is often used.

Fig. 7.8 $I–V$ and power curve of a PV cell

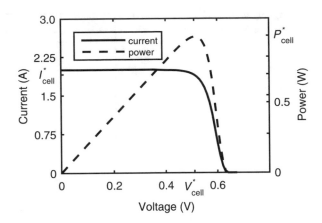

7.5.2 Maximum Power Point

As with all practical sources, there is a unique operating point that maximizes the power production from a PV cell. Consider a PV cell whose $I–V$ curve for a given irradiance is shown in Fig. 7.8. The power corresponding to each operating point is the product of the voltage and current

$$P_{\text{cell}} = I_{\text{cell}} V_{\text{cell}}. \tag{7.13}$$

The power associated with each voltage is also shown in Fig. 7.8. Importantly, we see that the operating point that maximizes power production is unique. This point is known as the "maximum power point" (MPP). The MPP is located near the knee of the $I–V$ curve. The voltage and current at the MPP are denoted V^*_{cell} and I^*_{cell}, respectively. Here the superscript "*" should not be confused with complex conjugate operator, which has no practical meaning in DC circuit analysis. The corresponding maximum power is P^* so that

$$P^*_{\text{cell}} = V^*_{\text{cell}} I^*_{\text{cell}}. \tag{7.14}$$

A PV cell will only produce its maximum power if it is connected to load whose resistance is $R^* = V^*_{\text{cell}}/I^*_{\text{cell}}$. Alternatively, if the PV cell is used to charge a battery, the battery's terminal voltage must equal V^*_{cell} for maximum power to be produced by the cell. We must not forget that the $I–V$ curve of PV cell depends on several factors, primarily the irradiance. We therefore rewrite (7.14) to show that maximum power and the voltage and current corresponding to the MPP are all functions of irradiance G:

$$P^*_{\text{cell}}(G) = V^*_{\text{cell}}(G) I^*_{\text{cell}}(G). \tag{7.15}$$

Recall that the illumination current is proportional to the irradiance received by the cell. From (7.10), the short-circuit current will then also be proportional

Fig. 7.9 The maximum power point of an example PV cell for different levels irradiance; the "x" mark the maximum power points

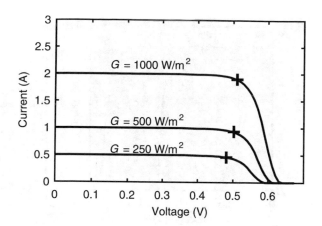

to the irradiance. Plotting the characteristic equation for different values of the illumination current results in a family $I–V$ curves, as shown in Fig. 7.9.

The resistance corresponding to the MPP will also depend on the irradiance $R^*(G)$. Therefore, as the irradiance changes, so must the resistance in order to maximize the power. The practical consequence of this is that any device powered by the PV cell must continually adjust its equivalent input resistance to track the MPP; otherwise, the power production will not be maximized. We shall see in Chap. 10 that specialized devices called "maximum power point trackers" are used to operate a PV module at its maximum power point.

7.6 PV Modules

PV cells are usually connected in series inside a module. The result is a stretching of the $I–V$ curve along the voltage axis, as shown in Fig. 7.10. The short-circuit current is not affected, and the open-circuit voltage increases in proportion to the number of cells. The characteristic equation becomes:

$$I_{module} = I_G - I_0 \left(e^{V_{module}/(V_T N_{ser})} - 1 \right) \tag{7.16}$$

where I_{module} is the module current and V_{module} is the module voltage. The module voltage is simply the cell voltage multiplied by the number of cells in series

$$V_{module} = N_{ser} V_{cell}. \tag{7.17}$$

PV cells can be connected in parallel to increase the current, although it is somewhat uncommon. This stretches the $I–V$ curve along the current axis as shown in Fig. 7.10. The characteristic equation for the module becomes:

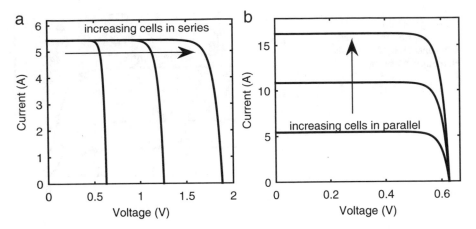

Fig. 7.10 (**a**) The I–V curve is changed as PV cells are connected in series and (**b**) parallel

Table 7.1 Example electrical characteristics of PV module

Characteristic at STC	Value
Maximum power	185.3 W
Optimum operating voltage (V^*)	36.4 V
Optimum operating current (I^*)	5.08 A
Open-circuit voltage (V_{OC})	45.0 V
Short-circuit current (I_{SC})	5.43 A
Short-circuit current temp. coeff. (α_i)	0.055 %/K
Open-circuit temp. coeff. (α_v)	−0.37%/K
Max. power temp. coeff. (α_P)	−0.48%/K
NOCT	45°C
Number of Cells	72 (series)

$$\frac{I_{\text{module}}}{N_{\text{par}}} = I_G - I_0 \left(e^{V_{\text{module}}/(V_T N_{\text{ser}})} - 1 \right) \tag{7.18}$$

where N_{par} is the number of parallel-connected cells.

The I–V curve of a PV module resembles that of a cell. The open-circuit voltage and short-circuit current are simply N_{ser} and N_{par} times greater than that of a single cell. Like a cell, a PV module has a unique maximum power point. In general, we are more interested in the behavior of a module or array than of a single cell. Hereafter, we will suppress the "module" subscript so that V, I, and P refer to the voltage, current, and power of a module.

Manufacturers provide information about the characteristics of a PV module in their data sheets. See Table 7.1 for example.

Example 7.2 Compute the open-circuit voltage if 60 of the cells described in the Example 7.1 are connected in series.

Solution Rearranging (7.16) and setting the module current to zero shows that

$$V_{OC} = N_s V_T \ln\left(\frac{I_G}{I_0} + 1\right) = 60 \times 0.028 \times \ln\left(\frac{8.46}{9^{-9}} + 1\right) = 36.81 \text{ V}.$$

More generally if the PV cells are connected in series within a module, the voltage of the module is equally divided among the cells. This holds under any loading conditions, including open and short circuit. Similarly, for parallel-connected cells, the module current is equally divided among the cells.

7.7 Standard Test Conditions

The power output of a PV module is dependent on both the irradiance and the resistance of the load. Given that both can vary, the power rating of a PV module must correspond to a set of specific operating conditions. The power rating, or *capacity,* of a PV module is the maximum power that can be produced under "standard test conditions" (STC). Standard test conditions are defined as:

- Irradiance of 1000 W/m^2
- Cell temperature of 25°C

In addition, STC also correspond to a specific distribution or spectrum of the photon wavelengths known as AM (Air Mass) 1.5.

The meaning of STC and how it relates to the rated power of a PV is especially important. For example, a PV panel rated at 100 W will produce 100 W if the panel operates under STC and is connected to a load whose input resistance results in maximum power point operation. In practice, PV modules deployed in actual power systems almost never operate at STC. It would not be surprising for a module to never produce its rated power.

We will use the subscript "STC" to refer to current, voltage, and power values referenced to STC. For example, P^*_{STC} is the maximum power output under STC—this is equivalent to the rated power of the PV module. The irradiance corresponding to STC, 1000 W/m^2, is hereafter denoted G_{STC}.

7.8   Correcting for Temperature

In practice, the temperature of a PV module T_C is rarely 25°C as specified in STC. In many locations, the ambient temperature routinely exceeds 25°C, and the exposure to direct sunlight further increases the module temperature. Figure 7.11 shows how the $I–V$ curve varies with temperature for a certain module.

The $I–V$ curve of a PV module, like that of a diode, is sensitive to temperature as expressed by the variable V_T in (7.2). From (7.3), as temperature increases, the current of the PV cell increases, but the voltage decreases. The decrease in voltage dominates the increase in current. Therefore, the maximum power from a PV cell is reduced as the temperature increases. PV module manufacturers provide information on the sensitivity of the PV module parameters to temperature. The temperature coefficients for the short-circuit current α_i, open-circuit voltage α_v, and maximum power α_p are used to correct for changes in temperature as follows:

$$I_{SC}(T_C) = I_{SC}(25° \text{ C}) \left(1 + \frac{\alpha_i}{100} \times (T_C - 25)\right) \tag{7.19}$$

$$V_{OC}(T_C) = V_{OC}(25° \text{ C}) \left(1 + \frac{\alpha_v}{100} \times (T_C - 25)\right) \tag{7.20}$$

$$P^*(T_C) = P^*(25° \text{ C}) \left(1 + \frac{\alpha_p}{100} \times (T_C - 25)\right). \tag{7.21}$$

We note that if the cell temperature is 25°C, which corresponds to STC, then no correction is made. The coefficients are expressed as percent per Kelvin. The open-circuit coefficient is sometimes expressed as millivolt per Kelvin. To convert this to percent per Kelvin, divide the coefficient by $V_{OC,STC}$.

Fig. 7.11 The effect of temperature on a PV $I–V$ curve for a typical 285 W module

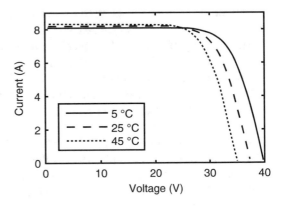

Example 7.3 The temperature of a PV module with characteristics in Table 7.1 is 47°C. Compute the short-circuit current, open-circuit voltage, and maximum power.

Solution We assume that the module is operating at STC with the exception of the temperature. We then apply (7.19)–(7.21) to compute the temperature-corrected short-circuit current, open-circuit voltage, and maximum power.

$$I_{SC}(T_C) = 5.43\,(1 + 0.00055 \times (47 - 25)) = 5.496\,\text{A}$$

$$V_{OC}(T_C) = 45.0\,(1 - 0.00370 \times (47 - 25)) = 41.337\,\text{V}$$

$$P^*(T_C) = 185.3\,(1 - 0.0048 \times (47 - 25)) = 165.732\,\text{W}$$

Notice that the magnitude of the percentage change in V_{OC} is greater than that of I_{SC} and the percent change in power is greater still.

7.8.1 Nominal Operating Cell Temperature

In order to apply the temperature corrections, we must know the temperature of the PV module. Similar to our discussion on distribution line thermal ratings, the temperature of the module depends on several factors, including the ambient temperature, wind speed, and irradiance. Because these conditions vary, manufacturers report the cell temperature, known as the Nominal Operating Cell Temperature (NOCT), when exposed to the following standard operating conditions (SOC):

- Irradiance: 800 W/m^2 (0.8 G_{STC})
- Ambient temperature: 20°C
- Wind speed: 1 m/s
- Spectral distribution: AM 1.5
- Power output: 0 W (no load)

Be careful not to confuse STC—the conditions in which PV electrical characteristics are referenced to—and SOC, the conditions in which the cell temperature are referenced to. Typical values of NOCT are 42°C to 50°C. Like STC, PV modules often operate in an environment that deviates from SOC. Most often, adjustments are only made for deviations in ambient temperature T_a and irradiance. In this case cell or module temperature is approximated as

$$T_C = T_a + (NOCT - 20)\,\frac{G}{800}. \tag{7.22}$$

It is common to assume that all cells in a module have the same temperature, so that (7.22) also applies to the module.

Example 7.4 Compute the module temperature of a PV module whose reported NOCT is 45°C when exposed to irradiance of 600 W/m^2 and an ambient temperature of 34° C.

Solution The PV module is not operating under SOC, and so (7.22) must be applied.

$$T_C = 34 + (45 - 20)\,\frac{600}{800} = 52.75°C.$$

7.9   Correcting for Irradiance

We now consider the common situation in which the irradiance does not correspond to G_{STC} (1000 W/m^2). A simple method of correcting for irradiance is to assume the maximum power changes in proportion to irradiance

$$P^*(G) = P^*_{STC} \times \frac{G}{G_{STC}}. \tag{7.23}$$

Using this method, the maximum power produced by a 100 W PV module exposed to irradiance of 500W/m^2 is 50 W. Underlying this model are the assumptions that

$$V^*(G) = V^*_{STC} \tag{7.24}$$

$$I^*(G) = I^*_{STC} \times \frac{G}{G_{STC}}. \tag{7.25}$$

That is, the voltage at the maximum power point is insensitive to changes in irradiance, and the current at the maximum power point scales proportionally to the irradiance. These are reasonable approximations in must circumstances.

Example 7.5 Compute the maximum power of the PV panel whose characteristics are described in Table 7.1 assuming the irradiance is 600 W/m^2.

Solution From (7.23):

$$P^*(G) = 185.3 \times \frac{600}{1000} = 111.18W.$$

7.10 Correcting for Temperature and Irradiance

We next consider the case in which both the irradiance and temperature deviate from STC. Several methods have been proposed for adjusting the maximum power output for varying temperature and irradiance [2]. One method that produces a reasonable balance of accuracy and simplicity is Osterwald's method [1]:

$$P = P_{STC}^* \times \frac{G}{1000} \times \left(1 + \frac{\alpha_p}{100} \times (T_C - 25)\right). \tag{7.26}$$

Osterwald's method corrects for irradiance using (7.23), and (7.21) for temperature.

Example 7.6 Compute the maximum power of the PV module whose characteristics are described in Table 7.1 assuming the irradiance is 600 W/m^2, and the ambient temperature is 34°C, using Osterwald's method.

Solution Osterwald's method requires the cell temperature to be computed, as done in Example 7.4. From (7.26), Osterwald's method yields a maximum power output of:

$$P = 185.3 \times \frac{600}{1000} \times \left(1 + \frac{-0.48}{100} \times (52.75 - 25)\right) = 96.37 \text{ W}.$$

In summary, the conditions most favorable for PV power production are when the PV modules are at a low temperature but high irradiance. These conditions do not often exist simultaneously, especially in regions of the world with low electricity access, with the exception of certain high-altitude locations.

In designing off-grid PV systems, we must be aware that the open-circuit voltage, as well as the short-circuit current provided by the manufacture can be exceeded when conditions deviate from STC, as they typically do.

7.11 PV Arrays

Commercially available PV panels rarely exceed 350 W in rated capacity. For mini-grids that require more capacity, multiple modules are combined to form a PV array. Modules can be combined in series, parallel, or a combination thereof, as shown in Fig. 7.12.

Series-connected PV modules are called "strings." If the modules are physically identical and exposed to the same conditions, then the output voltage of string is the voltage of an individual module multiplied by the number of modules. Strings with

Fig. 7.12 A PV array
consisting of three strings
with two modules in each
string

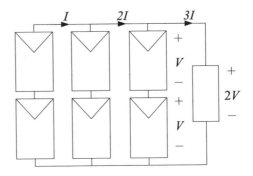

the same number of modules can be connected in parallel. The current to the load
increases in proportion to the number of strings. Mathematically

$$V_{\text{array}} = N_{\text{ser,str}} \times V \tag{7.27}$$

$$I_{\text{array}} = N_{\text{par,str}} \times I \tag{7.28}$$

where $N_{\text{ser,str}}$ is the number of series-connected modules in a string and $N_{\text{par,str}}$
is the number of strings. The subscript "array" has been used to identify variables
associated with the array.

The power produced by the array is:

$$P_{\text{array}} = V_{\text{array}} \times I_{\text{array}} = N_{\text{ser,str}} \times V \times N_{\text{par,str}} \times I = N_{\text{ser,str}} \times N_{\text{par,str}} \times P. \tag{7.29}$$

Note that $N_{\text{ser,str}} \times N_{\text{par,str}}$ is the total number of modules in the array and P is the
power output of one module. This tells us that the maximum power from an array
depends only on the total number of modules, not how they are connected. The
power from an array can be corrected for non-STC conditions in the same manner as
for individual modules. Hereafter, we will drop the "array" subscript. The voltage V,
current I, and power P for the remainder of this chapter will refer either to the cell,
module, or array quantities. This should be obvious by the context of the derivation
or problem.

The electrical characteristics of an array can be calculated from the characteristics of its constituent modules. Voltage characteristics are multiplied by the number
of modules in a string, and the current characteristics are multiplied by the number
of strings. Table 7.2 shows an example for an array with three strings and two
modules per string.

For a given number of modules, the selection of the number of strings and
number of modules per string is not arbitrary. Increasing the number of strings
increases the required ampacity of the cables connecting the array to the rest
of the system. However, increasing the number of modules per string increases
the operating voltage, which has safety and insulation consequences. Usually the

Table 7.2 Example
electrical characteristics of a
six-module array with
$N_{ser,str} = 2$, $N_{par,str} = 3$

Characteristic at STC	Module	Array
Maximum power	290.16 W	1740.96 W
Optimum operating voltage (V^*)	36 V	72 V
Optimum operating current (I^*)	8.06 A	24.18 A
Open circuit voltage (V_{OC})	45.5 V	91.0 V
Short circuit current (I_{SC})	8.56 A	25.68 A

voltage capability of a storage battery or charge controller is used to determine the
maximum voltage rating of a string.

Example 7.7 A mini-grid requires a PV array rated at least 2.5 kW. Each PV
module is rated at 190 W. Each has an open-circuit voltage of 43.2 V and
short-circuit current of 5.98 A (all under standard test conditions). The array
is connected to a battery charge controller. The input voltage of the charge
controller cannot exceed 150 V, and input current cannot exceed 85 A. The
maximum array power through the charge controller cannot exceed 3.2 kW.

Design a PV array that is compatible with the charge controller under STC.
Sketch the layout of the array.

Solution The minimum number of modules required is found from

$$\frac{\text{required array capacity}}{\text{capacity per module}} = \frac{2.5\,\text{kW}}{0.190\,\text{kW}} = 13.15$$

so at least 14 modules are needed.

The controller's input voltage limit is 150 V. The maximum number of
modules that can be placed in a string is found from

$$\frac{\text{max. string voltage}}{\text{module open} - \text{circuit voltage}} = \frac{150\,\text{V}}{43.2\,\text{V}} = 3.47$$

so at most three modules can be connected in a string.

Each string must have the same number of modules. Therefore, we can
select five strings of three modules each, for a total of 15 modules. The
maximum power produced by this array is 15 x 190 = 2850 W, which is under
the charge controller's power limit.

The current from each string during short circuit is 5.98 A, giving a total
current of $5 \times 5.98 = 29.9$ A, well below the charge controller's current limit.

The corresponding array is shown in Fig. 7.13. Note that a more economical solution would be to use seven strings of two modules, for a total of 14
modules. There are also other viable designs.

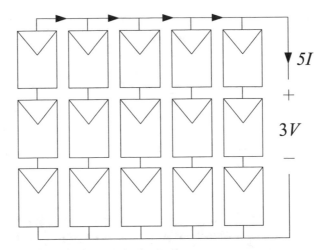

Fig. 7.13 A PV array with five strings of three modules per string

7.12   Effects of Shading

We have seen that the maximum power output from a PV array is approximately proportional to the irradiance it receives. What happens when a portion of the array is shaded? In many mini-grids, the PV array is affixed to an existing roof line. Trees and other building might shade the array during certain hours of the day or days of the year. In larger systems, there is often sufficient budget to clear trees and obstructions from the array, so this is less of a challenge.

Figure 7.14 shows the lossless circuit model of three series-connected ideal PV cells. Assume that the middle cell is completely shaded so that the illumination current is zero. Applying Kirchhoff's Current Law at the node above the shaded cell yields

$$I = I_{G,2} - I_0 \left(e^{V_2/V_T} - 1 \right) \tag{7.30}$$

$$I - I_{G,2} = -I_0 \left(e^{V_2/V_T} - 1 \right) \tag{7.31}$$

$$I = -I_0 \left(e^{V_2/V_T} - 1 \right). \tag{7.32}$$

The current I and I_0 are positive. The exponential term must be less than one for the right-hand side of (7.32) to be positive. The voltage of PV cell 2, V_2, must therefore be negative. The diode is therefore reverse biased, and the current I cannot exceed the reverse saturation current I_0. Recall that I_0 is typically several orders of magnitude smaller than the illumination current of the unshaded cells. We can therefore expect the power supplied by the PV module to be substantially decreased.

Fig. 7.14 Circuit model of
PV module with one cell
shaded

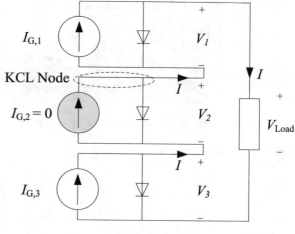

Fig. 7.15 *I–V* curve of a PV
module showing the effects of
shading

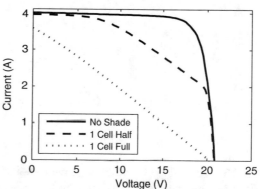

A memorable, but not entirely true, adage is that the current supplied by a PV
module with series-connected cells is limited to the current produced by its least
productive cell.

Figure 7.15 shows the effects of shading a single cell of a 66 W PV module with
40 series-connected cells. The shunt resistance of each cell has been modeled. The
maximum power drops to 42 W when one cell is shaded by 50% and to 19 W when
it is shaded by 100%. Shading is to be avoided not only because of the decreased
power output but also because the shaded cells dissipate power, which can lead to
local hot spots that permanently damage the module.

7.12.1   By-pass and Blocking Diodes

The substantial reduction in power and the potentially damaging effects of shading
can be mitigated by using by-pass diodes. If a diode is installed in parallel but in the
opposite direction as the diode in the PV cell model, then it will conduct whenever

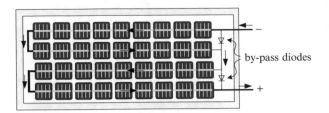

Fig. 7.16 PV modules usually contain several by-pass diodes

Fig. 7.17 Blocking diodes of
a PV array

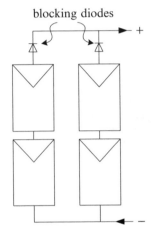

the PV cell voltage becomes negative (when the cell is shaded). The shaded cell does not meaningfully contribute to the power production, but it also does not substantially reduce it.

Adding by-pass diodes as described above to each cell is generally not done for economic reasons. Rather, a single diode is used to by-pass several cells, as shown in Fig. 7.16. In this figure, each diode serves as a by-pass for 18 of the 36 cells. Should one of the cells be shaded, then one half of the cells are by-passed. This strikes a reasonable balance between the cost of diodes and mitigating the effects of any shading. In larger arrays, external diodes are often used to by-pass entire modules in strings.

Figure 7.17 shows blocking diodes. The purpose of a blocking diode is to ensure that current only flows one way through the PV module or string. Blocking diodes should be included whenever a PV module or array is connected directly to a battery (in other words, in cases in which there is no charge controller). Otherwise, after sunset—or any time the module voltage is lower than the battery voltage— the battery will discharge through the array, draining the battery and potentially damaging the PV array.

7.13   Energy Production

The irradiance received by a PV array varies throughout the day. This is primarily caused by changing atmospheric conditions and the angle of incidence between the sun's rays and the face of the PV array. We can re-write (7.26) to show the dependency of irradiance and temperature on time:

$$P(t) = P_{STC}^* \times \frac{G(t)}{1000} \times \left(1 + \frac{\alpha_p}{100} \times (T_C(t) - 25)\right). \tag{7.33}$$

The plot of $P(t)$ depends on several factors, including the location and orientation of the PV array. A typical plot for an un-tilted (horizontal) 1 kW PV array located at the equator is shown in Fig. 7.18. Production begins in the morning, peaks at midday, and ends around sunset.

The production in Fig. 7.18a might appear low. After all, the PV array is rated at 1 kW, but its maximum production is less than 0.8 kW. Keep in mind that the rated power is based on STC, which rarely occurs in real-world conditions.

The energy produced by a PV array is the area under the curve in Fig. 7.18. It can be computed by integrating (7.33) between the time that irradiance is first and last received by the array t_{rise} and t_{set}

$$E = \int_{t_{rise}}^{t_{set}} P(t) dt. \tag{7.34}$$

The energy production corresponding to Fig. 7.18 is 5.1 kWh. During certain times of the year, it will be more, and other times it will be less. The factors influencing PV array energy production are discussed further in Chap. 12.

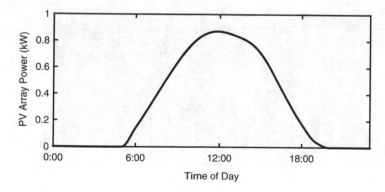

Fig. 7.18 Theoretical power production for a 1 kW PV array

Example 7.8 The irradiance received by a PV array can be modeled as

$$G(t) = 900 \sin \left(\frac{2\pi (t - 6)}{24} \right)$$

where t is in hours. The model is valid between sunrise $t = 6$ and sunset $t = 18$. The energy that the PV array is to provide each day to a mini-grid is 16 kWh. Compute the minimum capacity of the PV array to supply this energy. Ignore the effects of cell temperature. Compute the minimum capacity of a gen set needed to supply the required energy.

Solution From (7.33) and (7.34), the energy output of the PV array is expressed as:

$$E = \int_{t_{rise}}^{t_{set}} P(t) \mathrm{d}t = P_{STC}^* \frac{900}{1000} \int_6^{18} \sin \left(\frac{2\pi (t - 6)}{24} \right) \mathrm{d}t.$$

Setting the energy produced to the required 16 kWh and integrating using u substitution shows:

$$16 = P_{STC}^* \frac{900}{1000} \frac{24}{2\pi} 2 = 6.876 P_{STC}^*$$

and therefore the rating of the PV array must be at least $16/6.876 = 2.327$ kW.

$$P_{STC}^* = \frac{16}{6.876} = 2.327 \text{ kW}.$$

Assuming a gen set can be operated continuously, then the minimum required capacity is:

$$\frac{16 \text{ kWh/day}}{24 \text{h/day}} = 0.67 \text{ kw}.$$

For this simple scenario, the energy produced by a 2.327 kW PV array can be replaced by a 0.667 kW gen set. This example highlights the importance in considering energy production, not just power ratings, when comparing different energy conversion technologies.

7.14 Practical Considerations

PV arrays offer the following advantages in off-grid applications:

- Although the capital cost is higher than for gen sets, there are no fuel costs. In some cases, a PV system can provide energy at a lower cost than one using a gen set.
- Energy-impoverished communities are often in areas with abundant sunlight. Moreover, solar resource databases can be consulted to determine if a location is suitable for a PV-based system. Wind or hydro resources require very specific local conditions to be feasible, and on-site local measurements are typically required.
- The energy produced by PV modules correlates with daytime load.
- PV arrays are modular, which means that individual modules can easily be added (or removed) to achieve an appropriate capacity.
- PV arrays have no moving parts, which reduces maintenance (only periodic cleaning and inspection are required) and eliminates noise.
- PV arrays are considered to be environmentally benign as there is no particulate or carbon dioxide emissions associated with their operation.
- PV arrays and associated equipment such as the rackings upon which they are mounted, charge controllers, and batteries are widely available. The supply chain is maturing: licensed installers, original and replacement parts, and warranty service can be found in many developing countries.
- PV arrays do not directly consume water, although periodic washing is needed.

Balancing these advantages are the following considerations.

- The energy produced by PV arrays is variable and uncertain. PV array power production is driven by sunlight, which varies throughout the day and year. Cloud coverage is difficult to forecast, and production might be severely limited during rainy seasons. This adds uncertainty to the design process, leading to arrays that are larger than needed and consequentially more expensive, or smaller than needed causing the system to be unreliable.
- In certain locations, particularly those with perennial cloud coverage or at polar latitudes, the solar resource is inadequate for a PV array to be an economic and practical solution.
- Although PV array prices have fallen globally to less than US$1/W, energy storage, charge controllers, and other components are needed, increasing the cost and complexity.
- PV arrays have low power density, and so a large amount of roof space or land is needed. For example, a 5 kW system requires approximately 40 m^2 of surface area for the PV array. Further, the PV array must be tilted and oriented in a specific way to maximize power production. This often necessitates custom-made racking structures.

Despite these drawbacks, given the substantial solar resource in much of Africa and South Asia, many see PV-based mini-grids as a sustainable and scalable solution to off-grid electricity access.

7.15   Summary

This chapter provided the basic technical principles and characteristics of photovoltaic (PV) cells, modules, and arrays. PV modules are becoming the default choice for energy conversion technology in locations with suitable insolation. PV cells rely on the photovoltaic effect to produce DC current when illuminated. They can be modeled by a current source, representing the sunlight-driven current, in antiparallel with a diode, which models the p–n junction of the PV cell.

The I–V characteristic of a PV array is important in understanding its operation when connected to a load. In particular, there is a unique point of maximum power for a given irradiance and temperature. Maximum power production requires the load to be "matched" to the PV array's maximum power point.

PV modules are rated under Standard Test Conditions. These conditions rarely occur outside a laboratory setting. The electrical characteristics of a PV array are sensitive to irradiance and temperature. Osterwald's method can be used to estimate the power production for different conditions. Osterwald's method assumes the power is proportional to the irradiance, with the temperature effects modeled using the PV modules' temperature coefficient. Complete or partial shading is especially detrimental to PV power production. Shading of a single cell can dramatically reduce power produced by an entire module.

When PV modules are connected together, they form an array. The array voltage is proportional to the number of modules connected in series; the current is proportional to the number of parallel-connected module strings. The maximum power that can be produced by an array scales in proportion to the number of modules, regardless of how they are connected.

Problems

7.1 Consider PV module A whose characteristics are shown in Table 7.3. Compute:

- The cell temperature if the irradiance is 1000 W/m^2 and the ambient temperature is 31°C
- The short-circuit current, open-circuit voltage, and maximum power if the irradiance is 1000 W/m^2 and the cell temperature is as computed in the previous part of this problem

Table 7.3 Electrical characteristics of PV module A and module B

Characteristic at STC	Module A	Module B
Maximum power	350 W	20 W
Optimum operating voltage (V^*)	38.54 V	18 V
Optimum operating current (I^*)	9.08 A	1.11 A
Open-circuit voltage (V_{OC})	47.43 V	22.5 V
Short-circuit current (I_{SC})	9.49 A	1.23 A
Short-circuit current temp. coeff. (α_i)	0.040 %/K	0.045%/K
Open-circuit voltage temp. coeff. (α_v)	−0.29 %/K	−0.34 %/K
Max. power temp. coeff. (α_P)	−0.38 %/K	−0.47 %/K
NOCT	45°C	47°C
Number of cells	72 (series)	36 (series)

- The short-circuit current if the irradiance is 500 W/m^2
- The power produced using Osterwald's method if the irradiance is 650 W/m^2 and the ambient temperature is 20°C

7.2 Repeat Problem 7.1 but consider module B.

7.3 A mini-grid is supplied by a PV array consisting of 12 PV modules arranged in three strings. The module's parameters are provided in Table 7.3 (PV module A). Compute the total power output by the array under STC. Compute the corresponding voltage and current of the array under STC.

7.4 Consider the PV array in the previous problem. Compute the open-circuit voltage, short-circuit current and maximum power of the array if the irradiance is 1050 W/m^2, and the ambient temperature is 31°C using Osterwald's method. Assume the short-circuit current is proportional to the irradiance and the open-circuit voltage is only affected by the ambient temperature, not the irradiance (but its value must be adjusted according to the cell temperature).

7.5 A mini-grid is supplied by a PV array consisting of eight PV modules arranged in two strings. The module's parameters are provided in Table 7.3 (PV module B). Compute the total power output by the array under STC. Compute the corresponding voltage and current of the array under STC.

7.6 When designing a PV array for a mini-grid, the maximum open-circuit voltage of the array must be within the limits of the charge controller it is connected to. The maximum open-circuit voltage is usually based on the lowest expected daytime cell temperature, assuming the irradiance is also 1000 W/m^2. What is the maximum open-circuit voltage of the array in the previous problem if the minimum expected daytime cell temperature is 10°C?

7.7 A mini-grid is to be implemented for the community of Ababju. The community has 500 households. Each household is expected to consume 600 Wh per day. Assume the power consumption is constant throughout the day. Let the daily irradiance of Ababju be

$$G(t) = 700 \sin\left(\frac{2\pi(t-6)}{24}\right)$$

where t is in hours (valid between sunrise $t = 6$ and sunset $t = 18$). Determine how many of the modules described in Table 7.3 are needed to power Ababju. Assume that the mini-grid contains sufficient and lossless energy storage so that the excess energy produced during the day can be used to power the load in the evening.

7.8 The PV module of a solar home system is placed in a position such that one of its 36 cells is completely shaded. Explain, qualitatively, how this will affect the power output of the module.

7.9 Describe a condition in which a PV module rated at 30 W will produce 30 W of power.

7.10 A solar home system is powered by PV module B, whose parameters are provided in Table 7.3. Estimate the maximum power that can be supplied by the module when the irradiance is 600 W/m^2 and the ambient temperature is 24°C.

7.11 A small PV array is formed by connecting two PV modules in parallel. The modules' parameters are provided in Table 7.3 (module B). Compute the short-circuit current if the irradiance is 1250 W/m^2. Ignore the effects of temperature.

7.12 Use the typical $I–V$ curve of a PV module to explain why PV modules can be short-circuited without being damaged.

References

1. Fuentes, M., Nofuentes, G., Aguilera, J., Talavera, D., Castro, M.: Application and validation of algebraic methods to predict the behavior of crystalline silicon PV modules in Mediterranean climates. Solar Energy **81**, 1396–1408 (2007). DOI http://dx.doi.org/10.1016/j.solener.2006.12.008. URL http://www.sciencedirect.com/science/article/pii/S0038092X0700028X
2. Skoplaki, E., Palyvos, J.: On the temperature dependence of photovoltaic module electrical performance: A review of efficiency/power correlations. Solar Energy **83**(5), 614–624 (2009). DOI https://doi.org/10.1016/j.solener.2008.10.008. URL http://www.sciencedirect.com/science/article/pii/S0038092X08002788
3. Twidell, J., Weir, A.: Renewable Energy Resources, 2nd edn. Taylor & Francis (2006)

Part III
Energy Storage and Converters

Chapter 8
Battery Storage for Off-Grid Systems

8.1 Introduction

Energy storage devices are incorporated into off-grid systems to provide flexibility between when energy is produced and when it is consumed. The operation of a solar-powered mini-grid with a lead–acid battery illustrates this point. The battery is charged during the day when there is an abundance of input power and is discharged in the evening when the input power is lower than the output power required by the load (including losses). The battery can be thought of as "inhaling" during the day and "exhaling" in the evening, as shown in Fig. 8.1.

There are several energy storage technologies compatible with electrical systems. The most common type used in off-grid systems is the chemical battery, hereafter referred to simply as a battery. The basic concept of a battery is straightforward. A battery is a device that converts chemical energy into electrical energy. Batteries in which the conversion is reversible are referred to as "rechargeable" or "secondary batteries." Within this category, there are several chemistries as shown in Fig. 8.2. There are two general battery chemistries used in off-grid systems: lead–acid and lithium–ion (LI), which we discuss in this chapter.

Batteries are sometimes modeled as ideal voltage sources. While this might be reasonable under short-term low-loading conditions, in actuality, batteries exhibit nonlinear behavior. The characteristics are dependent primarily on the battery's temperature and chemical state. The chemicals in a battery are continuously reacting, and so their properties are always changing at some level. Although conceptually simple, the ideal voltage source model is inadequate.

Due to their complex internal behavior, batteries are often treated as black boxes, without regard to the mechanisms that give rise to their nonlinear characteristics. For most applications, the black-box treatment can be justified if the characteristics of interest—typically the voltage and energy availability—are documented in the

© Springer International Publishing AG, part of Springer Nature 2018
H. Louie, *Off-Grid Electrical Systems in Developing Countries*,
https://doi.org/10.1007/978-3-319-91890-7_8

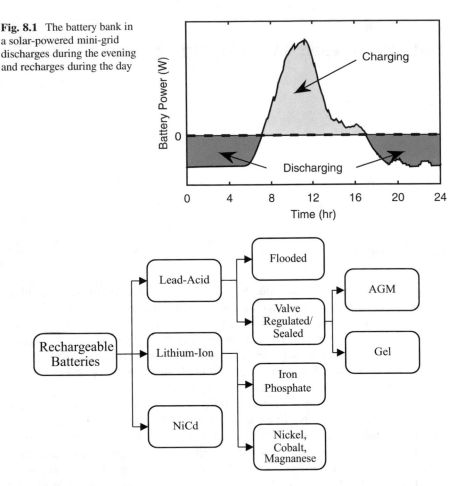

Fig. 8.1 The battery bank in a solar-powered mini-grid discharges during the evening and recharges during the day

Fig. 8.2 Types of rechargeable batteries

battery's specification sheet. In this chapter we discuss the external characteristics of batteries using a black-box approach. Coverage of the underlying electrochemical mechanisms is limited. However, understanding these mechanisms at least on a basic level is beneficial in de-mystifying the complex and nonlinear behavior exhibited by batteries. It is also important in the context of off-grid systems where the conditions encountered might vary considerably from the "typical" conditions found in specification sheets. A deeper treatment of the underlying electrochemistry in general or specifically for lead–acid and lithium–ion batteries is found in other textbooks [1, 2, 7, 9].

8.2 Basic Description

The first battery was developed by Alessandro Volta in 1799. A battery is comprised of one or more electrochemical cells. Each cell consists of two electrodes, which are usually strips or plates of different metals. One electrode is designated the positive electrode and the other the negative. By convention, the positive electrode is also referred to as the cathode and the negative electrode the anode.[1] A diagram of two cells connected in series is shown in Fig. 8.3. Actual batteries also include insulating separator is used to prevent the electrodes from touching.

Each electrode is partially submerged in a solution known as the "electrolyte." In some batteries, the electrolyte is a liquid; in others it resembles a paste (as used in so-called "dry cell" batteries) or gel. Traditional automotive batteries use liquid electrolytes. Batteries used to power consumer devices are often of the dry cell type. The primary purpose of the electrolyte is to allow ions to flow between the electrodes when the battery is charging or discharging. In a lead–acid battery, the electrolyte is comprised of sulfuric acid dissolved in pure water; in lithium–ion batteries, it is a lithium salt in an organic solvent.

The electrodes of several cells are often connected in series and encased into a single unit to form a battery. For example, an automotive battery consists of

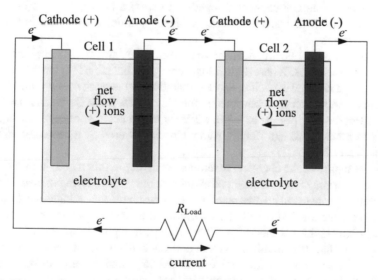

Fig. 8.3 Diagram of anodes, cathodes, and electrolytes of two cells in series and the flow of electrons and ions during discharge. Note that by convention, the direction of current is opposite to the flow of electrons

---

[1]Technically, the designation of anode and cathode depends on whether or not the cell is being charged or discharged, but this tends to add confusion, and so we will always refer to the positive electrode as the cathode and negative as the anode.

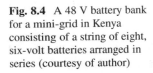

Fig. 8.4 A 48 V battery bank for a mini-grid in Kenya consisting of a string of eight, six-volt batteries arranged in series (courtesy of author)

six series-connected lead–acid cells. Only two electrodes are accessible. They are known as the "terminals" of a battery. One is positive and the other negative, just like an individual cell. The load or charging circuit is connected to the terminals. Several batteries can be connected together to form a battery bank as in Fig. 8.4. Series connections increase the overall voltage; parallel connections increase the available current.

A cell relies on electrochemical reactions to establish a charge imbalance between its electrodes. Wherever there is a charge imbalance, for example, between the plates of a charged capacitor, an electrostatic field is present and voltage exists. When an external circuit is connected to the electrodes, the voltage causes electrons to flow from the anode to the cathode, while there is a net flow of negative ions from cathode to anode. This is different from a circuit powered by a generator, in which only electrons flow.

It is the nature of the chemical reactions occurring within a cell, not its physical size, that determines its voltage. In other words, the voltage is an *intrinsic* property. The chemicals involved in the reactions that produce the voltage are known as the "active materials" of a cell. This includes the electrodes and, for some cells, including lead–acid, the electrolyte. The complete depletion of any of the active materials will halt the chemical reaction and the cell's ability to power an external circuit. Cells also contain non-active material to enhance their performance, for example, by improving the conductivity of the electrodes.

Different cell voltages can be achieved by using different combinations of active materials. A lead–acid cell has a typical open-circuit voltage of 2.10 V, whereas a lithium iron phosphate cell has an open-circuit voltage of approximately 3.2 V. But these voltages vary depending on the conditions. Alkaline batteries, such as those found in R6 (AA) batteries, use zinc and manganese oxide electrodes and have a typical open-circuit voltage of 1.43 V. These voltages are too low for high-power applications, and so several batteries or cells are connected together in series.

8.3 Lead–Acid Batteries

We begin by discussing the particularities of lead–acid batteries. Lead–acid batteries are commonly used in mini-grids due to their low cost, wide availability, and technical maturity. Because they are stationary once installed, their size and heavy weight are less of a consideration.

The lead–acid battery was developed by Gaston Planté in 1859. Lead–acid batteries have been continually refined to improve their efficiency and lifespan ever since. Lead–acid batteries can be designed for specialized uses, for example, in automobiles, forklifts, and backup energy supplies. For most off-grid electrification applications, the battery is designed to supply power over a long period of time and has a long cycle life. A "cycle" refers to a charge/discharge sequence. For most mini-grids a cycle is completed every day. Lead–acid batteries used in mini-grids are known as "deep-cycle" batteries. This is in contrast to automotive batteries— also called starting, lighting, and ignition (SLI) batteries—which are designed to supply a short burst of current to start the vehicle. An SLI battery used in an off-grid application will quickly fail. The internal makeup of a typical lead–acid battery is shown in Fig. 8.5.

8.3.1 Electrolyte

The electrolyte in a lead–acid battery is sulfuric acid (H_2SO_4). The sulfuric acid is mixed with pure water where it is dissociated into SO_4^{-2} ions and two H^+ ions. The concentration of a fully charged lead–acid battery is typically 6 moles per liter. A mole is the measurement of the amount of a substance. The number of particles of a substance in 1 mole is 6.022×10^{23}, Avogadro's constant. The concentration drops to around 1 to 2 moles per liter when completely discharged. As we will discuss later, a "fully discharged" battery still stores some energy; it just cannot be meaningfully used because the battery voltage is too low or its use might permanently damage the battery. The fact that the concentration of sulfuric acid in the electrolyte varies as the battery is discharged is important.

8.3.2 Electrodes

The lead at the anode resembles a sponge. This increases its surface area so that more of it can interact with the electrolyte. The electrodes are usually formed around a metallic grid, which also improves the conduction of the electrons to the cell's terminals. The electrodes are close together. A plastic or other insulating material is used as a spacer to prevent accidental contact (an internal short circuit). When fully charged, the anode of a lead–acid cell is lead (Pb), and the cathode is lead dioxide (PbO_2), as shown in Fig. 8.6.

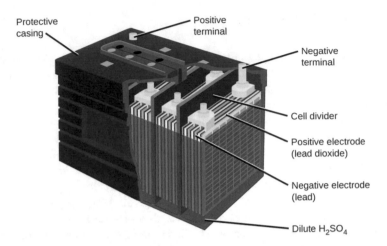

Fig. 8.5 Internal components of a typical multicell lead–acid battery (from https://opentextbc.
ca/chemistry/chapter/17-5-batteries-and-fuel-cells/ and is licensed under CC BY 4.0 https://
creativecommons.org/licenses/by/4.0/)

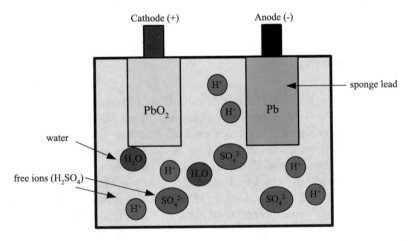

Fig. 8.6 A cell of a fully charged lead–acid battery consists of a lead anode, lead dioxide cathode,
and diluted sulfuric acid electrolyte

8.3.3   Discharging Reactions

When connected to an external circuit, as shown in Fig. 8.7, a charged battery
begins to discharge, and chemical reactions at both electrodes occur. There are two
reactions at the anode. The first is an oxidation reaction of lead which frees two
electrons:

$$Pb(s) \rightarrow Pb^{2+}(s) + 2e^- \tag{8.1}$$

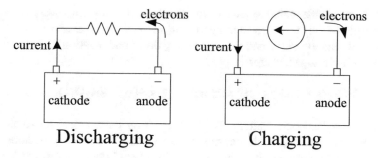

Fig. 8.7 Discharging and charging circuits of a battery

here the (s) indicates the substance is in solid form. The second is

$$Pb^{2+}(s) + SO_4^{-2}(aq) \rightarrow PbSO_4(s). \tag{8.2}$$

Here (aq) indicates that a substance is aqueous. The complete anode reaction is therefore written as

$$Pb(s) + SO_4^{-2}(aq) \rightarrow PbSO_4(s) + 2e^-. \tag{8.3}$$

The reaction converts the solid lead electrode into lead(II) sulfate ($PbSO_4$), which we will refer to as lead sulfate. The reaction consumes some of the SO_4^{-2} ions from the electrolyte. The reaction releases two electrons into the electrode which can flow through an external circuit, supplying power.

The electrons flow to the cathode where the following reduction reactions occur. The lead dioxide and liquid (ℓ) water react at the electrode, resulting in five ions

$$PbO_2(s) + 2H_2O(\ell) \rightarrow Pb^{+4}(s) + 4OH^-(aq). \tag{8.4}$$

The lead ion combines with two electrons from the external circuit (provided by the anode) to form

$$Pb^{+4}(s) + 2e^- \rightarrow Pb^{+2}(s). \tag{8.5}$$

As with the anode, the Pb^{+2} reacts with the SO_4^{-2} ions in the electrolyte to form

$$Pb^{2+}(s) + SO_4^{-2}(aq) \rightarrow PbSO_4(s). \tag{8.6}$$

The OH^- ions from (8.4) react with H^+ ions in the electrolyte to form water

$$4OH^-(aq) + 4H^+(aq) \rightarrow 4H_2O(\ell). \tag{8.7}$$

Written more concisely, the total cathode reaction during discharge is

$$PbO_2(s) + SO_4^{-2}(aq) + 4H^+(aq) + 2e^- \rightarrow PbSO_4(s) + 2H_2O(\ell). \tag{8.8}$$

The reaction at the cathode transforms the lead dioxide into lead sulfate and produces pure water while consuming the sulfuric acid in the electrolyte.

The total reaction of the battery considering the anode reaction (8.3) and cathode reaction (8.8) during discharge is

$$PbO_2(s) + Pb(s) + 2H_2SO_4(aq) \rightarrow 2PbSO_4 + 2H_2O(\ell). \tag{8.9}$$

Discharging a lead–acid battery therefore converts each electrode to the same substance, lead sulfate, while consuming the sulfuric acid but producing water. We now see why the concentration of the sulfuric acid decreases as the battery is discharged.

Example 8.1 The discharge reaction of a lead–acid battery yields 2 moles of electrons for every mole of lead. What is the mass of lead that is needed to provide 1 A of current for a period of 1 h? The molar mass of lead is 207.2 g/mole.

Solution A battery supplying 1 ampere of current for 1 h provides

$$1 \text{ amphour} = \frac{1 \text{ coulomb}}{\text{second}} \times \frac{3600 \text{ second}}{\text{hour}} = 3600 \text{ coulombs.}$$

The number of electrons that must be supplied is

$$1 \text{ amphour} = 3600 \text{ coulombs} \times 6.24 \times 10^{18} \text{ electrons/coulomb}$$
$$= 2.246 \times 10^{22} \text{ electrons.}$$

Each mole of lead provides 2 moles of electrons, which, per Avogadro's constant, is equal to

$$2 \text{ moles of electrons/mole of lead} \times 6.03 \times 10^{23} \text{electrons/mole of electrons}$$
$$= 12.06 \times 10^{23} \text{ electrons/mole of lead.}$$

Therefore, a total of $(2.246 \times 10^{22}) / (12.06 \times 10^{23}) = 0.0186$ moles of lead are needed to supply 1 ampere of current for 1 h. This weighs $0.0186 \times 207.2 = 3.8$ g. We could similarly show that the required lead dioxide and sulfuric acid bring the total mass of the active material required to supply 1 ampere for 1 h to 11.9 g. In practice, only a portion of the active material is accessible, and the added mass of non-active materials that increase the mechanical strength or conductivity of the battery makes the practical value closer to 50 g.

8.3.4 Charging Reactions

It is not possible to extract energy from a battery whose electrodes have each transformed into lead sulfate. However, it is possible to recharge the battery by connecting a battery charger, as shown in Fig. 8.7. The battery charger is modeled as an ideal current source. As we add electrons to the anode, the electrode will return to lead, and the SO_4^{-2} ions will return to the electrolyte

$$PbSO_4(s) + 2e^- \rightarrow Pb(s) + SO_4^{-2}. \tag{8.10}$$

At the cathode:

$$PbSO_4(s) + 2H_2O(\ell) \rightarrow PbO_2(s) + 4H^+(aq) + SO_4^{-2}(aq) + 2e^-. \tag{8.11}$$

We note that a fully charged lead–acid battery has an anode made of lead, and the cathode is lead dioxide. The electrolyte concentration is higher than in the discharged state as water is consumed and sulfuric acid (ions) is produced.

The complete chemical reaction for charging is thus

$$2PbSO_4(s) + 2H_2O(\ell) \rightarrow PbO_2(s) + Pb(s) + 2H_2SO_4(aq). \tag{8.12}$$

This is the same equation as (8.9), but the products and reactants have been reversed.

8.3.5 Other Reactions

The charge/discharge reactions enable the battery to store and supply energy. There are other reactions that also occur and affect the performance of the battery. These reactions are unwanted but inevitably occur at least to some degree.

8.3.5.1 Overdischarge

If a lead–acid battery is deeply discharged, then the lead sulfate can evolve into a dense crystalline form. This is known as sulfation. The crystals physically prevent the active material of the electrodes from interacting with the electrolyte. They also act as electrical insulators making electron exchange less likely. Sulfation reduces the energy that can be provided by a battery. The crystals can be difficult to remove. Once sulfation occurs, the battery can be permanently damaged or derated. When a component is "derated," it means that its effective rating is lower than its initial rating. Sulfation can be minimized by never letting the battery state-of-charge remain low for prolonged periods of time, by avoiding operation at high temperatures, and by preventing any deep discharge of the battery.

8.3.5.2 Self-Discharge

A charged lead–acid battery will discharge even in the absence of an external circuit. The discharge is internal to the battery and is called "self-discharge." Self-discharge occurs because a charged battery, be it fully or only partially charged, is not thermodynamically stable. In other words, there is a natural tendency for the lead and lead dioxide to react with electrolyte. The reaction occurring at the cathode is

$$PbO_2(s) + H_2SO_4(aq) \rightarrow PbSO_4(s) + H_2O(\ell) + 0.5O_2(g) \qquad (8.13)$$

and at the anode:

$$Pb(s) + H_2SO_4(aq) \rightarrow PbSO_4(s) + H_2(g). \qquad (8.14)$$

These reactions are different from the discharge reactions in (8.3)–(8.8). Nonetheless, the stored energy is decreased because the active material on both electrodes is converted to $PbSO_4$, and the electrolyte becomes depleted. We see that self-discharge also produces oxygen and hydrogen gas (g) which may be vented from the battery. This irreversibly reduces the water in the electrolyte. The anode and cathode reactions are independent of each other and occur at different rates, with the anode reaction being more rapid. Lead–acid batteries are designed to limit self-discharge. Most lose about 5% or their capacity each month. This is one reason why an automobile that has not been driven for a long period of time will have trouble starting.

8.3.5.3 Overcharging

When a battery is charged, its terminal voltage increases. As it approaches fully charged, the voltage is such that hydrogen and oxygen gas are rapidly evolved. Two overcharge reactions take place. At the anode, the electrons combine with the hydrogen ions in the electrolyte:

$$2H^+(aq) + 2e^- \rightarrow H_2(g) \qquad (8.15)$$

and in the cathode, the water in the electrolyte electrolyzes, splitting into oxygen and hydrogen:

$$H_2O(\ell) \rightarrow 0.5O_2(g) + 2H^+(aq) + 2e^-. \qquad (8.16)$$

The result is the production of hydrogen and oxygen gas. This is known as "gassing." Some gassing occurs even before the battery is fully charged. Gassing should be avoided for two important reasons. The first is that hydrogen gas is explosive, and therefore it poses a fire risk if allowed to escape the battery and accumulate in a

Fig. 8.8 Flooded lead–acid
batteries—from left to right:
6V, 290 Ah; 6V, 525 Ah; 2V,
1400 Ah (courtesy of
OutBack Power)

poorly ventilated space. The second is that water is consumed in the process. The
lost water must be replaced, which is a maintenance consideration.

8.3.6   Lead–Acid Battery Types

Nowadays there are several types of batteries designed for off-grid systems. They
come in different shapes, sizes, nominal voltages, and types. Several are shown in
Fig. 8.8. The choice of which to use is not arbitrary as each has different technical
performance, cost, safety, and maintenance considerations.

8.3.6.1   Flooded Lead–Acid

Flooded lead–acid batteries are the most common and mature type of lead–acid
battery. They are also referred to as "wet" cell batteries. The electrolyte is liquid
and the electrolyte and electrodes are at atmospheric pressure. The battery is not
permanently sealed—the interior is designed to be accessed. This allows pure
(distilled) water to be added to replace water lost to gassing, and the electrolyte
sampled to estimate the charge remaining, as discussed later in Sect. 8.5.1. Personnel
must be trained to safely maintain the battery. Care must be taken not to over- or
under-water the battery and to protect against any accidental spilling or splashing
of the electrolyte, which, because it is acidic, is potentially harmful. Flooded lead–
acid batteries are the mechanically weakest of the batteries and can leak. Precautions
should be made to contain any leak, for example, by installing a short curb around
the batteries. The open-circuit voltage for a given state-of-charge tends be somewhat
lower than the other types of lead–acid batteries.

8.3.6.2 Sealed Lead–Acid

Sealed lead–acid (SLA) batteries are sealed from the external environment under most operating conditions. Sealed lead–acid batteries as a class are also known as valve-regulated lead–acid (VRLA) batteries. This term stems from the one-way valve that separates the battery interior from the external environment. Only if the pressure inside the battery becomes high enough will the valve operate, emitting whatever gas caused the pressure to build, for example, the hydrogen or oxygen that forms during self-discharge or overcharge. This should not happen during normal operation.

An advantage of the sealed container is that any gases produced might recombine. For example, the hydrogen and oxygen can reform into water. The sealed container also prevents spillage, and so SLA batteries are common in non-stationary applications, for example, in solar home systems. SLA batteries do not require maintenance because any gassed hydrogen and oxygen can recombine to form water, eliminating the need for adding water (regardless, the electrolyte is inaccessible). For this reason, SLA batteries are sometimes marketed as "maintenance-free" batteries. SLA batteries tend to be more expensive than flooded. There are two general sub-types of SLA batteries: gel and absorbed glass mat.

The electrolyte in gel batteries has been thickened, for example, by introducing silica. The electrolyte does not move in bulk, but nonetheless the ions are able to travel within it. Some gel batteries use tubular plate cathodes. These batteries are known as OPzV—from the German phrase Ortsfest PanZerplatte Verschlossen (stationary tubular plate sealed). Tubular plate batteries are more common with European manufacturers. Absorbed glass mat (AGM) batteries use a fiberglass structure to absorb the electrolyte. Gel and AGM batteries offer similar advantages—they are maintenance-free, are leak-proof, and can be operated in any orientation. Presently, AGM seem to be preferred over gel for off-grid applications. However, the relative advantages of each largely depend on the specific model and manufacturer.

8.3.6.3 Comparison

Although there are many differences between flooded and sealed lead–acid batteries, in the context of off-grid systems, the important differences are cost, maintenance, lifespan, and efficiency. Flooded batteries tend to be less expensive—perhaps 50% less. They can have a somewhat longer lifespan, but this depends on the specific application. Sealed batteries are more efficient and are maintenance-free—an important feature if the off-grid system does not employ local staff. Another important distinction between flooded, gel, and AGM batteries is that their charging voltages and times are different. VRLA tend to charge somewhat faster. It is important that the battery charger set-points are matched to the particular battery used.

8.4   Basic Electrochemistry

The chemistry of batteries is not terribly complicated, but it does involve a number of intricacies that are not immediately obvious to the non-expert. These details are described in the following sections. One thing to keep in mind throughout is that the charging and discharging of a battery, although basically reverse actions, are not simply symmetrical opposites to one another.

You might wonder what causes a voltage to appear across the electrodes of a battery. Without going into extreme detail, in this section we describe the basics of the electrochemistry underlying battery operation. A functioning battery requires at least two features: voltage must exist between its terminals, and there must be a mechanism of continuously transporting charge within the battery. Transportation of charge within the battery is accomplished through the flow of charged ions in the electrolyte. A separation of charge can occur when two dissimilar substances with reasonable conductivity come into physical contact. This is true even if the substances, when apart, have no net charge. The voltage between two substances when in contact is known as the *Galvani potential*. The substances can be of any phase, for example, two metals, or in the case of many batteries, a metal and a liquid. In a lead–acid battery, there is voltage between the anode and the electrolyte and another, different, voltage between the cathode and electrolyte. It can be helpful to think of these voltages as being in series, with the voltage appearing across the electrodes the combined effect of the voltages.

When the electrode/electrolyte first comes into contact, some of the electrons from the electrode are accepted by the positive ions in the electrolyte, and some of the negative ions in electrolyte donate electrons to the electrode. Think of this as diffusion occurring. There are two flows of electrons, one to the electrode and one from it. If the two flows do not have the same rate, then the electrode and electrolyte become charged—one positive and one negative with respect to the other. Let us assume the electrode is positive. As the magnitude of the charge difference increases, it becomes harder for the electrons in the electrode to be accepted by the ions in the electrolyte. The reason is that the electrons have to "push" against the electric field caused by the net negative charge of the electrolyte. Therefore, the charge transfer in this direction slows. Eventually a steady-state equilibrium of charge transfer is reached where the diffusion tendency is balanced by the electric field. For lead–acid cells, the voltage corresponding to the equilibrium is approximately 2.04 V when measured across the cell's electrodes. But the exact cell voltage depends on several other factors, as discussed next.

8.4.1   Standard Cell Potential

It is possible to devise batteries using a variety of substances as electrodes and electrolytes. Although a voltage might be developed, it also might not be practical.

Electrochemists have measured and tabulated the equilibrium open-circuit voltages for a wide variety of substances and reactions. The measurements are conducted under *standard conditions*. Standard conditions are defined as reactions occurring at a pressure of 100 kPa and for the effective concentration of solutes to be 1 mole per liter. The temperature is not stipulated, but most tables report the measured voltage, known as the standard cell potential E^0_{cell}, at 25°C.

8.4.2   Nernst Equation

The open-circuit voltage of a lead–acid cell at equilibrium is dependent on the concentration of the electrolyte. Moreover, the concentration of the electrolyte, which is actively involved in the chemical reaction of a lead–acid battery, is almost never at the standard effective concentration of 1 mole per liter (it would be very deeply discharged). Therefore, the standard cell potentials tabulated by electrochemists cannot be directly used.

More generally, and for any cell, the open-circuit equilibrium voltage E_{cell} between its electrodes (terminals) can be adjusted for nonstandard concentrations by the use of the Nernst equation:

$$E_{cell} = E^0_{cell} - (RT/nF)\ln(Q_r) \tag{8.17}$$

where E_{cell} is the cell voltage under equilibrium conditions, R is the universal gas constant (8.314 J/mol/K), T is the temperature in Kelvin, n is the number of moles of electrons transferred in the reaction, F is the Faraday constant (96,485 C/mol), and Q_r is the reaction quotient of the reaction. The reaction quotient requires additional explanation. Consider a reaction with reactant chemicals generically named A and B and products C and D written as

$$a\text{A} + b\text{B} \rightarrow c\text{C} + d\text{D}. \tag{8.18}$$

The reaction quotient is

$$Q_r = \frac{\alpha_C^c \alpha_D^d}{\alpha_A^a \alpha_B^b} \tag{8.19}$$

where α_i is the "activity" of chemical i. The activity of a chemical can be thought of as the "effective" concentration.[2] Often, the concentration in moles per liter can be used as an approximation for the effective concentration. The activity of a solid is constant and equal to one, and the activity of pure water is often assumed to be one.

---

[2]The activity of a chemical is related to the concentration as $\alpha_i = \gamma_i c_i$ where γ_i is the activity coefficient and c_i is the concentration; often an activity coefficient of one is assumed.

Fig. 8.9 As the concentration of the electrolyte decreases during discharge, the open-circuit voltage also decreases according to the Nernst equation

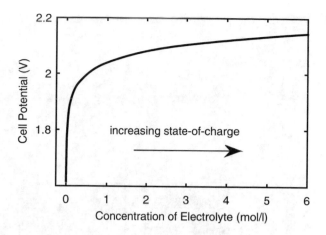

With these assumptions, the open-circuit equilibrium voltage of a lead–acid cell for different concentrations of sulfuric acid in a lead–acid battery is shown in Fig. 8.9. We would expect the actual voltage to be somewhat different due to the simplifying assumptions made [12].

Note that if (8.17) is applied to a reaction that occurs at standard state, then the reaction quotient is one, and $E = E^0_{cell}$ because $\ln(1) = 0$. The natural logarithm introduces nonlinearity in the relationship between the voltage and the concentration and state-of-charge in lead–acid batteries. The presence of temperature in (8.17) acts to scale the effect of activity-based deviations from the standard state. However, this does not capture the complete effect of temperature on battery voltage. These aspects are discussed further in Sect. 8.5.7.

Example 8.2 Assume that the standard cell voltage for a lead–acid battery is 2.04 V. The battery is fully charged, and so the concentration of the sulfuric acid in the electrolyte is 6 moles per liter. Compute the corresponding cell voltage. Approximate the activity of the sulfuric acid by its concentration. Assume the temperature of the battery is 25°C.

Solution The battery is not under standard conditions due to the increased concentration of the sulfuric acid. We can apply the Nernst equation (8.17) to adjust the open-circuit voltage. This requires the reaction quotient to be computed. For the reaction of a lead–acid battery (8.9), we let activities of the reactants be assigned as A→ PbO_2; B → Pb, and E → H_2SO_4. Note that we accommodated a third reactant, E, just like the first two. The products are assigned as C→ $PbSO_4$ and D→ H_2O so that a reasonable approximation of the reaction coefficient is

(continued)

$$Q_r = \frac{\alpha_C^c \alpha_D^d}{\alpha_A^a \alpha_B^b \alpha_E^e}$$

$$= \frac{1^2 \times 1^2}{1^1 \times 1^1 \times 6^2} = \frac{1}{36}.$$

We also see from (8.9) that the reaction yields 2 moles of electrons, so that $n = 2$. Applying the Nernst equation (8.17):

$$E_{cell} = E_{cell}^0 - (RT/nF)\ln(Q_r)$$

$$= 2.04 - ((8.314 \times 298.15) / (2 \times 96,485))\ln\left(\frac{1}{36}\right) = 2.086 \text{V}.$$

This roughly corresponds to the open-circuit voltage of a fully charged lead–acid cell.

8.5 Basic Characteristics

Now that we have a basic understanding of how a battery functions, we can discuss the characteristics that are especially relevant to off-grid systems. They are:

1. Open-circuit voltage
2. Current–voltage relationship
3. Battery resistance
4. Charge capacity and energy capacity
5. Cycle life
6. Efficiency
7. Self-discharge
8. Effect of temperature

8.5.1 Open-Circuit Voltage

The terminal voltage of a battery V_B, measured from the positive terminal to the negative terminal, along with the battery's charge or energy capacity, are the most important characteristics of a battery. The terminal voltage dictates what other components a battery is compatible with—for example, you cannot replace your laptop battery with your automobile's battery and expect it to function, even though your automobile's battery stores a greater quantity of energy. If the battery bank in

Table 8.1 Battery state-of-charge and corresponding open-circuit voltage and specific gravity for lead–acid batteries

Open-circuit cell voltage	Approximate state-of-charge (%)										
	0	10	20	30	40	50	60	70	80	90	100
Flooded (V/cell) (V/cell)	1.90	1.92	1.94	1.96	1.99	2.01	2.03	2.05	2.07	2.09	2.11
AGM (V/cell)	1.94	1.96	1.98	2.00	2.02	2.04	2.06	2.08	2.10	2.12	2.14
Specific gravity (g/ml)	1.048	1.072	1.096	1.119	1.142	1.165	1.187	1.207	1.227	1.246	1.260

an off-grid system is rated at 48 V, then every component connected to it must be compatible with this nominal voltage.

The Nernst equation tells us that the equilibrium open-circuit voltage of a battery depends on the temperature, chemical composition, and state of the battery's active materials. We are most interested in how the open-circuit voltage varies with the state-of-charge of the battery. A formal definition of *state-of-charge* (SoC) is given later, but for now it suffices to think of it as the percentage of the charge remaining in the cell or battery that can be meaningfully used. The open-circuit voltage as it relates to the SoC for flooded and AGM batteries is shown in Table 8.1. These values are approximate and representative—they vary somewhat from manufacturer to manufacturer and from one battery type to the next. For example, the open-circuit voltage of an SLI battery will be different from those shown. If the cells are connected in series, then multiply the values in the table by the number of series-connected cells.

You might have heard of the lead–acid battery in most automobiles as being a "12 V" battery. A 12 V lead–acid battery has six series-connected cells. According to Table 8.1, the fully charged open-circuit voltage for 12 V battery is 6 × 2.11 = 12.66 V not 12 V. In fact, if a 12 V battery has an open-circuit voltage of 12.0 V, it has less than 50% of its charge remaining. Despite the added confusion, we often describe batteries by their "nominal[3]" voltage, rather than their actual open-circuit voltage. If we know the nominal voltage of a cell and that of the battery, we easily can compute how many cells are connected in series and then can apply a table such as Table 8.1. For example, certain lithium–ion chemistries have a nominal cell voltage of 3.6 V. Therefore, we refer to a battery with five of these cells connected in series as an "18 V" battery (3.6 × 5), which are commonly used in cordless power tools. Note that some manufacturers do not adhere to this convention. They inflate their nominal voltages slightly so that their batteries seem superior.

We emphasize that the values in Table 8.1 and those reported by manufacturers are the *open-circuit voltage* as measured at the battery terminals during *equilibrium conditions* and at a certain temperature, usually 25°C. If the battery is being charged and discharged or has recently been charged or discharged, the table should not be

---

[3]The nominal voltage itself approximately refers to the average terminal voltage when discharged.

used. The battery must be isolated from other components and allowed sufficient time—usually several hours—to rest before the SoC can be inferred from the battery terminal voltage V_B. It is often impractical to disconnect the battery from the system to measure the open-circuit voltage.

An alternative way of estimating the SoC is by measuring the concentration of the electrolyte. This can only be done in flooded lead–acid batteries because the electrolyte is accessible in these types of lead–acid batteries. The concentration of sulfuric acid in the electrolyte of a fully charged lead–acid battery is usually 6 moles per liter. The specific gravity of pure water is 1.0 g/ml and 1.85 g/ml for sulfuric acid. Since sulfuric acid is denser than water, the specific gravity of the electrolyte will increase from unity as the concentration of sulfuric acid increases. The specific gravity is measured using a hydrometer. The measured value can be compared to a table that relates specific gravity to state-of-charge, as shown in Table 8.1. The reading should be adjusted for the temperature of battery, typically by 0.00075 g/ml for every degree increase above 25°C. Note that stratification of concentration does occur in the electrolyte, and so samples taken at the top of the electrolyte might not accurately measure the average specific gravity.

8.5.2   I–V Characteristic

We next consider the electrical characteristics of a battery or cell when connected to an external circuit. The $I-V$ curve of a typical battery when charging and discharging is shown in Fig. 8.10. This curve is also known as a "polarization curve." We see that as the discharge current increases, the terminal voltage decreases. The decrease is not linear—there are steep drops at low current and high current. The current–voltage relationship is influenced by several factors, including the temperature, SoC, and the shape and separation distance of the electrodes. An $I-V$ curve therefore is valid only for a given set of parameters and conditions. In particular, the curve can change dramatically for the same battery as the SoC increases or decreases.

The $I-V$ curve of an idealized battery would be a horizontal line located at the battery's open-circuit voltage, with no variation in terminal voltage as the discharge current increases. That of course would be impossible because it suggests infinite power capability. The downward sloping, nonlinear curve of a practical battery is caused by three types of "polarizations": ohmic, activation, and concentration [2, 7]. Each of these can be thought of as a voltage drop that causes the $I-V$ curve to deviate from the ideal.

Ohmic polarization is the voltage drop caused by the battery current passing through the resistance of the electrodes and the electrolyte (the resistance of the electrolyte can be thought of as how easily the ions can be transported through it). The magnitude of the resistance varies somewhat with the SoC and whether the battery is charging or discharging. Most notably, there tends to be a sharp increase when the battery is discharging and at a low SoC. For example, the resistance of lead sulfate (0.3×10^{10} Ω cm) is several orders of magnitude greater than for lead or lead

Fig. 8.10 *I–V* characteristic of a typical battery during charging and discharging

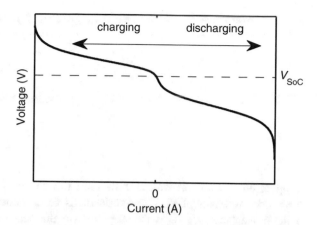

dioxide; and the resistivity of the sulfuric acid electrolyte increases by about 60% from full SoC to empty. Ohmic polarization is largely responsible for the nearly linear middle portions of the curve while charging or discharging in Fig. 8.10.

The activation polarization is related to the kinetics of the chemical reactions. For there to be a net flow of current at the electrode, as is the case whenever the battery is charged or discharged, the equilibrium of the charge exchange described in Sect. 8.4 must be upset. When charging, a voltage higher than the equilibrium is needed. When discharging the voltage is lower. Conceptually, activation polarization can be thought of as a voltage drop (when discharging) or voltage rise (when charging). The voltage drop/rise is generally not the same for the same discharge/charge current. In other words, the polarity of the current matters. The activation polarization increases very rapidly as the battery is charging and approaches a full SoC. Activation polarization is responsible for the steep drop (rise) in voltage in the *I–V* curve at low discharge (charge) current.

The concentration polarization reflects that at high current, there can be a localized reduction in concentration of the electrolyte—caused by a mismatch in the ion transportation rate in the electrolyte with those being transformed at the electrodes. From the Nernst equation, we expect any change in concentration to affect the developed voltage. The concentration polarization is responsible for the large voltage drop (rise) at high discharge (charge) current. Like the activation polarization, the voltage drop (rise) associated with concentration polarization increases rapidly when charging at a near full SoC.

How might we model the effects of polarization using circuit elements? A simple model of a cell is shown in Fig. 8.11 where we find a voltage source connected in series with two resistances [14]. The voltage source itself is shown to be a function of the cell's SoC and temperature. Under open-circuit conditions:

$$V_{cell} = V_{SoC,cell}. \tag{8.20}$$

The resistance $R_{\Omega,cell}$ represents the ohmic polarization. As discussed, this is dependent on the SoC but also the temperature. The other resistance $R_{\eta,cell}$ models the concentration and activation polarizations. Its value is a function of SoC,

Fig. 8.11 Polarization
resistance model of a battery

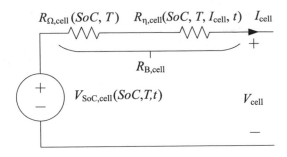

temperature, current as well as how long the current has been applied and how long since it was applied. These resistances can be combined into a hypothetical battery resistance $R_{B,cell}$ which depends on the same parameters as $R_{\Omega,cell}$ and $R_{\eta,cell}$. Therefore

$$V_{cell} = V_{SoC,cell} - I_{cell} R_{B,cell}. \tag{8.21}$$

Using this model, the terminal voltage of a battery with N_{cells} in series is

$$R_B = N_{cells} \times R_{B,cell} \tag{8.22}$$

$$I_B = I_{cell} \tag{8.23}$$

$$V_{SoC} = N_{cells} \times V_{SoC,cell} \tag{8.24}$$

$$V_B = V_{SoC} - I_B R_B \tag{8.25}$$

In interpreting R_B we must always remember it is not a resistor, but it is resistance. It is variable with the state and characteristics of the battery and reflects a number of physical phenomena, including the battery's age [5]. Further, R_B should not be confused with the "internal" resistance reported by manufacturers. The reported resistance is usually the resistance associated with the ohmic polarization only. Hereafter, whenever the resistance of a battery is discussed, it is in reference to the battery resistance R_B, not the "internal" ohmic resistance. The model also does not represent all of the dynamic characteristics of a battery. More sophisticated models[4] include a capacitive element [8]. This models a battery's "inertia"—that a battery's terminal voltage takes some time to return to V_{SoC} after the sudden disconnection of the charging circuit or load.

The value of the battery resistance for a larger flooded lead–acid battery—one that could supply 1 kWh of energy—typically ranges from about 1 mΩ/cell to about 25 mΩ/cell during charging. Smaller capacity batteries tend to have larger battery resistances.

---

[4]See, for example, the widely used Randle's model.

Example 8.3 The equilibrium open-circuit cell voltage of a 24 V lead–acid AGM battery is 24.96 V. An external circuit is connected to the battery that draws 21 A. Assume the battery resistance R_B under the these conditions is 0.060 Ω. Compute the terminal voltage. Estimate the SoC using Table 8.1. How does this estimate compare to SoC if the open-circuit voltage is used? Which is more accurate?

Solution The terminal voltage will be less than the equilibrium voltage because the battery is discharging. Applying (8.22)

$$V_B = 24.96 - 21.0 \times 0.060 = 23.70 \text{ V}.$$

Normally, we do not know R_B, but if we do we can compute V_{SoC} and estimate the SoC. The corresponding cell voltage is found by dividing the voltage by the number of cells: 23.70/12 = 1.975 V/cell. If this is wrongly used to estimate the SoC according to Table 8.1, then the estimated SoC would be approximately 17.5%. However, the SoC should be estimated using the equilibrium open-circuit cell voltage of 24.96/12 = 2.08 V/cell, corresponding to an SoC of 70%. This illustrates why the terminal voltage is a poor estimator of SoC unless the battery is open-circuited.

Figure 8.12 is a typical plot of the terminal voltage of a lead–acid battery during a discharge and charge cycle if the magnitude of the current is constant throughout. As the battery is discharged, the voltage drops immediately from the open-circuit value (approximately 2.1 V). This is due to the voltage drop associated with the battery resistance. Over time, the terminal voltage decreases, primarily due to the decrease in V_{SoC} brought about by the dilution of the electrolyte concentration and the increasing value of R_B at low SoC while discharging. When the battery transitions from discharging to charging, the terminal voltage immediately and abruptly increases. This is primarily due to the voltage drop across the battery resistance now having the opposite polarity. As the battery is recharged, its terminal voltage rises nonlinearly due to the increasing V_{SoC}. The sharp rise near the far right of the figure is caused by the rapid increase in R_B—primarily due to increases in the activation and concentration polarization—that occurs when a battery with a full SoC continues to be charged.

8.5.3 Battery Capacity

The capacity of a battery tells us how long it can meaningfully supply a constant current load. Although this is conceptually simple, there is some nuance to its

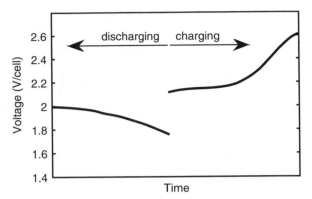

interpretation, and the reader should pay particular attention to how capacity and the related state-of-charge (SoC) and depth-of-discharge (DoD) are defined. Capacity refers to either the amount of *charge* or the amount of *energy* a battery can provide to an external load.

8.5.3.1 Charge Capacity

The most common method of quantifying the capacity of a battery is the charge capacity. The charge capacity is the amount of electric charge that can be supplied to an external load. Recall that the relationship between current and charge is

$$1 \text{ Ampere} = 1 \text{ coulomb/second.} \tag{8.26}$$

Out of convenience, the charge capacity of a battery is specified in terms of amphour (ampere-hour, Ah), rather than coulombs. One amphour is equivalent to 3600 coulombs. It follows that since it takes the charge of 1.602×10^{19} electrons to equal 1 coulomb, 1 amphour is also equivalent to 2.246×10^{22} electrons. The number of electrons that a battery can supply is directly related to the mass of the active material (see Example 8.1). This explains why higher-capacity batteries are necessarily physically larger than smaller-capacity batteries.

We can devise a simple test to determine the charge capacity of a battery. We do this by discharging a fully charged battery through a variable resistor. The resistance is varied so that the discharge current has constant value of x as R_B and V_{SoC} change. The charge supplied by the battery in amphours during constant current discharge is

$$c_x = x T_{d,x} \tag{8.27}$$

where $T_{d,x}$ is the discharge time.

If we were also measuring the terminal voltage, we would notice that the terminal voltage decreases over time, for example, as shown in Fig. 8.13. As can be expected, the terminal voltage decreases nonlinearly as the discharge progresses, even as the

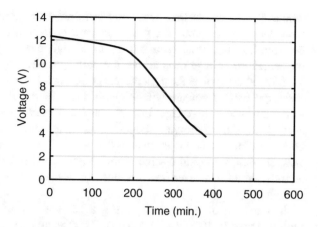

Fig. 8.13 Discharge curve of a 12 V, 40 Ah, gel battery at a rate of $x = 8$ A

current is held constant. Of particular importance is that the decrease is first gradual and then rapid. The battery is unable to supply $x = 8$ A after about 390 min ($T_{d,x} = 6.5$ h). The charge supplied by the battery shown in Fig. 8.13 is therefore approximately $8 \times 6.5 = 52$ Ah. The discharge ceases when the battery is no longer able to supply 8 A. However, we note that at the end of the discharge cycle, the voltage of the battery was far below its nominal value of 12 V. In practice, it is unlikely that the load connected to the battery would properly function across the range of voltages exhibited. Moreover, the battery is almost surely permanently damaged due to overdischarge. In other words, we are not able to meaningfully extract all 52 Ah of charge. But at what point should the test be terminated and the charge capacity calculated?

The standard practice is to consider a battery "fully" discharged when the terminal voltage reaches a predefined *cut-off voltage*. The cut-off voltage is selected so that, when reached, the load will likely still function and the battery will not be seriously damaged. The cut-off voltage typically corresponds to the point at which the battery voltage sharply decreases (slightly above 10 V for the battery in Fig. 8.13). For lead–acid batteries this is usually 1.75 V/cell; for lithium–ion batteries it is typically between 2.0 and 3.0 V/cell.

The presence of the battery resistance makes it so that the amount of charge provided before the cut-off voltage is reached will vary with the magnitude of the discharge current. In other words, a battery discharged at a current of x will provide somewhat less charge before the terminal voltage reaches the cut-off voltage than if it were discharged at $0.5x$. This is because the voltage drop across R_B will be approximately twice as large when discharged at x. There might also be an increase in unwanted side reactions. We therefore must associate the charge capacity rating of a battery with the current it was discharged at. Mathematically, the capacity c_x at discharge rate of x amperes is

$$c_x = \int_0^{T_{\text{cut-off}}} x \, dt \qquad (8.28)$$

where $T_{\text{cut-off}}$ is the time when the terminal voltage equals the cut-off voltage. Note that it is common for manufacturers to decrease the cut-off voltage when determining the rating of a battery at a high discharge current to account for the increased voltage drop across the battery resistance. Values of 1.67 V/cell or 1.33 V/cell are typical for high discharge current. Note that as soon as the discharge ends, the battery voltage will rebound past the cut-off value. This is partly due to the absence of a voltage drop across the battery resistance when the current is zero. Also, the electrolyte gradually mixes, removing any localized areas of low concentration around the electrodes that might have lowered the voltage. If we wait long enough, we could discharge the battery further before the cut-off voltage is again reached. However, rest periods are not used when determining a battery's capacity.

In the case of the battery in Fig. 8.13, the cut-off voltage specified by the manufacturer is 10.2 V (1.7 V/cell). This occurs after approximately 3.75 h of discharge. The corresponding capacity is 8 A × 3.75 h = 30 Ah. This is considerably less than the charge extracted if the cut-off voltage is ignored.

We emphasize that the capacity rating of a battery depends on its terminal voltage during discharge. The capacity is therefore related to, but not a direct measurement of, the accessible active material that can supply charge. Regardless of the discharge rate, the same mass of active material is used to supply each electron to the external circuit [6]. Thus, the same active material will be consumed if a battery is discharged at 10 A for 1 h or 1 A for 10 h.

8.5.3.2 State-of-Charge

We know that we can extract charge in excess of a battery's capacity rating if we are willing to discharge it past the specified cut-off voltage. However, we do this to the detriment of the battery, and so this additional capacity should be considered "off-limits." We often describe the charge that a battery can meaningfully provide in its present condition as its state-of-charge (SoC). The SoC is always in reference to the capacity rating.

$$SoC = 100 \times \frac{c_x - \int_0^{T_{\text{use}}} i_B(t)\,dt}{c_x} \tag{8.29}$$

where SoC is the state-of-charge, in percent, $i_B(t)$ is the battery current—which is positive when discharging and negative when charging—and T_{use} is the time the battery has been in use. Despite the mathematical exactness of this definition, there are several practical limitations to its use. First, it requires that the current be measured and integrated throughout the life of the battery. This is known as "coulomb counting." Because integration is involved, any measurement bias in the current sensor will accumulate over time leading to large errors in the SoC. Second, this definition does not account for the effects of self-discharge. The SoC of a fully charged battery that has never supplied current will decrease over time. Third, it is possible in some situations for the SoC to be negative or exceed 100%. A negative

SoC occurs when the battery has been discharged past the cut-off voltage. An SoC in excess of 100% can occur in lithium–ion batteries that are overcharged. Lastly, the selection of the capacity c_x upon which to calculate the SoC is somewhat arbitrary, and the discharge current will likely differ from x. For example, if the SoC calculation is based on a capacity of c_{10} (10 A discharge), but the battery is actually discharged at 30 A, then the cut-off voltage will be reached before the SoC is zero. In practice then, the SoC should not be viewed as a precise indicator of how much charge is remaining in a battery. Rather, it is a general indicator. Despite these drawbacks, the SoC is very useful because several characteristics of the battery can be related to the SoC, as we have seen with the open-circuit voltage.

8.5.3.3 Depth-of-Discharge

The complementary measure to the SoC is the depth-of-discharge (DoD). The DoD is the percent of rated capacity that has been discharged:

$$DoD = 100 \times \frac{\int_0^{T_{use}} i_B(t)dt}{c_x}. \tag{8.30}$$

The DoD and SoC are related by

$$DOD = 100 - SoC. \tag{8.31}$$

Example 8.4 A nominal 6 V battery is discharged at a constant current of 15 A. After 20 h the terminal voltage reaches the cut-off voltage (5.25 V). The battery is continued to be discharged for another 2 h at the same current. What is the c_{15} capacity of the battery? Compute the SoC and DoD at 6 h and at 20 h.

Solution Since the battery reaches the cut-off voltage after 20 h, its c_{15} capacity is $20 \times 15 = 300$ Ah. That the battery is discharged for another 2 h is irrelevant to the calculation. The SoC and DoD at 6 h are found using (8.29) and (8.30):

$$SoC = 100 \times \frac{c_x - \int_0^{T_{use}} i_B(t)dt}{c_x} = 100 \times \frac{300 - 6 \times 15}{300} = 70\%$$

$$DoD = 100 \times \frac{\int_0^{T_{use}} i_B(t)dt}{c_x} = DoD = 100 \times \frac{6 \times 15}{300} = 30\%.$$

A similar approach shows that after 20 h, the SoC is 0% and the DoD is 100%.

8.5.3.4 Charge Rate (C-Rate)

To have meaning, the charge capacity c_x of a battery must be referenced to a specific discharge current x. In addition to expressing the current directly in amps, the *charge rate* (C-rate) is used. The C-rate standardizes the current in terms of the cell capacity. The C-rate τ is defined as

$$\tau = \frac{\text{discharge current (A)}}{\text{capacity when discharged at } x \text{ amps (Ah)}} = \frac{x}{c_x}. \tag{8.32}$$

The units of C-rate are therefore h^{-1} but are frequently dropped. Rather, the letter C is used to indicate that the value is a C-rate. For example, it is common to write C-rates as 0.05C, 0.1C, 1C, and so on rather than 0.05/h, 0.1/h, and 1/h. A C-rate of τ indicates that the capacity is referenced to an amount of current that will discharge the battery in $1/\tau$ h. Equivalently, the C-rate tells us what portion of the charge capacity is discharged (or charged) each hour. For example, when a battery is discharged at 0.25C, it loses 25% of its charge each hour and therefore will only be able to do so for 4 h. As another example, if the battery is discharged at a C-rate of 2C, then the battery will discharge 200% of its charge per hour, and so it can only discharge at this rate for 30 min (half an hour).

The numerical value of the discharge or charge current is found by multiplying the C-rate by the stated capacity, per (8.32). For example, a battery rated at 2 Ah at a C-rate of 0.20C means the battery is capable of supplying $0.20 \times 2 = 0.4$A continuously for 5 h until its terminal voltage falls to the cut-off voltage. The inverse of the C-rate $1/\tau$ is known as the *hour rate*. The hour rate has units of time (hours) and is also commonly used to describe the discharge conditions in which a capacity is referenced to. A battery rated 2 Ah at a C-rate of 0.20C could equivalently be described as having a capacity of 2 Ah at the *5-hour rate*. If a battery is rated at 220 Ah at 0.05C, then the proper interpretation is that "the battery can supply 11 A (220×0.05) continuously for 20 h ($1/0.05$) until the terminal voltage drops below the manufacturer prescribed cut-off voltage."

Example 8.5 A wind energy conversion system is used to charge a battery. The battery supplies electricity to a school. The battery is discharged at 50 A for 10.5 h before the terminal voltage drops below the cut-off voltage of 1.75 V/cell. Compute the capacity, in amphours, of the battery. What C-rate does this capacity correspond to? What is the hour rate?

Solution The capacity of the battery when discharged at 50 A is

$$c_{50} = 50 \times 10.5 = 525 \text{ Ah}$$

The C-rate is $\tau = 50/525 = 0.0952/\text{hr}$ (an hour rate of 10.5 h).

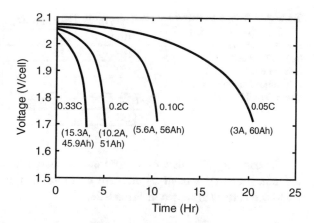

Fig. 8.14 Capacity curves of a battery at different discharge rates for a 60 Ah (at 0.05C) lead–acid battery

Manufacturers will sometimes provide the capacity of a battery without explicit presentation of the C-rate or hour rate. For example, you might encounter a battery rated at 100 Ah. Without more information, you cannot confidently know if, for example, this battery can supply 100 A for 1 h or 1 A for 100 h. While this adds confusion, the C-rate to which the capacity is referenced is implied by its application or chemistry. For example, the rating of a AA-cell battery is given assuming a C-rate of 1; lead–acid batteries in mini-grids use a C-rate of 0.05 (20 h); batteries used in forklifts might use a C-rate of 0.16C (6 h). If unsure, the battery vendor should be consulted. The remainder of this book assumes lead–acid batteries are referenced to a C-rate of 0.05C (20 h) unless otherwise noted. This is also the capacity which the SoC and DoD are referenced to.

8.5.3.5 Capacity Curves

Manufacturers often provide a battery's "capacity curves." Capacity curves are a family of plots of voltage versus discharge time for different C-rates. An example is shown in Fig. 8.14. As we expect, at higher C-rates, the battery voltage drops sooner and more rapidly. The total amphours provided is found by multiplying the current corresponding to the C-rate by the time until the cut-off voltage is reached, as in (8.27). As the C-rate increases, the rated capacity decreases. For example, if the battery in Fig. 8.14 is discharged at 3 A, it will last 20 h (60 Ah). But if it is discharged at 10.2 A, it will only last 5 h (51 Ah). Note that the cut-off voltage for the lower C-rates is somewhat higher.

From Fig 8.14, we can deduce that a battery's capacity is inversely related to the discharge current. Thus, no single value can describe the capacity of battery; it must always be referenced to a discharge rate. The dependence of capacity on discharge current is largely due to the cut-off voltage being reached more quickly at higher current. This is because the voltage drop across R_B increases, not only because the current is greater, but also because R_B itself increases with current.

Table 8.2 Charge capacity of a 2V, 1000 Ah flooded lead–acid battery. The battery's capacity is in reference to its 0.05C discharge characteristics, shown in bold

C-rate (h^{-1})	Hour rate (h)	Current (A)	Capacity (Ah)
0.01C	100	14.10	1410
0.02C	50	24.60	1230
0.05C	**20**	**50.00**	**1000**
0.10C	10	83.00	830
0.5C	2	240	480
1.0C	1	340	340

The capacity of a battery at different discharge rates is often provided by the manufacturer in a tabular format. An example of a 1000 Ah (at 0.05C) flooded lead–acid battery is shown in Table 8.2.

Example 8.6 Consider the battery whose discharge curve is plotted in Fig. 8.13. Estimate the charge capacity if a cut-off voltage of 1.33 V/cell is used.

Solution The battery has a nominal voltage of 12 V, so there are six cells. The cut-off voltage therefore is $6 \times 1.33 \approx 8$ V. Approximately 260 min (4.33 h) have elapsed when the terminal voltage reaches 8 V. Since the battery is discharged at a constant current of 8 A, the corresponding capacity would be $4.33 \times 8 = 34.67$ Ah.

8.5.3.6   Peukert's Equation

The relationship between capacity and discharge current can be approximated using Peukert's equation.[5] It relates the known capacity at a discharge rate of c_{x_r} at current x_r to the capacity c_x if the battery is discharged at some other constant rate x:

$$c_x = c_{x_r} \left(\frac{x_r}{x}\right)^{k-1} \tag{8.33}$$

where k is the experimentally determined and battery-specific Peukert exponent.[6] The Peukert exponent typically is between 1.0 and 1.3. Larger values of k are less desirable as they indicate that the battery capacity is more severely reduced by high discharge rates. This effect is seen in Fig. 8.15 for a hypothetical battery whose rating is 65 Ah at 1 A (0.0154C). If the battery's Peukert coefficient is 1.15, then the capacity at C-rate of 0.05C (20-h discharge time) is approximately 55.7 Ah

---

[5] Peukert's equation is sometimes known as "Peukert's law," but this is a misnomer.
[6] Peukert's exponent is often, but erroneously, referred to as "Peukert's coefficient."

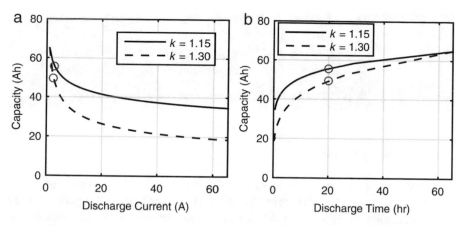

Fig. 8.15 The effect of discharge current and time on battery capacity for batteries with different Peukert exponents

(2.875 A $\times$ 20). If the Peukert coefficient is 1.30, then the capacity at a 0.05C is reduced to 49.5 Ah (2.475A $\times$ 20). These points are shown as the circles in Fig. 8.15. We must apply Peukert's equation with care. If the battery discharge current varies during the discharge or the temperature changes, then the capacity suggested by Peukert's equation will not be accurate.

Example 8.7 A battery in an off-grid system is rated at 105 Ah when discharged at 5.25 A. What is the capacity when discharged at 11.7 A? What is the corresponding C-rate? Let $k = 1.18$.

Solution Setting $x = 11.7$A, $c_{x_r} = 105$ Ah and $x_r = 5.25$ A and using (8.33):

$$c_{11.7} = c_{x_r}\left(\frac{x_r}{x}\right)^{k-1} = 105\left(\frac{5.25}{11.7}\right)^{1.18-1} = 90.90 \text{ Ah.}$$

The C-rate is found by dividing the discharge current by the corresponding capacity $c_{11.7}$: $11.7/90.90 = 0.129$C. The battery can supply a current of 11.7 A for $1/0.129 = 7.77$ h.

Example 8.8 A hypothetical battery has a capacity of 5 Ah at 1C and $k = 1.0$. How many hours will the battery be able to supply a current of 15 A? What C-rate does this discharge current correspond to?

(continued)

Solution The Peukert exponent is 1.0, which means there is no "penalty" for fast discharge. Therefore, the capacity is 5 Ah regardless of the discharge current. From (8.32), the battery can supply 15 A for 5/15 = 0.333 h. The corresponding C-rate is 1/0.333 = 3C.

8.5.3.7 Energy Capacity

Another way of expressing the capacity of a battery is by its energy. Unlike the charge capacity, the energy capacity considers the voltage that the charge is supplied at. Determining the energy capacity therefore requires simultaneously monitoring voltage and current during discharge. The energy capacity e_x of a battery when discharged at a constant current of x is expressed in watthours or kilowatthours and is calculated as

$$e_x = x \int_0^{T_{d,x}} v_B(t)dt. \tag{8.34}$$

The terminal voltage $v_B(t)$ will vary with time, and so for the discharge current x to be constant, the discharge resistance will need to be varied with time. The discharge time ends when the terminal voltage drops below a specified cut-off voltage, just like for the charge capacity. The energy capacity will depend on the discharge rate, with less energy capacity associated with higher discharge. Energy capacity values are rarely provided by the manufacturer. Instead, it is common to approximate the energy capacity by multiplying the charge capacity by the battery's nominal voltage V_{nom}

$$\hat{e}_x = c_x \times V_{nom}. \tag{8.35}$$

The energy capacity of a battery is useful because its units are watthours. Many loads are rated in watts, and so the duration that a battery can supply a given load is found by dividing its energy capacity by the power of the load.

Example 8.9 Consider the 12 V battery discharged at 8 A whose voltage is shown in Fig. 8.13. The voltage can be modeled as a piecewise linear function:

$$v_B(t) = \begin{cases} -0.006088t + 12.39 : 0 \le t \le 180 \\ -0.02769t + 16.11 \ : 180 < t \le 213.5 \end{cases}$$

(continued)

where t is the time in minutes. Compute the energy capacity of the battery if the cut-off voltage is 10.2 V. Compare this to the energy capacity estimated by the battery's nominal voltage.

Solution We know the battery has been discharged at a constant current of $x = 8$ A. From (8.34), to compute the energy capacity, we need to integrate the battery voltage.

$$e_x = 8 \int_0^{180} (-0.006088t + 12.39)\, dt + 8 \int_{180}^{213.5} (-0.02769t + 16.11)\, dt$$

$$= 8 \times (2131.6 - 0) + 8 \times (2808.4 - 2451.2) = 19,910.0 \text{ wattminutes}$$

Converting to watthours: 19,910.0/60 = 331.84 Wh. The energy capacity when estimated from the nominal is $12 \times (8 \times 213.5/60) = 341.6$ Wh. The two values are fairly close.

8.5.4 Efficiency

Just as with energy conversion technologies, no form of energy storage is 100% efficient. The efficiency of a battery is found by dividing the energy that can be meaningfully output by the energy that is input:

$$\eta_B = \frac{\int_0^{T_{B,D}} i_{B,D}(t) v_{B,D}(t)\, dt}{\int_{T_D}^{T_C + T_D} i_{B,C}(t) v_{B,C}(t)\, dt} \tag{8.36}$$

where the subscripts C and D denote whether the battery is being charged or discharged, we have assumed that charging immediately follows discharging, and the current is not necessarily constant.

For most charge and discharge scenarios, the efficiency can be approximated by

$$\eta_B \approx \frac{\bar{V}_{B,D} \int_0^{T_D} i_{B,D}(t)\, dt}{\bar{V}_{B,C} \int_{T_D}^{T_C + T_D} i_{B,C}(t)\, dt} \tag{8.37}$$

were $\bar{V}_{B,D}$ and $\bar{V}_{B,C}$ are the average terminal voltage during discharge and charge, respectively.

The energy efficiency can then be split into two parts: the voltage efficiency η_V and the coulombic efficiency η_C. Thus, the battery efficiency can be approximated as

$$\eta_B \approx \eta_V \times \eta_C. \tag{8.38}$$

The energy efficiency is often expressed as a percent.

8.5.4.1 Voltage Efficiency

The voltage efficiency is the ratio of the average voltage during discharge to the average voltage during charge:

$$\eta_V = \frac{\bar{V}_{B,D}}{\bar{V}_{B,C}}. \tag{8.39}$$

The voltage efficiency is often expressed as a percentage. The average voltage of a discharge period T_D is computed as any other average:

$$\bar{V}_{B,D} = \frac{1}{T_D} \int_t^{t+T_D} v_B(t)dt. \tag{8.40}$$

The average voltage during a charge period is computed in a similar fashion. The average discharge voltage is always less than the average charging voltage due to the battery resistance R_B. The voltage efficiency will therefore be less than 100%. The voltage efficiency depends upon the charge and discharge rate, as well as the other factors that influence battery resistance in general. Voltage efficiencies generally range between 80 and 90%.

8.5.4.2 Coulombic Efficiency

The coulombic efficiency or "Faraday efficiency" of a battery is the ratio of charge input to the battery during charging to the charge that is output from the battery during discharge

$$\eta_C = \frac{\int_0^{T_D} i_{B,D}(t)dt}{\int_{T_D}^{T_C+T_D} i_{B,C}(t)dt}. \tag{8.41}$$

The coulombic efficiency is also often expressed as a percent. The coulombic efficiency tends to be high, between 90 and 95%. A coulombic efficiency less than 100% might seem strange—what happens to the electrons that enter the battery but never leave? These electrons are consumed in "side reactions" inside the battery, for example, producing hydrogen and oxygen gases in a lead–acid battery. The "loss" of electrons primarily occurs during charging when voltages tend to be higher and in particular when a battery is nearly fully charged [6].

8.5.4.3   Energy Efficiency

The energy efficiency η_B of the battery is typically in the range of 70% to 85%. Lower efficiencies are associated with rapid charging and discharging and in general any condition that increases R_B or side reactions. A practical consequence of this is that to supply 1 kWh of energy, the battery might require 1.4 kWh of input energy. It is sometimes convenient to define a charging efficiency η_C and a discharging efficiency η_D of the battery, so that

$$\eta_B = \eta_C \times \eta_D \tag{8.42}$$

This is useful in simulations. It is common to set the charging and discharging efficiencies equal to each other, so that

$$\eta_B = \sqrt{\eta_C} = \sqrt{\eta_D} \tag{8.43}$$

This is a simplification, again in part to dependency of charging efficiency on SoC. The battery efficiency must be accounted for in designing an off-grid system, which we discuss in Chap. 12.

8.5.5   Cycle Life

The rated charge or energy capacity of a battery assumes the battery is in new condition. Unfortunately, as a battery degrades over time and with use, its capacity diminishes. The cycle life of a battery is the number of charge and discharge cycles the battery can experience before its capacity drops below a certain value—typically 60 or 80% of the initial rated value. In other words, if a battery is rated at 100 Ah at 0.05C, after a certain number of charge and discharge cycles, its effective rating will only be 80 Ah at the same C-rate. At this point, the battery is said to have reached its end of life. The battery is still functional and so it does not necessarily need to be immediately replaced, but its capacity has been appreciably diminished.

Figure 8.16 shows the typical cycle life of flooded and AGM batteries with respect to the DoD of each cycle. The number of cycles decreases as the DoD increases. Each cycle causes some irreversible damage to the cell's electrode, which reduces the amount of active material that can be used in the chemical reactions. Some battery types, like lithium iron phosphate (LiFePO$_4$) batteries, have a cycle life far exceeding lead–acid. The cycle life is also dependent on the calendar age of the battery. The active materials will corrode over time, leading to a decrease in capacity. This effect is accelerated with temperature.

This is an important concern. To prolong battery life, many systems limit the DoD to 40 to 60%. Each cycle can be taken as a day so that most lead–acid batteries will last between 4 and 6 years if the DoD is limited. However, other factors, primarily elevated temperature, can shorten the cycle life.

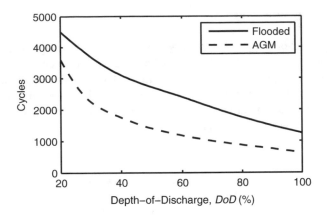

Fig. 8.16 Cycle life of lead–acid batteries decreases with the depth-of-discharge

Example 8.10 A 220 Ah flooded lead–acid battery is discharged at a constant rate of 11.0A for 8 h each day. Estimate the number of days the battery will last before it reaches the end of its cycle life. Use the curve in Fig. 8.16 to estimate the cycle life.

Solution To use the curve in Fig. 8.16, we must know the depth-of-discharge. The battery capacity is given as 220 Ah, which we assume is the 20-h (0.05C) rate. The corresponding current is $220/20 = 11$ A, which conveniently matches the discharge rate. The charge supplied by the battery each day is:

$$8\,\text{h} \times 11\,\text{A} = 88\,\text{Ah}.$$

Using (8.30), this corresponds to 40% DoD. Consulting Fig. 8.16, 40% DoD corresponds to approximately 3000 cycles, or $3000/365 = 8.2$ years.

8.5.6 Self-Discharge

A battery that is not connected to an external circuit will discharge internally, causing it to lose a portion of the energy it stores over time. The self-discharge is not caused by internal current; rather, it is caused by chemical "side" reactions. A typical self-discharge rate is 2.5% to 5% per month for lead–acid batteries designed for off-grid applications. The rate is lower for lithium–ion batteries. However, many lithium–ion batteries are packaged with integrated protection circuits, which might also consume another 1 to 2% per month. Self-discharge is accelerated at high temperatures, approximately doubling for every 10°C increase in temperature.

Depending on the discharge rate, it is advisable to periodically charge batteries, for example, every 3 months, when not in use. Batteries can be stored at lower temperatures to minimize self-discharge.

8.5.7 Temperature Dependence

The performance and characteristics of all batteries depend on the temperature of operation, which is often near the ambient temperature. Qualitatively, the following effects are observed for lead–acid batteries as temperature increases past about 25°C.

- Decreased internal impedance
- Increased capacity
- Increased open-circuit voltage
- Increased self-discharge
- Decreased cycle life

In many mini-grids, the batteries will not be in temperature-controlled environments, and thus the influence of temperature should be considered in the design phase. While the first three effects can be considered beneficial, they are typically outweighed by the last two—the decreased cycle life in particular.

The voltage of a fully charged lead–acid cell will increase by approximately 0.2 mV/cell/°C. But this rate depends on the SoC, as seen in Fig. 8.17. Thus at a very low or negative SoC, an increase in temperature leads to a *decrease* in voltage. The voltage increases because temperature increases the rate of all chemical reactions, including those in the battery. The battery capacity is effectively increased with temperature, as shown in Fig. 8.18. At very high temperatures, the battery capacity increases by perhaps 10%, and at very low temperatures, it is perhaps just 60% of its rated value.

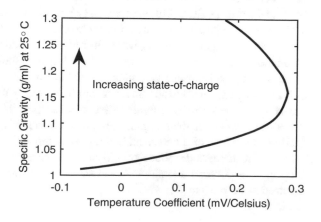

Fig. 8.17 The temperature coefficient of a lead–acid battery depends on the specific gravity of the electrolyte [10]

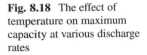

Fig. 8.18 The effect of temperature on maximum capacity at various discharge rates

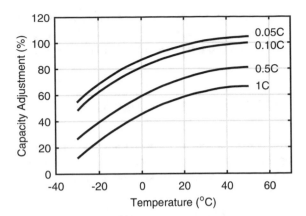

High temperature also promotes the unwanted side reactions and self-discharge in a battery. The rate approximately doubles for every 10°C increase in temperature. Gassing can also increase, and the voltage that the battery is charged at should be reduced. This will be discussed in more detail in Chap. 10.

8.6   Lithium–Ion Batteries

Lithium–ion (LI) batteries are a relatively new technology, only becoming commercially available since the 1990s. They generally offer higher energy density, longer cycle life, a flatter $I–V$ curve, less severe high-discharge penalty (Peukert exponent close to 1.0), and slower self-discharge rate than lead–acid batteries. Balancing these benefits are higher cost and some safety concerns. We will cover the fundamentals of LI batteries; other resources may be consulted for additional information [3, 11, 13].

There are two general types of LI battery chemistries: lithium iron phosphate (LiFePO$_4$) and those using lithium nickel, cobalt, or manganese. LI batteries operate by transferring positive lithium ions between electrodes through the electrolyte while electrons pass through an external circuit. The anode is typically a graphite carbon and the cathode one of several lithium metal oxide materials. The electrodes also incorporate copper or other materials to improve their conductivity.

In a completely discharged state, all the lithium is in the cathode. When the cell is charging, lithium ions travel through the electrolyte to the anode; the ions are shuttled back to the cathode when discharging. The electrolyte is usually a lithium salt dissolved in an organic solvent such as LiPF$_6$ or LiBF$_4$. The electrolyte facilitates the flow of the ions but is not part of the active material as in a lead–acid battery. The anode and cathode are physically separated by an electrically insulating microporous polymer membrane. The ions can pass through the membrane, but the electrodes are prevented from touching each other, as this would cause a damaging short-circuit.

The anode must have a physical structure that allows lithium ions to be inserted and extracted (to electrochemists this is known as "intercalation" and "de-intercalation"). The more lithium ions that can be inserted into the anode, the greater the charge capacity of the cell. The most common anode material is graphite carbon. Its layered molecular structure has space for one lithium ion for every six carbon atoms. As the cell becomes charged, the anode evolves from C_6 to Li_yC_6 where $0 \leq y \leq 1$. Li_yC_6 is expressed as an empirical formula, which is different from the structural molecular formula most often used in introductory chemistry classes. An empirical chemical formula expresses the ratio of the elements in a compound, and so the subscripts need not be whole numbers. For our purposes, we should be comfortable with the concept that as y increases, the proportion of lithium in the anode increases, and consequentially the SoC of the battery increases. We can write a similar empirical formula for the cathode in which the portion of lithium decreases as the SoC increases.

The basic reaction for an LI cell using a lithium cobalt oxide ($LiCoO_2$) cathode as an example during charging (insertion of the lithium–ion) is, for the anode:

$$yLi^+ + C_6 + ye^- \rightarrow Li_yC_6 \tag{8.44}$$

and for the cathode

$$LiCoO_2 \rightarrow Li_{1-z}CoO_2 + yLi^+ + ye^-. \tag{8.45}$$

The reactions are reversed during discharge. Most $LiCoO_2$ cells are charged so that $y = 1$ and $z \approx 0.5$; the overall reaction can be expressed as

$$2LiCoO_2 + C_6 \leftrightarrow 2Li_{0.5}CoO_2 + LiC_6. \tag{8.46}$$

8.6.1   Open-Circuit Voltage

The open-circuit voltage for a $LiCoO_2$ battery is shown in Fig. 8.19. In this figure, the horizontal axis is the fraction of lithium ions in the anode z. The right side of the figure is when all of the lithium ions have left the cathode and are in the anode (fully charged). The left side of the figure is when there are no lithium–ions in the anode and they are in the cathode (fully discharged). As with lead–acid batteries, the voltage developed depends on the chemical composition of the electrodes, and so this varies with the amount of lithium in the anode or cathode. The amount of lithium directly relates to the SoC, and so the horizontal axis can also be thought of as a SoC of sorts.

An important aspect of LI batteries is what happens near either end of the curves. When z approaches 1.0, the energy stored in the cell is the highest. The voltage is also at its highest level, in excess of 4.40 V. The voltage is so high in fact that the

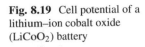

Fig. 8.19 Cell potential of a
lithium–ion cobalt oxide
($LiCoO_2$) battery

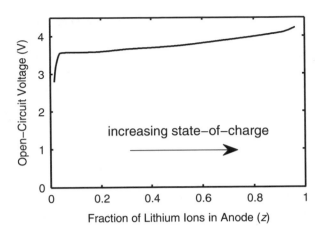

electrolyte begins to irreversibly degrade. The lithium becomes plated on the surface on the anode. Remember, we want the lithium ions to be inserted into the anode, not plated onto it. The plated lithium can no longer be removed during discharge.

Low voltages are also harmful to the electrodes. The copper conductor at the anode begins to irreversibly dissolve. When the battery is recharged, the copper is precipitated wherever it might be. This can cause an internal short-circuit. The cathode gradually breaks down, which reduces its capacity. If these two harmful extremes are avoided, though, the voltage of an LI battery remains fairly constant over a wide range of lithium–ion concentration.

Just like lead–acid batteries, we define the SoC as being the range of charge or energy that can be meaningfully provide without damaging the battery or the battery voltage becoming too low. The cut-off voltage varies depending on the type of LI battery, but it is often 2.5 to 3.0 V/cell. This roughly corresponds to the steep drop in the cell potential curve. For reasons previously described, the maximum voltage is limited to around 4.0 V/cell. The 100% SoC is in reference to the stored charge at this voltage. Note that the point at which 100% SoC is defined is not necessarily at $z = 1.0$. For example, z is limited to less than 0.5 for $LiCoO_2$ batteries in order to prolong the life of the cathode.

An LI battery can be charged in excess of 4.0 V/cell, in which case the SoC would exceed 100%. The battery lifespan and capacity would be reduced, however due to lithium plating and other reactions. Selecting the charging voltage of an LI battery therefore presents a trade-off—a higher voltage increases the energy that is stored (and consequentially can be extracted) but shortens its lifespan. As a familiar example, if you want your mobile phone battery to last for several years, then you should not fully charge it. By prematurely ending the charging, for example, at 70%, the battery voltage will be lower, and the battery's lifespan will be longer. However, the trade-off is of course that you will need to charge the battery more frequently because it is never full.

Fig. 8.20 Capacity curve of
a LiFePO$_4$ battery

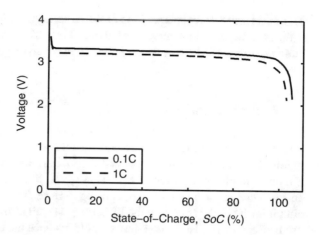

We should also expect the voltage to decrease with state of charge. However, the decrease is nonlinear as there are many reactions that are occurring. This will be explored in the capacity curve of the battery.

8.6.2 Capacity Curve

The typical capacity curve for a LiFePO$_4$ battery is shown in Fig. 8.20. Instead of discharge time, the horizontal axis is the SoC referenced to 1.0C. LI batteries are often referenced to 1C instead of 0.05C. Although the capacity of a lead–acid battery is lower at 1.0C than 0.05C, the capacity of an LI battery is less sensitive to the magnitude of the discharge current. In other words, its Peukert exponent is near 1.0. LiFePO$_4$ batteries in particular exhibit flat curves. This is desirable because it means their output voltage is less variable while discharging.

8.6.3 Safety

LI batteries have special safety concerns that must be acknowledged. In particular, a failure mode known as *thermal runaway* can cause LI batteries to ignite [4]. This has led to special handling procedures when LI batteries are transported on aircraft. The risk generally stems from the heat caused when the battery develops an internal short-circuit. The membrane separating the anode and cathode is very thin. If accidentally punctured, the anode and cathode could come into contact. The heat from the resulting short-circuit will cause the battery's internal temperature to rise. This can cause the separator to shrink, exacerbating the problem. At high temperatures, around 80°C, a series of unwanted heat-producing reactions occur. Among them is the decomposition of the cathode, which releases gaseous oxygen.

This accelerates the decomposition of the electrolyte which releases more gas. This is the runaway condition mentioned above. The cell temperature can exceed several hundred degrees Celsius at which point the electrolyte ignites.

8.6.4 Protection

The sensitivity and safety risk associated with over- and undervoltage conditions require LI batteries to include specialized battery protection circuits. The cells in a battery might have different SoC at a given time. This can cause one cell to be overcharged or undercharged, causing failure. To avoid this, a cell balancing circuit is often integrated into the LI battery as well as overcharge and undercharge protection circuits. There are a few ways to balance the cells. Most involve circuits that discharge the cells that have a higher voltage than the rest through a resistor. This, however, is inefficient. More sophisticated designs will transfer this energy to the cells that have a lower charge, thus reducing losses. Pressure release valves are sometimes included.

8.6.5 Cycle Life

LI batteries tend to have a longer cycle life than lead–acid batteries. For $LiCoO_2$, the increase is more modest, perhaps lasting 500 cycles at a DoD of 80%. However, $LiFePO_4$ batteries can withstand several thousand cycles.

8.6.6 Charging

An advantage of LI batteries is that they can be charged more rapidly than lead–acid batteries. The rate varies with the chemistry, but it is usually less than 4 h and for $LiFePO_4$, and for LI manganese batteries as few as 2 h. Of course the charger must be capable of supplying this high rate of charge.

8.6.7 Comparison

The different choices of electrodes and electrolytes change the characteristics of the LI battery. Some general comparisons are shown in Table 8.3. These are approximate only as the performance of any battery depends on many factors. In the context of off-grid systems, $LiFePO_4$ batteries are increasingly common in solar home systems (SHS) and solar lanterns (SL). They offer the highest cycle

Table 8.3 Lithium–ion characteristics

Characteristic	Lead-acid	Lithium cobalt oxide	Lithium nickel manganese cobalt oxide	Lithium iron phosphate
Specific energy (Wh/kg)	40	150–200	150–220	90–120
Nominal voltage/cell (V)	2.04	3.60–3.70	3.60–3.70	3.2
Operating range/cell (V)	1.75–2.50	3.0–4.2	3.0–4.2	2.5–3.65
Cycles (80% DOD)	1000–2000	500–1000	500–1000	2000–5000
Efficiency (%)	70–90	85–95	85–95	85–95
Cost (USD/kWh)	150–500	2000–3000	500–1500	750–1250
Toxicity	High	Low	Low	Low
Self-discharge (%/month)	5–8	2–5	2–5	2–5

life, even up to 10,000 if not deeply discharged or overcharged. They offer good thermal stability and so are less prone to thermal runaway and are more tolerant to overcharging. Their discharge curve is flat and so the voltage regulation is good. However, $LiFePO_4$ have lower nominal voltage, and so their specific energy is lower than other LI batteries.

8.7   Battery Banks

The largest individual batteries used in off-grid applications rarely exceed a few kilowatthours of capacity, and most have a nominal voltage of no more than 12 V (or 12.8 V for $LiFePO_4$). In applications requiring more energy storage or greater voltage, the batteries are connected in series, parallel, or combination thereof to form a battery bank. The battery banks for larger mini-grids can be substantial. A 48 V, 6800 Ah bank is shown in Fig. 8.21. This battery bank fills an entire room. The batteries should be as similar as possible, preferably the same model, with the same parameters, from the same supplier and manufactured at the same time.

As with a PV array, the batteries connected in series are a "string". The string voltage is equal to the sum of the individual voltages in that string:

$$V_{bank} = \sum_{i=1}^{N_{series}} V_{B,i} \qquad (8.47)$$

where N_{series} is the number of batteries in the string. During normal operation, the batteries all have the same voltage, V_B:

$$V_{bank} = N_{series} \times V_B. \qquad (8.48)$$

Battery strings can be connected in parallel, so long as the voltage of each string is equal or nearly equal. We expect the battery bank current I_{Bank} to be divided equally by the strings.

Fig. 8.21 Battery bank from
a mini-grid in Nigeria; the
battery bank has not been
connected to the DC bus
(courtesy of GVE Projects)

Fig. 8.22 A diagram of a
battery bank consisting of
two strings each with three
batteries

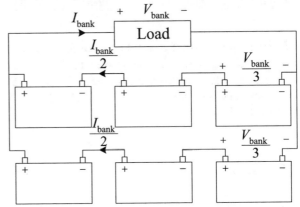

$$I_{\text{string}} = \frac{I_{\text{Bank}}}{N_{\text{string}}}. \qquad (8.49)$$

In practice, the number of strings is often limited to about four. The resistance of the
cables connecting each string to the DC bus should be kept as uniform as possible.
This ensures the batteries are charged and discharged evenly. Should a battery in
a string fail, for example, by an internal short circuit, then the other batteries will
discharge through the string with the faulty battery. This can cause a catastrophic
failure of the entire battery bank.

Figure 8.22 shows a battery bank consisting of two strings of three batteries
each. The discharge current is evenly divided among the strings. Similarly, the
battery bank voltage is equally divided among the batteries in each string. This
is an idealization of course. There will be slight variations in individual batteries.
The energy capacity of a battery bank is the sum of the capacities of the individual
batteries, which should all be equal

$$E_{\text{bank},x} = N_{\text{string}} \times N_{\text{series}} \times e_{\text{B},x}. \qquad (8.50)$$

In other words, the battery bank capacity is independent of how the individual batteries are connected. We must be careful when computing the charge capacity rating of a battery bank. For example, when two, 12V, 100 Ah, are connected in series, the charge capacity remains at 100 Ah (the current through each battery is the same). The voltage, however, has been increased to 24 V and so the energy capacity has doubled.

We often refer to the battery bank as simply the "battery" without providing specific information about how the bank is configured. In this case, V_B and I_B refer to the terminal voltage of and current from the bank, not the individual batteries. The nominal voltage of the battery bank is the sum of the nominal voltages of the series-connected batteries.

The terminals of a battery bank are connected to the positive and negative DC bus. The DC bus voltage therefore is the same as the battery bank voltage.

Example 8.11 The battery bank of a mini-grid consists of 32 lead–acid batteries. Each battery has a nominal voltage of 6 V and a capacity of 375 Ah at the 20-h rate. The nominal voltage of the battery bank is 48 V. Compute the charge capacity of the battery bank, and estimate the energy capacity of each battery at the 20-h rate. According to the manufacturer, the maximum charging current is limited to 18% of the 20-h capacity. What is the maximum charging current that can be supplied to the battery bank without violating this limit?

Solution Each string of the battery bank consists of

$$\frac{\text{battery bank nominal voltage}}{\text{individual battery voltage}} = \frac{48}{6} = 8 \text{ batteries.}$$

Therefore, the battery bank has

$$\frac{\text{total batteries}}{\text{batteries per string}} = \frac{32}{8} = 4 \text{ strings.}$$

Each string has a capacity of 375 Ah, and so the battery bank has a total charge capacity of

$$\text{Ah per string} \times \text{strings} = 375 \times 4 = 1500 \text{ Ah.}$$

The charge capacity is 1500 Ah at the 20-h rate. The energy of each battery is

$$\text{Ah capacity} \times \text{nominal voltage} = 375 \times 6 = 2250 \text{ Wh.}$$

(continued)

Therefore, the energy capacity of the battery bank is $2250 \times 32 = 72.0$kWh. We could also arrive at this value by multiplying the amphour rating of the bank by its nominal voltage: $1500 \times 48 = 72$ kWh.

The maximum charging current per string is 18% of the 20-h capacity: $0.18 \times 375 = 67.5$ A. Since there are four strings, and assuming the current divides equally into each string, the maximum total charging current is $67.5 \times 4 = 270$ A.

8.8 Summary

Electrochemical batteries are an important component in many off-grid systems, particularly those that rely on PV and WECSs. Lead–acid and lithium–ion (LI) batteries are most common in off-grid applications. These, like all batteries, rely on electrochemical reactions to develop a voltage between their terminals and to provide power when connected to a load. The reactions are decidedly nonlinear and temperature sensitive. Because of this complexity, we must be careful in interpreting the ratings and other information provided by battery manufacturers, and when analyzing the operation of a battery-based system. Among the most important to understand is that the battery capacity is based on the charge or energy that can meaningfully extracted without the voltage becoming too low or the battery being damaged. This capacity depends on the rate of discharge, temperature, and the specified cut-off voltage. The battery can be modeled as voltage source V_{SoC} in series with a resistance R_B—but both of these are dependent on several factors, and the battery resistance itself represents three polarization affects, only one of which is actually ohmic in nature.

When using a battery, you must be careful not to overcharge or overdischarge it. This is especially true for lithium–ion (LI) batteries as their failure can present a safety hazard. Lead–acid batteries can last for several hundred up to a few thousand cycles. If one assumes a daily charge/discharge cycle, that suggests a battery life of several years. Some LI batteries, $LiFePO_4$, in particular, can last for several thousand cycles. Limiting the depth-of-discharge and avoiding high operating temperatures promote longer battery life.

The choice between lead–acid and LI depends on the application. LI tend to be used in solar lanterns and solar home system as they are lighter. Lead–acid batteries, flooded, in particular, are less expensive but require regular maintenance in the form of watering.

Problems

8.1 The standard cell voltage for a lead–acid battery is 2.04 V. Compute the corresponding open-circuit voltage of a nominal 48 V battery bank if the concentration of the sulfuric acid in the electrolyte is 2 moles/liter. Assume the temperature is 25°C. Approximate the activity of the sulfuric acid by its concentration. Repeat the calculation with a concentration of 5 moles/liter.

8.2 A flooded 12 V lead–acid battery is charged at a current of 10 A. The terminal voltage is measured to be 11.9 V. The battery resistance is 0.032 Ω. Compute the voltage V_{SoC} and estimate the battery's SoC using Table 8.1.

8.3 A battery is charged in two sequential stages. In the first stage, a constant current of 20 A is applied for 2 h. In the second, the current decreases according to the equation:

$$i_B(t) = 20e^{\frac{-(t-2)}{1.25}}$$

where t is the time since charging began, in hours. The second stage lasts for 4 h. Compute the total charge, in amphours, provided to the battery.

8.4 The capacity of a battery at its 72-h rate is 394 Ah and is 375 Ah at its 20 hour-rate. Compute the current corresponding to the 72- and the 20-h rates. What are the corresponding C-rates?

8.5 A 1100 Ah AGM battery (20-h rate) is used in a mini-grid. The mini-grid supplies a critical load whose current draw is 25 A. Estimate the number of hours the battery can supply this load if the Peukert exponent is 1.046.

8.6 Repeat the previous problem but if a 1100 Ah flooded lead–acid battery with a Peukert exponent of 1.09 is used.

8.7 The capacity of a flooded lead–acid battery at 0.05C is 1547 Ah and at 0.01C is 1990Ah. Compute the Peukert exponent.

8.8 The average voltage of a battery is 13.56 V when charged at 5 A, and the average voltage is 11.46 V when discharged at 5 A. Compute the voltage efficiency of the battery.

8.9 A battery bank consists of three strings of eight 6 V lead–acid batteries. Each battery is rated at 440 Ah. Estimate how long the battery bank can supply a load of 120 A. Assume the Peukert exponent is 1.0. Estimate the energy capacity of the battery bank, in kilowatthours.

8.10 A 48 V battery bank has one string of 2 V lead–acid batteries. Each battery is rated at 1160 Ah. Estimate how long the battery bank can supply a load of 80 A. Assume the Peukert exponent is 1.08. Estimate the energy capacity of the battery bank, in kilowatthours.

8.11 Consider the batteries in the previous problem. Instead of a 48 V battery bank, consider an alternate design where the same number of batteries are arranged in a 24 V battery bank. Since the DC bus voltage has been reduced by one half, the current doubles to 160 A so that the same power to the load is provided. Estimate how long the battery bank can supply the load. Assume the Peukert exponent is 1.08. Estimate the energy capacity of the battery bank, in kilowatthours. Compare the results to the previous problem.

8.12 A 340 Ah battery supplies a constant current load of 17 A for 8 h each day. Compute the battery's state-of-charge and depth-of-discharge at the end of each day.

8.13 Estimate the number of cycles the battery in the previous problem will last if its life cycle characteristic is that of Fig. 8.16. Estimate the number of cycles the battery will last if the load is reduced to 9 A for 8 h each day. Assume the battery is a flooded lead–acid battery.

8.14 A large mini-grid requires its battery to supply 80,000 Wh of energy at 48 V each day. The charge capacity characteristics of the particular 2 V nominal battery type to be used are provided in Table 8.2. Design a battery bank (provide the number of strings and number of batteries per string) to supply this load while discharging the battery no deeper than 60%. Assume the battery is discharged at a constant current over a 10-h period each day.

References

1. Awode, M.R.: Introduction to Electrochemistry, 2nd edn. Himalaya Publishing House (2010)
2. Barak, M. (ed.): Electrochemical Power Sources: Primary and Secondary Batteries. The Institution of Electrical Engineers (1980)
3. Chagnes, A.: Chapter 2 - fundamentals in electrochemistry and hydrometallurgy. In: Chagnes, A., Wiatowska, J. (eds.) Lithium Process Chemistry, pp. 41–80. Elsevier, Amsterdam (2015). DOI https://doi.org/10.1016/B978-0-12-801417-2.00002-5. URL https://www.sciencedirect.com/science/article/pii/B9780128014172000025
4. Feng, X., Ouyang, M., Liu, X., Lu, L., Xia, Y., He, X.: Thermal runaway mechanism of lithium ion battery for electric vehicles: A review. Energy Storage Mater. **10**, 246–267 (2018). DOI https://doi.org/10.1016/j.ensm.2017.05.013. URL http://www.sciencedirect.com/science/article/pii/S2405829716303464
5. Franke, M., Kowal, J.: Empirical sulfation model for valve-regulated lead-acid batteries under cycling operation. J. Power Sources **380**, 76–82 (2018). DOI https://doi.org/10.1016/j.jpowsour.2018.01.053. URL https://www.sciencedirect.com/science/article/pii/S0378775318300533
6. Kaushik, R., Mawston, I.: Coulombic efficiency of lead/acid batteries, particularly in remote-area power-supply (RAPS) systems. J. Power Sources **35**(4), 377–383 (1991). DOI https://doi.org/10.1016/0378-7753(91)80055-3. URL http://www.sciencedirect.com/science/article/pii/0378775391800553
7. Mantell, C.: Batteries and Energy Systems, 2nd edn. McGraw-Hill (1983)
8. Mauracher, P., Karden, E.: Dynamic modelling of lead/acid batteries using impedance spectroscopy for parameter identification. J. Power Sources **67**(1), 69–84 (1997). DOI https://doi.org/10.1016/S0378-7753(97)02498-1. URL http://www.sciencedirect.com/science/article/pii/S0378775397024981. Proceedings of the Fifth European Lead Battery Conference

9. Rand, D., Moseley, P.: Secondary batteries – lead – acid systems overview. In: Garche, J. (ed.) Encyclopedia of Electrochemical Power Sources, pp. 550–575. Elsevier, Amsterdam (2009). DOI https://doi.org/10.1016/B978-044452745-5.00126-X. URL https://www.sciencedirect.com/science/article/pii/B978044452745500126X

10. Reddy, T.B. (ed.): Linden's Handbook of Batteries. McGraw Hill, New York (2011)

11. Spataru, C., Bouffaron, P.: Chapter 22 - off-grid energy storage. In: Letcher, T.M. (ed.) Storing Energy, pp. 477–497. Elsevier, Oxford (2016). DOI https://doi.org/10.1016/B978-0-12-803440-8.00022-1. URL https://www.sciencedirect.com/science/article/pii/B9780128034408000221

12. Treptow, R.S.: The lead-acid battery: Its voltage in theory and in practice. J. Chem. Educ. 79(3), 334 (2002). DOI 10.1021/ed079p334. URL https://doi.org/10.1021/ed079p334

13. Vetter, M., Lux, S.: Chapter 11 - rechargeable batteries with special reference to lithium-ion batteries. In: Letcher, T.M. (ed.) Storing Energy, pp. 205–225. Elsevier, Oxford (2016). DOI https://doi.org/10.1016/B978-0-12-803440-8.00011-7. URL https://www.sciencedirect.com/science/article/pii/B9780128034408000117

14. White, C., Deveau, J., Swan, L.G.: Evolution of internal resistance during formation of flooded lead-acid batteries. J. Power Sources 327, 160–170 (2016). DOI https://doi.org/10.1016/j.jpowsour.2016.07.020. URL http://www.sciencedirect.com/science/article/pii/S037877531630876X

Chapter 9
Off-Grid System Converters and Controllers

9.1 Introduction

It is possible to create an off-grid system consisting of only an energy source, a battery, and a load. These "improvised" or "non-engineered" systems are very common in rural households (see Fig. 9.1). As you might expect, they are limited in size and function and are often short-lived, inefficient, and unreliable. In this book we are considering more formally engineered systems. They involve the use of energy sources, generators, loads, and, pertinent to this chapter, control devices.

In earlier chapters, references have been made to controllers that are used to maintain mini-grids at their desired states of operation. These controllers can be either mechanical or electrical. Many such electrical control systems are categorized as *converters*. Converters are a broad class of power electronic devices that are designed to perform a wide range of functions within DC and AC systems. Converters add cost and complexity to the system but also offer several benefits. They can act as an interface, enabling otherwise incompatible components to work together. For example, converters make it possible for a DC-coupled system to serve an AC load. Converters facilitate the flow of power between the buses in AC–DC coupled systems and regulate battery charging and discharging. Converters are used in maximum power point trackers that increase the production from PV arrays and wind turbines.

The study of converters lies within the subfield of electrical engineering known as power electronics. There are numerous textbooks devoted to the subject [6, 10]. We cannot cover converters in detail in a single chapter. Instead, this chapter presents the basic function of each converter, its principle circuit, and application considerations. Many converters require control circuitry and logic to properly function. We do not dwell on how the control signals are generated. Rather, we consider how the signals affect the operation of the converter.

From the outside, converters appear as metal enclosures with indicator lights and switches on the surface. Inside are the power conditioning and control circuitry, as

© Springer International Publishing AG, part of Springer Nature 2018
H. Louie, *Off-Grid Electrical Systems in Developing Countries*,
https://doi.org/10.1007/978-3-319-91890-7_9

Fig. 9.1 This improvised
solar-powered system has no
controllers (courtesy of
author)

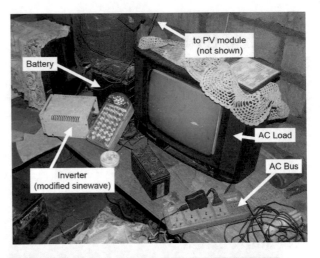

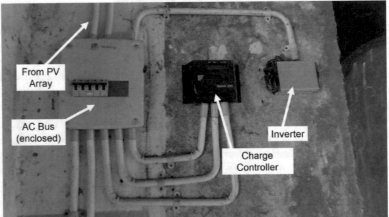

Fig. 9.2 Converters and controllers of a small PV system installed in a school in Malawi (courtesy of author)

well as fans, fuses, and connection points for other mini-grid components. Most converters are no larger than a microwave oven. A recent trend by manufacturers is to combine several converters into a single unit. For example, one can readily purchase a unit that combines maximum power point tracking, charge control, and inverter functionality. Integrated units can reduce the space requirements and simplify the wiring. We shall generally consider single-purpose converters.

Small systems use only a few converters, as shown in Fig. 9.2. Larger systems will require dedicated control rooms, as in Fig. 9.3.

Table 9.1 shows some of the commonly-available converter types and their applications in off-grid systems. These will be discussed in the following sections.

Fig. 9.3 The control room of a large mini-grid in Nigeria, consisting of inverters, charge controllers, and system monitoring equipment (courtesy of GVE Projects)

Table 9.1 Function and applications of converters

Converter	Basic function	Section
DC–DC converter	Increases or decreases output voltage	9.3
Maximum power point tracker	Increases power produced by PV arrays or WECS	9.4
Solar battery charger	Charges batteries directly from PV sources	9.5
AC battery charger	Converts AC produced by generators or other sources to DC and manages battery charging	9.6
Rectifier	Converts AC to DC	9.6.1–9.6.2
Automatic voltage regulator	Adjusts excitation to synchronous generators	9.7
Electronic load controller	Controls power to ballast load to regulate frequency	9.8
Inverter	Converts DC to AC	9.9
Solar inverter	Converts DC from PV sources to AC	9.10
Grid-tied inverter	Converts DC to AC and synchronizes with AC bus	9.11
Bi-directional converter	Allows power to be exchanged between the DC and AC buses	9.12

9.2 Basic Concepts

We begin by covering basic concepts. To readers with a background in semiconductor devices or power electronics, much of this will be a review.

9.2.1 Solid-State Switching Elements

The primary circuit elements used in converters are solid-state switches, inductors, and capacitors. Inductors and capacitors provide small amounts of short-term energy storage and are useful as filters. Solid-state switches operate like mechanical switches but at very high speeds, usually tens of thousands of times per second. The circuit symbols for several solid-state switches are shown in Fig. 9.4.

Some switches are operated by a control signal, which we generically denote by the time-dependent binary or logic variable $q(t)$. We use the convention that $q(t) = 1$ is a "close" signal and $q(t) = 0$ is an "open" signal. In practice, the signal must have the appropriate electrical characteristics to initiate conduction, but we will ignore this detail.

Power MOSFETs
Power MOSFETs (metal–oxide–semiconductor field-effect transistors) are a type of transistor designed to withstand the high voltages and currents present in power electronic circuits. A MOSFET is a three-terminal device. Conduction between two of the terminals, the drain and the source, is controlled by a control signal $q(t)$ applied to the gate. Conduction continues between the source and the drain as long the signal is applied and the current has positive polarity. If current in the negative direction is required, for example, in a bi-directional converter or in rectifiers with inductive load, an antiparallel diode is placed across the drain and source, as shown in Fig. 9.4. The diode current, however, is not controlled by the gate.

Thyristors
Thyristors, also known as silicon-controlled rectifiers (SCRs), are controllable switches. Like MOSFETs, thyristors are three-terminal devices. Conduction begins when the thyristor is forward biased and when the signal $q(t) = 1$ is applied to its

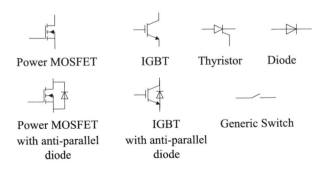

Fig. 9.4 Circuit symbols for different electronic switch components

gate. Conduction continues even after the close signal to the gate returns to $q(t) = 0$, so long as thyristor is forward biased. Thyristors therefore act like a latched switch. Conduction ends when the current drops below a threshold known as the "holding current," which we will assume to be zero. The thyristor current cannot be negative. In order to allow current to flow in the opposite direction, a diode can be placed in antiparallel. Thyristors are commonly used in phase-controlled rectifiers and some inverters.

IGBTs

IGBTs (insulated-gate bipolar transistor) are three-terminal devices designed for use in higher-power applications. IGBTs conduct when $q(t) = 1$. The voltage drop associated with an IGBT when conducting is approximately 2 V. For this reason, it is mostly used in applications with higher voltages, for example, above 96 V.

9.2.2 Distortion and Filtering

Many converters have nonlinear characteristics that result in nonconstant, non-sinusoidal current and voltage waveforms. These distorted waveforms can damage or cause malfunction in certain loads. Distortion can also cause generators to overheat, and some devices will emit an annoying humming sound or visible flicker. Obviously, distortion should be minimized.

AC circuit analysis is more complicated when the waveforms are not sinusoidal because phasor-based analysis cannot be directly used. Instead, we use harmonic analysis. The basic idea is to first decompose the distorted waveform into its harmonic components using the Fourier Series [10]. The harmonic components are sinusoids with different amplitudes and phase angles and whose frequencies are at integer multiples of the frequency of the distorted waveform. If the distorted waveform does not have a zero mean, then a constant term is also present in the series. Phasor-based circuit analysis is performed on each harmonic separately (or often only those harmonics that have large enough amplitude to be of consequence), and the results recombined to complete the analysis.

Any periodic zero-average waveform $f(t)$ can be decomposed into its harmonic components by applying the Fourier Series. The Fourier Series is

$$f(t) = \sum_{k=1}^{\infty} F_k \sin(k\omega_0 t + \delta_k) \tag{9.1}$$

where ω_0 is the fundamental frequency of $f(t)$, k is the harmonic number, $k\omega_0$ is the frequency of the kth harmonic, and F_k and δ_k are the magnitude and phase shift associated with the kth harmonic. Equation (9.1) applies equally to voltage and current. The magnitude of F_k indicates the "strength" or contribution of a particular harmonic to the total distorted waveform. Waveforms with little distortion will feature a fundamental component F_1 much greater than the other components.

The distortion of a waveform is measured by the Total Harmonic Distortion (THD)[7]:

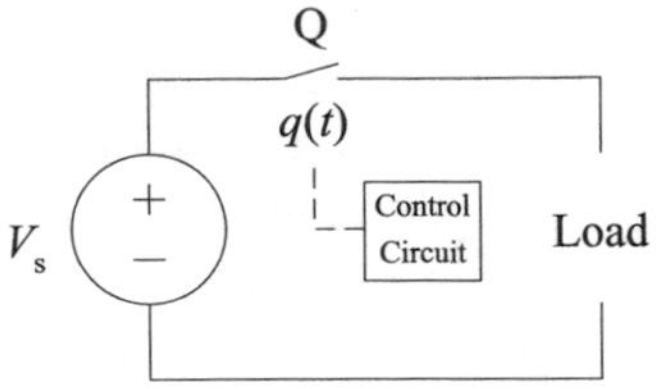

Fig. 9.5 Pulse width
modulation can be used to
control the average voltage to
the load

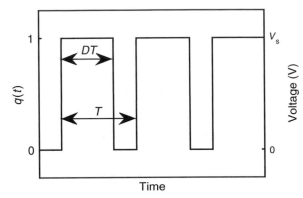

Fig. 9.6 Pulse train with
duty cycle D and period T

$$THD = \frac{\sqrt{\sum_{k=2}^{\infty} F_k^2}}{F_1}. \tag{9.2}$$

Keep in mind that distortion is a measure of the signals' deviation from the desired
pure sinusoid. The THD is commonly expressed as a percentage. The THD is zero
when there is no distortion. In general, we should select gen sets and inverters whose
output voltage has low THD. The THD is often reported by manufacturers. THD
values less than 5% can often be used, but this depends on the application.

A common way to reduce distortion is by using passive filters at the output or
input of device. Such filters attenuate higher harmonics, leaving the fundamental.

9.2.3   Pulse Width Modulation (PWM)

Pulse width modulation (PWM) is a technique in which the width of a series of
pulses is adjusted to obtain a desired average value. The signals that control the
switching elements in converters often use PWM.

As an example of how PWM can be used in converters, consider the circuit shown
in Fig. 9.5. This common circuit is known as a "chopper." The switch Q is controlled
by a signal $q(t)$. Let $q(t)$ be the periodic train pulses as shown in Fig. 9.6. The period
between pulses is some time T. Within each period, the duration when the signal
$q(t) = 1$ is DT, and the duration when it $q(t) = 0$ is $(1-D)T$. The variable D is the
duty cycle or *duty ratio*. The duty cycle is often expressed as a percentage, in which
case it varies between 0% (constant minimum) and 100% (constant maximum). In
PWM, the duty cycle is variable.

When the switch in Fig. 9.5 is closed ($q(t) = 1$), the voltage across the load is V_s; otherwise, it is zero. The average value of the voltage across the load $\bar{V}_{Load}$ is found through integration:

$$\bar{V}_{Load} = \frac{1}{T} \int_0^T v(t)\mathrm{d}t = \frac{1}{T} \int_0^{DT} V_s \mathrm{d}t = DV_s \tag{9.3}$$

We can therefore adjust the duty cycle to control the average value of the voltage applied to the load.

For a general pulse train whose maximum and minimum values are V_{max} and V_{min}, the average value is:

$$\bar{V}_{Load} = \frac{1}{T} \int_0^T v(t)\mathrm{d}t = \frac{1}{T} \left(\int_0^{DT} V_{max}\mathrm{d}t + \int_{DT}^T V_{min}\mathrm{d}t \right) = DV_{max} + (1-D)V_{min}. \tag{9.4}$$

We are mostly concerned with time-averaged rather than instantaneous values in DC switching circuits. Hereafter, variables in uppercase will denote constant or time-average values.

9.3 DC–DC Converters

DC–DC converters are conceptually similar to transformers in AC circuits in that they are used to change voltage levels. This capability is extremely useful. For example, a DC–DC converter can be used to boost the voltage from a nominal 96 V battery bank to 220 V for distribution, and another converter can be used at the consumer's house to reduce the voltage to a nominal 12 V—a voltage that many DC appliances are compatible with.

DC–DC circuits are used within many converters, including battery charge controllers, maximum power point trackers (MPPTs), and inverters. They are also used in solar home systems and solar lanterns, especially those that have USB ports for device charging.

A DC–DC converter has the following property that relates the input and output voltage and current with the power through the converter and its efficiency:

$$P = \eta_{DC–DC} V_{in} I_{in} = V_{out} I_{out}. \tag{9.5}$$

Equation (9.5) is nothing more than the conservation of power. Converters that increase the output voltage necessarily decrease the output current and vice versa. This can be done very efficiently. DC–DC converters tend to have efficiencies above 90%.

DC–DC converters incorporate high-frequency solid-state switching and energy storage components—inductors and capacitors—to change the output voltage. The difference between the input and the output voltage is controlled by adjusting the duty cycle of the switch. A controller is used to adjust the duty cycle to achieve a targeted output voltage.

Table 9.2 DC–DC converter
input/output relationships

Converter	Relationship
Boost	$V_{out} = \frac{1}{1-D} V_{in}$
Buck	$V_{out} = D V_{in}$
Buck–boost	$V_{out} = \frac{-D}{1-D} V_{in}$

Fig. 9.7 Boost converter
circuit

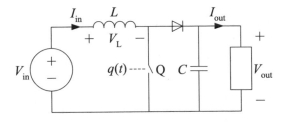

There are several types of DC–DC converters. The most popular are the boost, buck, and buck–boost. Boost converters increase the output voltage; buck converters decrease the output voltage; and buck–boost either increase or decrease the output voltage. Note that the polarity of the output voltage of a buck-boost converter is negative. The circuit topology of each converter is different, as is the effect of adjusting the duty cycle, as shown in Table 9.2.

To understand the basic principles of how DC–DC converters function, we will briefly discuss the boost converter. The basic circuit is provided in Fig. 9.7. When the switch signal is $q(t) = 1$, the switch Q conducts. The diode is not forward biased and so it does not conduct. The voltage across the inductor V_L is then equal to V_{in}. The current through the inductor consequentially increases according to

$$I_{in}(t) = \frac{1}{L} \int V_L dt. \tag{9.6}$$

As the current increases, so does the energy stored in the magnetic field of the inductor.

When $q(t) = 0$, Q opens. The inductor current cannot instantly change, and so the diode must be forward biased to provide a pathway for the current. The inductor current is therefore supplied to the load. Applying Kirchhoff's Voltage Law and ignoring the often-negligible voltage drop across the diode:

$$V_{in} = V_L + V_{out}. \tag{9.7}$$

The inductor current decreases as energy is transferred to the load. From (9.6), when the inductor current decreases, the voltage across it is negative. Therefore, from (9.7), the output voltage is *greater* than the input, and the basic function of the boost converter has been accomplished.

The input and output voltages are related by the duty cycle, which is derived as follows. The average voltage across the inductor during a switching period must be zero. If it were non-zero, then the inductor current would increase or decrease from one period to the next—this cannot continue indefinitely.

When Q is closed, the voltage across the inductor is V_{in}. When Q is open, from (9.6), the voltage is $V_L = V_{in} - V_{out}$. The switch is closed for the period D and open for $1 - D$. Therefore, setting the average voltage to zero shows that

$$DV_{in} + (1 - D)(V_{in} - V_{out}) = 0 \qquad (9.8)$$

where theoretically $0 \leq D \leq 1$. In practice D must be somewhat lower than 1 so that the output voltage cannot be arbitrarily large. Rearranging (9.8):

$$V_{out} = \frac{1}{1 - D} V_{in}. \qquad (9.9)$$

This equation, as well as those in Table 9.2, is valid so long as the inductor and switching frequency are such that the inductor current does not drop to zero.

A few notes. A capacitor is usually placed in parallel to the output so that the voltage is stable. The capacitor will charge when Q is open and discharge when Q is closed so that the load current is nearly constant. The inductor on the input side is usually sized so that energy is supplied by the input source, regardless of the state of the switch, and therefore makes good use of the power supply capability of the input voltage source.

9.4 Maximum Power Point Tracker

Maximum power point trackers (MPPTs) are used to increase the power produced by a PV module or array. Although controlled differently, MPPTs can also be used with WECS to provide the same function. However, most MPPTs are used with PV arrays, and so we focus our discussion on this application.

The basic premise of an MPPT is to decouple the load voltage from the PV array voltage, allowing the PV array to operate at $V^*(G)$—the voltage corresponding to its maximum power point. The basic scheme is shown in Fig. 9.8. The PV array is connected to a DC–DC converter; the output of the converter is connected directly to the load, battery, or charge controller. The type of DC–DC converter used depends on the maximum power point voltage of the array relative to the battery or load voltage. In any case, the duty cycle of the converter is controlled to track $V^*(G)$ as irradiance, temperature, and shading changes. Most MPPT uses digital control, but analog control is also possible. The algorithms used to track the maximum power point (MPP) are discussed in the next chapter.

As an example, consider a PV module connected to a battery through a MPPT, as shown in Fig. 9.9. The $I-V$ curve of the module is shown in Fig. 9.10. The MPP is at "A." Without an MPPT, the PV module voltage is equal to the battery terminal voltage V_B since they would be directly connected to each other. The operating point of the PV module is therefore point "B." The voltage is lower than $V^*(G)$, and the power corresponding to point B is less than A.

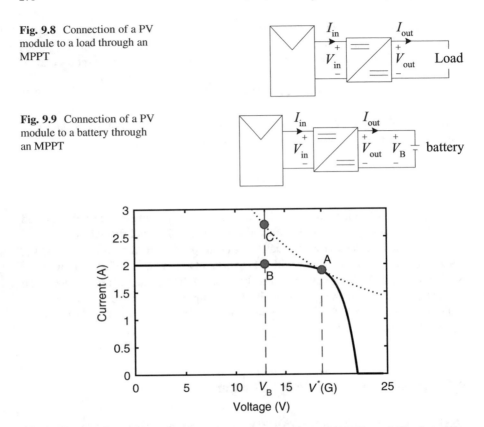

Fig. 9.8 Connection of a PV module to a load through an MPPT

Fig. 9.9 Connection of a PV module to a battery through an MPPT

Fig. 9.10 Operating points of the PV module and battery when an MPPT is used. The dashed line is the equi-power curve of the maximum power point. The product of voltage and power at each point along this curve is equal to $P^*(G)$

When an MPPT is used, the module's voltage is increased so that it operates at $V^*(G)$ (point "A") and therefore produces the maximum power $P^*(G)$. The battery's operating voltage remains at V_B. However, assuming the MPPT is lossless, the power to the battery must also be $P^*(G)$. The current corresponding to this operating point is $P^*(G)/V_B$, so that the operating point of the battery is point "C." This point lies along the same equi-power curve as the maximum power point. The MPPT is operating as a buck converter since the module voltage is greater than the battery voltage.

MPPTs can increase the total energy production by about 10 to 15%. This increase must be weighed against the cost of the MPPT unit. MPPTs are almost always used in larger-capacity off-grid systems, for example, above 2 kW. MPPT units are now commonly integrated with solar battery chargers. Manufacturers tend to brand units without MPPT functionality as "PWM" charge controllers; but this is confusing, as MPPT charge controllers also use PWM to control their internal DC–DC converter.

Example 9.1 An off-grid house has an improvised system consisting of a PV module that is directly connected to a battery. The *I*–*V* curve of a PV module under the present irradiance and temperature conditions is shown in Fig. 9.10. The PV module is used to charge a battery whose terminal voltage is 13 V. Estimate the power produced by the PV module. Next, consider the scenario in which the house has an MPPT (buck converter) that is connected between the module and the battery. The voltage and current corresponding to the maximum power point are 18.7 V and 1.89 A. Compute the duty cycle so that the PV module operates at its maximum power point and the corresponding power and current into the battery.

Solution By inspection of Fig. 9.10, the current when the PV module voltage is 13 V is approximately 2 A. Therefore the PV module produces $13 \times 2 = 26$ W when connected directly to the battery.

The duty cycle of the buck converter is found by arranging the corresponding equation in Table 9.2:

$$D = \frac{V_{\text{out}}}{V_{\text{in}}} = \frac{13}{18.7} = 0.695.$$

The corresponding power and current are:

$$P = V^* I^* = 18.7 \times 1.89 = 35.34 \text{ W}$$

$$I_{\text{out}} = \frac{P}{V_{\text{out}}} = \frac{P}{V_{\text{B}}} = \frac{35.34}{13.0} = 2.72 \text{ A}$$

The MPPT increased the PV power by 35% (35.34 W versus 26 W).

9.5 Solar Battery Charger

Solar battery chargers, also known as solar charge controllers, regulate the charging of a battery from a PV array or module. They are sometimes integrated with an MPPT, but we will not consider this scenario for the sake of clarity. They are used to prevent the battery from being damaged or degraded during charging. They are used in mini-grids and most solar home systems and solar lanterns. There are three general types of solar charge controllers: shunt, series, and pulse width modulation (PWM) [12].

Fig. 9.11 Shunt-type battery
charger

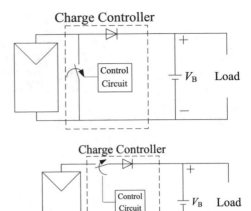

Fig. 9.12 Series-type battery
charger

9.5.1   Shunt-Type Solar Battery Charger

Shunt-type charge controllers are the simplest type and the first that were developed. A schematic is shown in Fig. 9.11. The controller functions by operating a solid-state switch or relay that is in parallel (shunt) with the PV module. When closed, the switch short-circuits the module and prevents the battery from being charged. The diode between the switch and the battery prevents the battery from discharging back through the shunt. The simplicity and low cost of the design make shunt controllers viable for low-end off-grid systems. Shunt controllers can also be used with other energy conversion technologies such as a wind energy conversion system (WECS) that can be shorted without damage to the source.

9.5.2   Series-Type Solar Battery Charger

A series-type solar battery charger is a chopper whose input is a PV module, as shown in Fig. 9.12. When the switch is opened, the energy source is open-circuited, isolating it from the battery. A series-type charge controller can also be used with energy conversion technologies that can be safely open-circuited: fossil fuel or biomass gen sets and micro hydro power systems with speed control.

9.5.3   Series-Type Solar Battery Charger with PWM

The series-type solar battery charger can be improved by using pulse width modulation to control the solid-state switch. The duty cycle is adjusted to vary the average current to the battery. The duty cycle is increased as greater current is desired. When the battery voltage reaches a predefined upper limit, the duty cycle is set to zero, thus open-circuiting the PV module. The greater control of the current in PWM type controllers allows more sophisticated charging algorithms to be used such as the three-stage algorithm discussed in Sect. 10.2.1.

Example 9.2 Consider a series-type charger with PWM. The PV array is modeled as a constant voltage source V_{PV} whose value is 16 V. The battery state-of-charge (SoC) voltage is $V_{SoC} = 12.8$ V and its resistance $R_B = 0.1\ \Omega$. Determine the duty cycle D needed for the battery current to have an average value of 1.5 A.

Solution The current to the battery at any time t is:

$$i_B(t) = \left(\frac{V_{PV} - V_{SOC}}{R_B} \right) q(t).$$

Applying (9.4) to the battery current pulse train:

$$I_B = D i_{max}(t) + (1 - D) i_{min}(t) = \left(\frac{V_{PV} - V_{SOC}}{R_B} \right) D + (1 - D) 0$$

$$= \left(\frac{V_{PV} - V_{SOC}}{R_B} \right) D.$$

Solving for D yields

$$D = I_B \left(\frac{R_B}{V_{PV} - V_{SOC}} \right) = 1.5 \left(\frac{0.1}{16 - 12.8} \right) = 0.0469.$$

9.6 AC Battery Charger

Batteries can also be charged from AC sources such as gen sets, WECS, MHP systems, and, in general, an AC bus. The basic principles of charging the large stationary batteries in mini-grids also apply to the smaller-capacity batteries found in solar home systems and other electronic devices such as mobile phones and laptop computers. Within the charging circuit of all of these devices is a rectifier, which converts the input AC voltage to the DC voltage used to charge a battery.

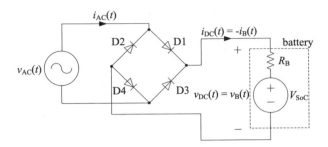

Fig. 9.13 Single-phase full-bridge rectifier connected to a battery

9.6.1   Single-Phase Full-Bridge Rectifier

A battery charger whose input is single-phase AC uses a single-phase full-bridge rectifier to convert AC to DC. The circuit consists of four diodes as shown in Fig. 9.13.

The diodes conduct when they are forward biased. This occurs in pairs, so that D1 and D4 are conducting together while D2 and D3 are blocking and vice versa. The state of the diodes changes automatically, without the need for external control. For this reason, diode-based rectifiers are known as "uncontrolled" rectifiers.

For simplicity, the input is modeled as an AC voltage source. We will assume that the rectifier output is connected to a battery. We will sometimes refer to the input side of the rectifier as the AC side and the output as the DC side. It should be noted that the output side is not a constant voltage, but it does have a positive average value. The voltage at the AC side and DC side is denoted as $v_{AC}(t)$ and $v_{DC}(t)$, respectively. Since the DC side of the rectifier is connected to the battery

$$v_{DC}(t) = v_B(t) = R_B i_{DC}(t) + V_{SoC} \tag{9.10}$$

where $i_{DC}(t)$ is the current output by the rectifier, which is also the negative of the battery current. (Recall from Chap. 8 that the battery current is defined as positive when discharging.)

Let the voltage on the AC side of the rectifier have amplitude v_{AC}^{max} and operate at a frequency of ω so that

$$v_{AC}(t) = v_{AC}^{max} \sin(\omega t). \tag{9.11}$$

When the diodes are not conducting, $i_{DC}(t) = 0$ and from (9.10) $V_{DC}(t) = V_{SoC}$. A pair of diodes will conduct when they are forward biased. This occurs when $|v_{AC}(t) - 2V_D| > V_{SoC}$ where V_D is the voltage drop associated with a single diode. The phase angle at which conduction begins varies with the battery voltage. The conduction angle θ_{on} is

$$\theta_{on} = \sin^{-1}\left(\frac{V_{SoC} + 2V_D}{v_{AC}^{max}}\right). \tag{9.12}$$

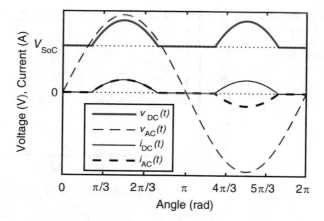

Fig. 9.14 Waveforms of a single-phase full-bridge rectifier connected to a battery

Due to symmetry, the angle at which conduction ends for each half cycle is

$$\theta_{\text{off}} = \pi - \theta_{\text{on}}. \tag{9.13}$$

The voltage on the DC side is therefore

$$v_{\text{DC}}(t) = \begin{cases} v_{\text{AC}}^{\max} \sin(\omega t) - 2V_{\text{D}} : \theta_{\text{on}} < \theta < \theta_{\text{off}} \\ V_{\text{SoC}} \qquad\qquad\qquad : \text{else} \end{cases} \tag{9.14}$$

The current to the battery is found by rearranging (9.10)

$$i_{\text{DC}}(t) = \frac{v_{\text{DC}}(t) - V_{\text{SoC}}}{R_{\text{B}}}. \tag{9.15}$$

The resulting waveforms are shown in Fig. 9.14. The reader should confirm that the waveforms are as expected. The battery voltage pulsates at twice the frequency of the AC source. When the diodes are not conducting, the battery voltage is V_{SoC}. The current from the AC source is only non-zero when the diodes are conducting. The AC current is not sinusoidal and so there is some level of harmonic distortion. We note that the voltage and current from the AC source peak at the same time (angle), and both are symmetric around the peak for each half of the waveform. They are in phase.

The instantaneous power to the battery $p(t)$ is the product of the instantaneous voltage and current on the DC side of the rectifier:

$$p(t) = v_{\text{DC}}(t)i_{\text{DC}}(t) \tag{9.16}$$

The real power P is defined as the average of the instantaneous power. The power delivered to the battery is found by substituting (9.14) and (9.15) into (9.16) and averaging over one half cycle:

$$P_{\text{B}} = \frac{1}{\pi} \int_{\theta_{\text{on}}}^{\theta_{\text{off}}} \left(v_{\text{AC}}^{\max} \sin(\theta) - 2V_{\text{D}}\right) \frac{v_{\text{AC}}^{\max} \sin(\theta) - 2V_{\text{D}} - V_{\text{SoC}}}{R_{\text{B}}} d\theta. \tag{9.17}$$

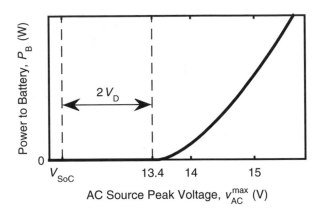

Fig. 9.15 The power to the battery increases nonlinearly with the voltage of the AC source. The battery begins charging once the AC voltage is sufficient to bias the diodes in the rectifier

From this we see that the power delivered to the battery is dependent on the amplitude of the voltage on the AC side of the rectifier, the battery SoC (which affects V_{SOC}, θ_{on} and θ_{off}), and the battery resistance. Figure 9.15 shows an example of how the power to the battery changes with the voltage of the AC source. This could, for example, be a WECS. Notice that no current is delivered to the battery until the AC-side voltage exceeds the battery voltage by $2V_D$. If the battery terminal voltage is much greater than the diode voltage drops, they can be ignored, and (9.17) is reduced to

$$P_B = \frac{v_{AC}^{max}}{\pi R_B} \left(v_{AC}^{max} \int_{\theta_{on}}^{\theta_{off}} \sin^2(\theta)\, d\theta - V_{SoC} \int_{\theta_{on}}^{\theta_{off}} \sin\theta\, d\theta \right) \qquad (9.18)$$

where θ_{on} and θ_{off} are calculated from (9.12) and (9.13) with $V_D = 0$.

9.6.2 Three-Phase Rectifier

The basic principles of a single-phase rectifier apply to a three-phase rectifier. The principle circuit is shown in Fig. 9.16. There are now six diodes instead of two. Hereafter, we assume that the AC voltage is such that there is always a conduction path between the AC side and DC side, and that the voltage drop associated with the diodes is negligible. In other words, the battery is always being charged. In this case, the voltage on the DC side of the rectifier is as shown in Fig. 9.17. The peak of $v_{DC}(t)$ is the peak of line–line voltage on the AC-side $V_{\ell\ell}^{max}$. The output voltage stays close to this average value, but pulsates with a period of $\pi/3$ radians (60°). By symmetry, we can analyze a single 60° segment and extend the results to the entire period. For convenience, we will select a period centered around the peak of one of the humps:

$$v_{DC}(t) = V_{\ell\ell}^{max} \cos(\theta) : -\pi/6 \leq \theta \leq \pi/6 \qquad (9.19)$$

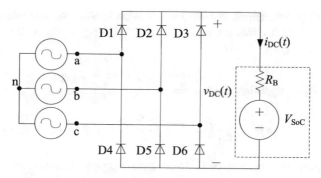

Fig. 9.16 Three-phase full-bridge rectifier

Fig. 9.17 Voltage output waveform of an ideal three-phase rectifier. The dashed lines are the line–line voltages of the AC source

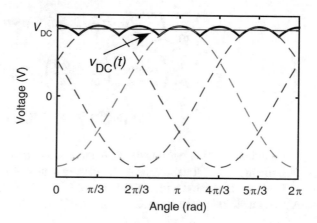

The DC-side current during this period is:

$$i_{DC}(t) = \frac{V_{DC}(t) - V_{SoC}}{R_B} = \frac{V_{\ell\ell}^{max} \cos(\theta) - V_{SoC}}{R_B}. \qquad (9.20)$$

The *average* value of the voltage and current on the DC side of the rectifier is

$$V_{DC} = \frac{3}{\pi} \int_{-\pi/6}^{\pi/6} V_{\ell\ell}^{max} \cos(\theta) d\theta = \frac{3}{\pi} V_{\ell\ell}^{max} \approx 0.955 V_{\ell\ell}^{max} \qquad (9.21)$$

$$I_{DC} = \frac{V_{DC} - V_{SoC}}{R_B}. \qquad (9.22)$$

In other words, the average voltage on the DC side of the rectifier is approximately 95.5% of the peak line–line voltage on the AC side. We cannot simply multiply the DC-side average voltage and current to find the power to the battery because the voltage and current are not constant. Instead, we must take the average of the instantaneous power, which repeats with a period $\pi/3$. The instantaneous power into the battery during this period is found by multiplying the voltage (9.19) and current (9.20):

$$p(t) = v_{DC}(t)i_{DC}(t) = \frac{V_{\ell\ell}^{max}\cos(\theta)\left(V_{\ell\ell}^{max}\cos(\theta) - V_{SoC}\right)}{R_B} \tag{9.23}$$

$$p(t) = \frac{1}{R_B}\left(\left(V_{\ell\ell}^{max}\right)^2\cos^2(\theta) - V_{\ell\ell}^{max}V_{SoC}\cos(\theta)\right) \tag{9.24}$$

The real power into the battery is found by averaging $p(t)$ over the period from $-\pi/6$ to $\pi/6$:

$$P_B = \frac{3}{\pi}\frac{1}{R_B}\int_{-\pi/6}^{\pi/6}\left(V_{\ell\ell}^{max}\right)^2\cos^2(\theta) - V_{\ell\ell}^{max}V_{SoC}\cos(\theta)d\theta \tag{9.25}$$

$$= \frac{3}{\pi}\frac{1}{R_B}\left(\left(V_{\ell\ell}^{max}\right)^2\left(\frac{\pi}{6} + \frac{\sqrt{3}}{4}\right) - V_{\ell\ell}^{max}V_{SoC}\right) \tag{9.26}$$

$$= \frac{3}{\pi}\frac{1}{R_B}\left(0.957\left(V_{\ell\ell}^{max}\right)^2 - V_{\ell\ell}^{max}V_{SoC}\right). \tag{9.27}$$

Equation (9.27) shows that the power to the battery increases as the voltage on the AC side increases. It decreases as the battery state-of-charge increases. The current and voltage associated with a single phase of the AC source are shown in Fig. 9.18. As with a single-phase rectifier, the current is distorted, but it is in phase with the voltage.

The assumption that the voltage on the AC side is unchanging in steady state is not always valid. It is a reasonable approximation for systems where the AC bus voltage is formed and controlled by an inverter or synchronous generator with an automatic voltage regulator. However, in some mini-grids this will not be the case.

Fig. 9.18 Phase voltage (line-to-neutral) and current from a single phase of a three-phase AC source charging a battery through a rectifier

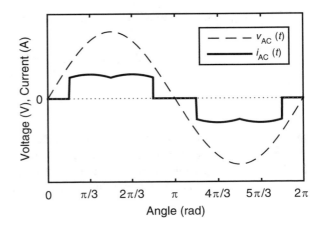

Fig. 9.19 Single-phase
phasor-domain model of a
three-phase AC generator
charging a battery through a
three-phase rectifier

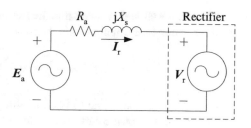

An example is a mini-grid powered only by WECS. In this case, terminal voltage
of the WECS and hence the voltage on the AC side of the rectifier will vary with
current due to the voltage drop associated with the winding impedance. It will also
vary with rotational speed under changing wind conditions.

Calculating the power to the battery under these conditions requires reconciling
the AC (generator side) and DC (battery side) circuits. Our approach is to model the
DC side as an equivalent AC load and use phasors to analyze the circuit [11]. This
is useful in understanding how the variables affect the complete system. A more
accurate, but less insightful, analysis can be obtained using a circuit model-based
simulation.

We begin by making the assumption that the voltage drop across the diodes is
negligible compared to the amplitude of the AC input. We next develop a model
that is useful in analyzing the AC side of the circuit. The model, which represents
one of the three phases of a synchronous AC generator connected to the rectifier,
is shown in Fig. 9.19. The generator and its internal impedance are on the left side
of the circuit. The rectifier and battery cannot be represented in the phasor domain.
However, we can approximate them as a voltage source V_r if we are careful in
understanding the limitations and conditions of doing so. The model is only valid
when the battery is continuously being charged; therefore, the current I_r must be
positive in the direction shown. We noted that in Fig. 9.18, the phase voltage and
current are in phase with each other. Therefore, the power supplied to the battery
must only have a real component. This means that I_r and V_r must be in phase. For
convenience, we set the phase angle of both to zero (reference) so that

$$V_r = |V_r| \angle 0° \tag{9.28}$$

$$I_r = |I_r| \angle 0°. \tag{9.29}$$

Lastly, since the rectifier is modeled as a lossless component, the power into the AC
side must equal the power into the battery:

$$P_B = 3\text{Re}\left\{V_r I_r^*\right\} = 3|V_r||I_r|. \tag{9.30}$$

The factor of three is needed because the model pertains to a single phase of the
three-phase source.

What value should $|V_r|$ be? It is the line-to-neutral voltage on the AC side of the rectifier. It can be computed from the maximum line–line voltage:

$$|V_r| = \frac{V_\phi^{\max}}{\sqrt{2}} = \frac{V_{\ell\ell}^{\max}}{\sqrt{2}\sqrt{3}} = \frac{V_{\ell\ell}^{\max}}{\sqrt{6}} \tag{9.31}$$

where $V_\phi^{\max}$ is the peak line-to-neutral voltage.

The magnitude of the current I_r is found by rearranging (9.30) using (9.27) and using (9.31) to replace $V_{\ell\ell}^{\max}$ with $|V_r|$:

$$|I_r| = \frac{P_B}{3|V_r|} = \frac{6}{\pi}\frac{1}{R_B}\left(0.957|V_r| - \frac{1}{\sqrt{6}}V_{SoC}\right) \tag{9.32}$$

Simplifying further:

$$|I_r| = \frac{1.8270}{R_B}(|V_r| - 0.4268 V_{SoC}) \tag{9.33}$$

Assuming the generator resistance to be negligible ($R_a = 0$), and noting that I_r and V_r have an angle of zero degrees, we have:

$$E = jX_s I_r + V_r = jX_s|I_r| + |V_r| \tag{9.34}$$

Equation (9.33) and (9.34) are sufficient to analyze the circuit, as shown in the following example.

Example 9.3 A three-phase WECS is used to charge a battery through a rectifier. The per-phase RMS induced voltage at a certain rotational speed is 14 V at 10 Hz. The per-phase synchronous reactance at this frequency is 0.19 Ω. The battery internal SoC voltage is 12.3 V, and the corresponding battery resistance is 0.25 Ω. Determine the power to the battery.

Solution To determine the power to the battery, we must first compute I_r and V_r. Only the magnitude is needed as the angle of each of these phasors is set to zero degrees. Rewriting (9.34)

$$|E|\angle\delta = jX_s|I_r| + |V_r|.$$

Next, splitting this into real and imaginary parts using Euler's Identity:

$$|E|\cos\delta = 14\cos\delta = |V_r| \tag{9.35}$$

$$|E|\sin\delta = 14\sin\delta = 0.19|I_r| \tag{9.36}$$

(continued)

The current $|I_r|$ is related to the voltage $|V_r|$ from (9.33)

$$|I_r| = \frac{1.8270}{0.25}(|V_r| - 0.4268V_{\text{SOC}}) \tag{9.37}$$

$$= 7.308\,(|V_r| - 5.250) \tag{9.38}$$

$$= 7.308|V_r| - 38.364. \tag{9.39}$$

Using

$$\sin^2\delta + \cos^2\delta = 1$$

on (9.35) and (9.36) and substituting for $|I_r|$ using (9.37) yield

$$0.19^2\frac{(7.308|V_r| - 38.364)^2}{14^2} + \frac{|V_r|^2}{14^2} = 1$$

This can be solved through the quadratic equation to yield $|V_r| = 11.25\text{V}$. Using (9.33), the current is $|I_r| = 43.85$ A. The power to the battery is found from (9.30):

$$P_B = 3|V_r||I_r| = 1480.1 \text{ W}$$

Properly understanding I_r is not completely straightforward. Although I_r is modeled as a phasor in the circuit in Fig. 9.20, its waveform is not sinusoidal, as shown in Fig. 9.18. However, the calculated value of I_r is close, within 5% of the RMS value of the single-phase current.

The relationship between the generator speed and power delivered to the battery is as follows. The battery does not charge until the voltage is sufficient to bias the diodes. As speed increases, the induced voltage in the generator E_a increases in proportion. Once the diodes are biased, the power to the battery increases approximately linearly. This continues over a range of speeds and induced voltages. However, as the speed further increases, the power levels off. This is evident in Fig. 9.20. The nonlinearity is attributed to the inductive reactance, which increases with speed. The voltage drop across jX_s then increases, reducing the terminal voltage of the generator.

9.6.3 Charging Circuit

A rectifier is a necessary component in an AC battery charger, but it is usually only part of the total circuit. Most AC battery chargers use a series of stages to

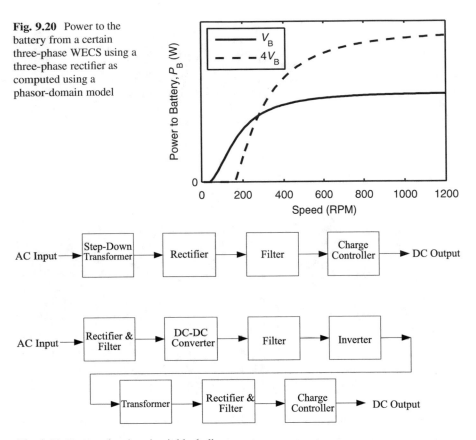

Fig. 9.20 Power to the battery from a certain three-phase WECS using a three-phase rectifier as computed using a phasor-domain model

Fig. 9.21 Battery charging circuit block diagram

convert the input AC to a stable DC voltage. Two such approaches are shown in Fig. 9.21 [4, 13, 15]. The different stages are usually combined into a single unit. The approaches assume the input AC has a higher peak voltage than the DC voltage required to charge the battery, as is often the case.

In the first approach, a transformer steps down the AC voltage to slightly above the maximum voltage the battery should be charged at. This lower-voltage AC is rectified and filtered so that it is a stable DC. The DC is input into the charge controller stage which is often a chopper circuit.

Another approach to producing DC is used in switch-mode power supplies. Using this approach the input voltage undergoes several conversions. The first stage is rectification and filtering to produce a stable DC voltage. The voltage is likely higher than needed. The next stages are a buck converter and filter, which reduce the voltage to the required magnitude. An inverter is then used to convert the DC to high-frequency, low-amplitude AC. The details of inverter circuits will be provided later. The high-frequency AC then is input into a transformer. Although this may further adjust the voltage magnitude, its primary purpose is to provide isolation between the input and output. An output rectifier and filter are used once again to produce stable DC voltage. This is fed into the charge controller stage.

9.7 Automatic Voltage Regulator

Automatic voltage regulators (AVR) perform the important function of holding the output voltage of a generator steady and at the desired value. Voltage control is needed in any generator forming the AC bus voltage in an AC-coupled system. Most synchronous generators, even those in small-capacity portable gen sets, have an AVR. Without an AVR, the terminal voltage will vary with the load and power factor, as discussed in Sect. 5.2.3. AVRs are also used to synchronize and share the reactive power load among generators in systems with multiple gen sets, as will be discussed in the next chapter.

AVRs function by controlling the generator's excitation system, adjusting the field current to the rotor of a synchronous generator. A control system continually monitors the output voltage and compares it against a target value, increasing or decreasing the field current accordingly. This can be done either with an analog circuit or with a digital controller.

Recall that a field winding requires DC current. One way of supplying the DC current is to rectify the AC output of generator. However, the current output by a diode-based rectifier is not controllable. A simple chopper circuit is not used due to the large inductance of field winding, which would result in a large voltage each time the switch was opened. Instead, a buck converter can be placed at the output of the rectifier.

Another option is to use a phase-controlled rectifier. The basic circuit for a phase-controlled rectifier is shown in Fig. 9.22. The topology is the same as for the 3-phase rectifier in Fig. 9.16, except that thyristors have replaced the diodes. Control signals are sent to the thyristors to control the angle that conduction begins. More specifically, the conduction angle is delayed by some angle α, known as the "firing angle." The resulting waveforms are shown in Fig. 9.23 for a firing angle of 30°. For simplicity, the load is assumed to be purely resistive. Delaying the conduction angle reduces the average voltage appearing on the DC side of the inverter. The average DC-side voltage can be derived from (9.21), accounting for the delay introduced by the firing angle:

$$V_{DC} = \frac{3V_{\ell\ell}^{max}}{\pi} \left(\sin(\pi/6 + \alpha) - \sin(-\pi/6 + \alpha) \right) \tag{9.40}$$

$$V_{DC} \frac{\pi}{3V_{\ell\ell}^{max}} = \sin(\pi/6)\cos(\alpha) + \cos(\pi/6)\sin(\alpha)$$

$$- \sin(-\pi/6)\cos(\alpha) - \cos(-\pi/6)\sin(\alpha)$$

$$\tag{9.41}$$

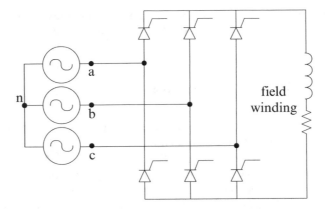

Fig. 9.22 Circuit of a three-phase phase-controlled rectifier connected to the field winding of a synchronous generator

Fig. 9.23 Voltage output waveform of a phase-controlled rectifier

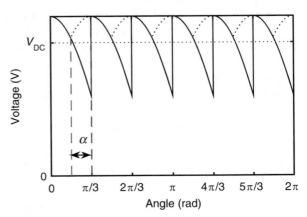

$$V_{DC} = \frac{3V_{\ell\ell}^{max}}{\pi} \left(0.5\cos(\alpha) + \frac{\sqrt{3}}{2}\sin(\alpha) + 0.5\cos(\alpha) - \frac{\sqrt{3}}{2}\sin(\alpha) \right) \quad (9.42)$$

$$V_{DC} \approx 0.955 V_{\ell\ell}^{max} \cos(\alpha) \quad (9.43)$$

We see that (9.43) reconciles with (9.21) when the firing angle is zero. The firing angle should not exceed 90°as the voltage would become negative. The average current to the field winding is simply the average voltage divided by the field winding resistance.

Figure 9.24 also shows the relationship between firing angle and power to a resistive load, which decreases as the firing angle increases. The values are in reference to $\alpha = 0°$.

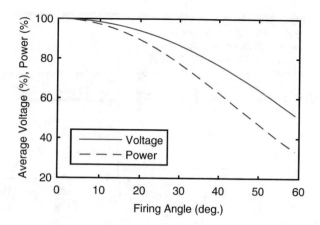

Fig. 9.24 Voltage and power output of a phase-controlled rectifier as a function of the firing angle

9.8 Electronic Load Controller

Recall from Sect. 6.3.7.3 that electronic load controllers (ELCs) are used in MHP systems as a method of demand-side frequency control on the AC bus. They allow the generator to operate at a constant speed [8, 14]. More generally, electronic load control is a method of balancing the net torque on the turbine shaft by adjusting the electromagnetic torque. ELCs are sometimes referred to as "electric load governors."

ELC's are preferred over mechanical speed control systems such as spear valves and deflectors because:

- they are able to quickly respond to rapid changes in load;
- they are less complex;
- the dissipated power can be put to productive use, for example, heating water;
- they require little maintenance;
- they can be less expensive.

An ELC maintains constant torque by adjusting the real power to a diversion load (also referred to as a "ballast load") so that the real power supplied by the generator is constant. We will consider the ELC and ballast load as a collective unit. Keep in mind that when we refer to the "ELC power" the power is actually being consumed by the ballast load. The ballast load itself is simply a resistor with a high power rating.

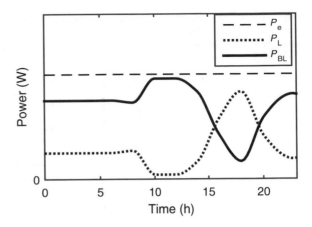

Fig. 9.25 Power consumed by the ELC P_{BL} is adjusted to maintain constant generator power P_e as the load P_L changes

The operation of an ELC is illustrated in Fig. 9.25. The real power from the generator P_e is the sum of the power to the load P_L and, to the ELC (ballast load), P_{BL}

$$P_e = P_L + P_{BL}. \tag{9.44}$$

To keep the generator power constant, the power to the ELC must equally balance any change in load. As the load increases, the ELC power decreases and vice versa. A frequency-based control scheme is usually used by the ELC. Any increase in electrical frequency above the targeted 50 Hz or 60 Hz leads to a decrease in P_{BL} and vice versa. The ballast load is typically sized to match the generator's power rating so that even if P_L drops to zero, constant speed can be maintained.

Electronic load control can be accomplished in several ways, including binary-weighted resistor network, phase-angle-controlled rectification, and impedance control.

9.8.1   Binary-Weighted Resistor Network

In the binary-weighted approach, the ELC controls relays that connect and disconnect fixed-value parallel resistors to the generator to balance changes in user load. The resistors are "binary weighted" in that their values follow the pattern of 2^n, for example, R, $2R$, $4R$, $8R$, and so on. An example of a single phase of a binary-weighted resistor network is shown in Fig. 9.26. The switches can be closed in different combinations to achieve several discrete steps of equivalent resistance and power consumption. The resistors should be sized so that when all are connected, the power consumed equals the rated power from the generator. The binary-weighted method is conceptually simple and easy to implement, but the power to the diversion load cannot be finely controlled, and so the frequency is not tightly regulated.

Fig. 9.26 A single phase of
an ELC ballast load using
binary-weighted resistor
network

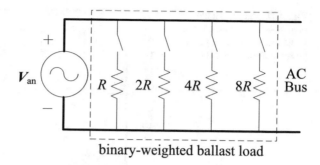

binary-weighted ballast load

9.8.2   Phase-Angle-Controlled Rectification: Revisited

An alternative approach to controlling the power to the ballast load is to use a
phase-angle-controlled rectifier. This allows for more precise control than a binary-
weighted resistor network. However, as the firing angle increases, the current
from the AC source begins to lag the voltage. When the current lags voltage,
reactive power is consumed. This might seem strange, since the load is purely
resistive. However, the power factor associated with any periodic but not necessarily
sinusoidal current $i(t)$ and voltage $v(t)$, both with period T, is

$$PF = \frac{P}{I_{RMS} V_{RMS}} = \frac{\frac{1}{T} \int_0^T i(t)v(t)\,dt}{\sqrt{\frac{1}{T}\int_0^T i^2(t)\,dt}\sqrt{\frac{1}{T}\int_0^T v^2(t)\,dt}} \qquad (9.45)$$

where I_{RMS} and V_{RMS} are the RMS values of the current and voltage. There is
nothing about this definition that requires resistive loads to have unity power factor.
The notion that resistive loads operate at unity power factor is only valid if the
circuit consists entirely of linear elements. This is a common assumption and one
that makes phasor-based circuit analysis possible. However, it is not valid when
power electronic converters with nonlinear characteristics such as a phase-controlled
rectifier are in the circuit.

As an example, the current and voltage of a single phase of a three-phase AC
source connected to a phase-angle-controlled rectifier are shown in Fig. 9.27. The
firing angle is 30°. The waveforms are not symmetric in phase as they were for
uncontrolled rectifiers (see Fig. 9.18). It can be shown that the power factor of a
three-phase phase-controlled rectifier is

$$PF = \frac{I_{RMS,1}}{I_{RMS}} \cos(\alpha) \qquad (9.46)$$

where $I_{RMS,1}$ is the RMS value of the first harmonic of the current from a phase of
the generator. From (9.46), we see that the power factor is affected by both the firing
angle and the distortion of the current waveform.

Fig. 9.27 Phase voltage (line-to-neutral) and current waveforms of a three-phase phase-controlled rectifier

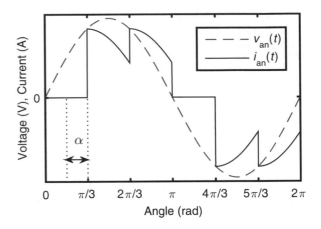

The firing angle therefore affects both the real power and reactive power consumed by the ELC. One cannot be controlled independent of the other. This can be a problem. Reactive power consumption increases the overall losses in a system and can cause the terminal voltage of the generators to be reduced, particularly with induction generators.

Example 9.4 A mini-grid consists of an AC-coupled MHP system with a phase-angle-controlled ELC. The mini-grid supplies power to a three-phase AC load. The peak line–line voltage is 560 V. The MHP system produces 18 kW of real power. At a certain moment, the power to the load and ELC are 15 kW and 3 kW, respectively. If the load decreases to 13 kW, determine the power that should be consumed by the ELC and the corresponding firing angle assuming the ballast load resistance is 15 Ω.

Solution The purpose of the ELC is to maintain the power output by the MHP generator at a constant value, in this case 18 kW. Therefore, if the load decreases by $15 - 13 = 2$ kW, then P_{BL} should increase to $3 + 2 = 5$ kW. The average voltage required to consume this power is

$$V_{DC} = \sqrt{R_{BL} P_{BL}} = \sqrt{15 \times 5000} = 273.9 \text{ V}.$$

The corresponding firing angle is computed by rearranging (9.43)

$$\bar{V}_{DC} \approx 0.9546 V_{ll}^{max} \cos(\alpha)$$

$$273.9 \approx 0.9546 \times 560 \times \cos(\alpha)$$

$$\alpha = \cos^{-1}(0.5121) = 1.033 \text{ rad} = 59.20°.$$

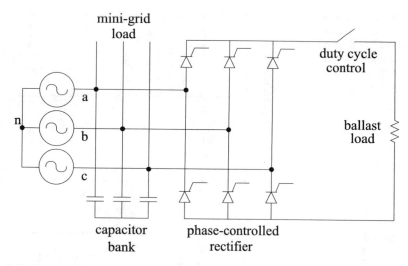

Fig. 9.28 Impedance controller circuit for electronic load control with capacitor bank

9.8.3 Impedance Controller

The phase-angle-controlled rectifier can be modified to allow greater control of the real and reactive power consumed by the ELC. The DC output of the phase-angle-controlled rectifier is connected to a chopper and then to the load. In parallel with the ballast load are capacitors, as shown in Fig. 9.28. The combined circuit is known as an ELC with impedance control or simply as an "impedance controller" [1, 2]. The capacitors can be omitted if the ability of the impedance controller to supply reactive power is not necessary. The impedance controller has two control variables: the firing angle and the duty cycle of the chopper. This allows the ELC to adjust the reactive power and real power consumed. Although the real and reactive power are not entirely independent of each other, better control is offered than from a phase-controlled rectifier. The capacitors allow the controller to supply reactive power as well as consume it.

An example waveform of the current of a single phase of a three-phase ELC with impedance control is shown in Fig. 9.29. The phase voltage is shown for reference. The reader should compare the current waveform with that in Fig. 9.27. Figure 9.30 shows the real and reactive power for different firing angles and duty cycles. The effect of the capacitors has been excluded so that the effects of the firing angle and duty cycle are highlighted.

The capacitors have a fixed value, and so the reactive power they supply is also fixed. The total reactive power consumed by the impedance controller Q_{IC} is:

$$Q_{IC} = Q_{BL}(\alpha, D) - Q_{cap} \tag{9.47}$$

Fig. 9.29 Phase voltage
(line-to-neutral) and current
waveforms of a three-phase
phase-controlled rectifier with
firing angle of 30° and a duty
cycle of 25%

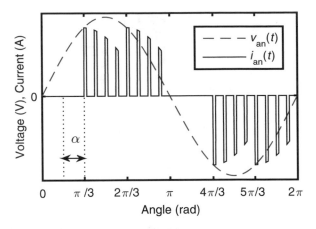

Fig. 9.30 Family of P–Q
curves of an impedance
controller without capacitors.
Increasing the duty cycle
increases the real power, and
increasing the firing angle
increases the reactive power
consumption. The duty cycle
and firing angle must be
coordinated to achieve the
desired operating point

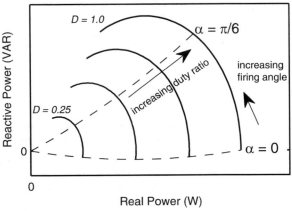

where $Q_{BL}(\alpha, D)$ is the reactive power consumed by the ballast load portion of the
impedance controller. The real power of the impedance controller is the same as
the ELC without fixed capacitance: $P_{IC} = P_{BL}$. The impedance controller reactive
power can be positive or negative. When it is negative, it is supplying reactive power.
Impedance controllers are often used when in MHP systems with induction rather
than synchronous generators.

Induction generators do not have a controllable excitation system, and so an AVR
cannot be used. The generator voltage therefore changes with the current it provides. It
varies with the load and the power factor of the load. This is problematic. An
impedance controller can be used to set the speed of the generator by adjusting
P_{ELC} and the voltage by controlling Q_{ELC}.

9.9 Inverters

Inverters perform the important function of converting DC voltage to AC voltage [5, 6]. They are used in DC-coupled systems to supply an AC load and in AC–DC systems to facilitate the flow of power from the DC bus to the AC bus. An example of inverters used in a mini-grid is shown in Fig. 9.31. Some inverters are capable of forming the AC bus voltage and synchronizing with other AC sources. Inverters are a critical component in many off-grid systems, including mini-grids and some solar home systems. Keep in mind though that inverters are not necessary in systems that either do not have a DC bus or only have DC loads.

9.9.1 Principle Circuit

The basic premise of an inverter is to use solid-state switching to alternate the polarity of the voltage applied to a load. A schematic of a simple inverter is shown in Fig. 9.32. The control circuitry is not shown. A DC voltage source is used to represent the battery, but this could also be a large charged capacitor. For inverters rated up to 5 kVA and 96 V, power MOSFETs are most often used as the switching elements. Above these ratings IGBTs are used.

A controller, not shown, is used to control the state of the switches. The switches are operated in pairs: Q1 and Q3 are always in the same state, as are Q2 and Q4. When Q1 and Q3 are conducting, Q2 and Q4 are not conducting. As the switching alternates, so does the polarity of the applied voltage and current. The resulting voltage waveform is shown in Fig. 9.33. The time each switch spends conducting and not conducting is the same so that the applied voltage waveform has an average of zero. The basic goal of achieving an alternating voltage output has been realized.

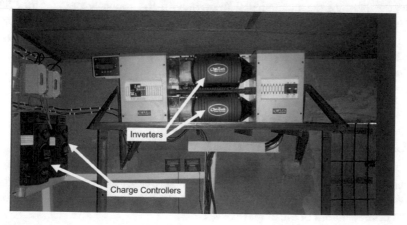

Fig. 9.31 A pair of inverters provide AC to a mini-grid in Nigeria (courtesy GVE Projects)

Fig. 9.32 Basic circuit of single-phase inverter

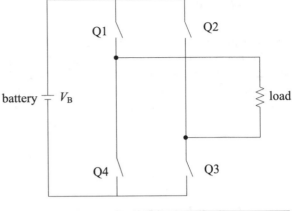

Fig. 9.33 Voltage output of a simple inverter circuit

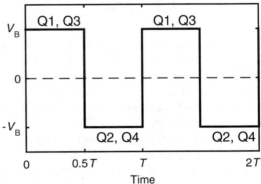

The waveform, although no longer DC, is also not sinusoidal. It is a square wave. Square waves have high harmonic distortion relative to the desired pure sinusoidal waveform. A low-pass filter can be placed at the output to reduce the distortion. However, because the fundamental frequency of the square wave is either 50 or 60 Hz, the filter capacitor must be large and therefore expensive. The output voltage varies between $\pm V_B$, which is typically lower than the 120 VAC or 230 VAC (RMS) that is desired. An output transformer can be used to increase the voltage to the desired level. Alternatively, a boost DC–DC converter can be used to increase the DC voltage input to the inverter.

9.9.2   Sinusoidal Pulse Width Modulation Switching

We can reduce the distortion of the output voltage by using pulse width modulation. However, rather than using a fixed duty cycle, the duty cycle varies over the course of the desired period of the output voltage.

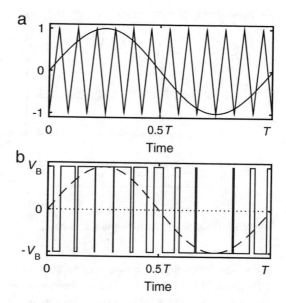

Fig. 9.34 (**a**) Carrier (triangular) and modulation (sinusoidal) signals used in sine wave pulse width modulation. The carrier frequency shown is much lower than in practice. (**b**) Output voltage form and its fundamental component (dashed)

The switching pattern is realized using a carrier-based PWM technique known as sinusoidal PWM (SPWM). In SPWM, the modulating signal is a sinusoid $f_m(t)$ with frequency ω_m and amplitude a_m. The carrier is a triangle-shaped waveform $f_\Delta(t)$ with amplitude a_Δ and frequency ω_Δ. The modulating frequency is much lower than the carrier $\omega_m \ll \omega_\Delta$. For example, the modular frequency is 50 or 60 Hz, but the carrier frequency is several kilohertz. The switches are controlled such that when $f_m(t) > f_\Delta(t)$, Q1 and Q3 are on and Q2 and Q4 are off; when $f_m(t) < f_\Delta(t)$, Q1 and Q3 are off and Q2 and Q4 are on. The resulting waveforms are shown in Fig. 9.34.

Although the output waveform in Fig. 9.34b (the variable duty cycle waveform) does not appear to be sinusoidal, it in fact has a large Fourier Series component at the fundamental frequency. This is shown as the dashed line in Fig. 9.34b. The distortion is largely due to harmonics near the carrier frequency. This higher frequency distortion can be filtered using a small, inexpensive capacitor so that the output is nearly sinusoidal.

SPWM offers a way to control the amplitude of the filtered output voltage. The amplitude is changed by adjusting the modulation index. The modulation index is defined as the ratio of the amplitude of the modulating signal to the carrier signal.

$$m_a = \frac{a_m}{a_\Delta}. \tag{9.48}$$

We can show that the output voltage at the fundamental frequency is:

$$V_{AC}^{max} = m_a V_{DC} \tag{9.49}$$

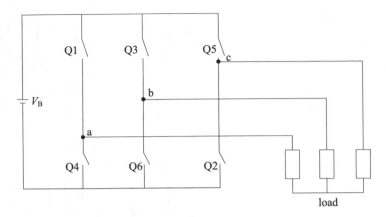

Fig. 9.35 Basic circuit of a three-phase inverter

for modulation indexes between zero and one. The ability to control the voltage magnitude is important as it allows the output voltage to be tightly regulated, even if the load or input voltage source changes. It also allows the reactive power to be shared between multiple synchronized inverters, which is discussed in Sect. 10.10.2. The voltage can be boosted using a transformer at the output or a boost converter at the inverter's input.

9.9.3   Three-Phase Inverters

Some mini-grid users, especially those with large motors or high-power loads, require three-phase power. It is technically possible for three single-phase inverters to be connected together to create a three-phase supply. However, not all inverters have this functionality. Instead, a three-phase inverter can be used.

Three-phase inverters rely on the same principles as a single-phase inverter with SPWM. The circuit however now contains six switches, as shown in Fig. 9.35. The output line–line voltage waveforms are shown in Fig. 9.36 for a single period T. The switches that are conducting for each sixth of a cycle are shown in Fig. 9.36a. At any moment, three of the MOSFETS are conducting. The line–line voltage waveforms are each shifted by $0.33T$ (120°), as in a three-phase supply. The waveforms approximate sine waves but are heavily distorted but can be filtered.

Some mini-grids might require the use of more than one inverter. The implementation is straightforward if each inverter supplies separate, isolated AC circuits. However, if the inverters are to be connected in parallel anywhere in the system, most likely the AC breaker box, then care must be taken to synchronize their outputs.

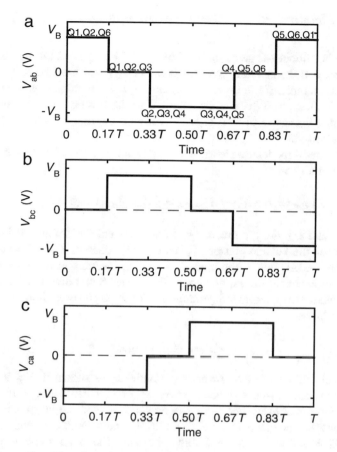

Fig. 9.36 Output line–line voltage for a three-phase inverter

9.9.4 Modified Sine Wave Inverters

In applications in which extreme affordability is a requirement, *modified sine wave* inverters can be considered. As the name implies, the output voltage waveform is not sinusoidal. Most often it is a square wave.

Modified sine wave inverters are less expensive and so are popular in improvised systems often found in rural areas. The distortion is high. The THD for a square wave is about 48%. Modified sine wave inverters should only be used to supply loads that can tolerate heavily distorted voltage, such as heaters, power tools, and incandescent bulbs. Modified sine wave inverters typically have poor voltage regulation—as the battery voltage or load varies, the output voltage changes in proportion. This means that the output voltage can change by more than 10% as the a battery discharges. As a general rule, "pure sine wave" inverters—those with low harmonic distortion—should be used.

9.9.5 Efficiency

In mini-grids without AC generators, all of the AC load passes through an inverter.
The efficiency of the inverter then affects the energy that must be supplied by the
system. For example, if the inverter is operating at 80% efficiency, then for every
1 kWh of load, 1.25 kWh must be supplied to the inverter. This means that the
battery bank, PV array, or other energy sources must be larger than if a more efficient
inverter were used.

If the inverter is modeled as lossless, then power is conserved and the input power
equals the output power.

$$P_{\text{inv,in}} = V_{\text{DC}} I_{\text{DC}} = P_{\text{inv,out}} = |V_{\text{inv}}||I_{\text{inv}}| \cos(\Phi) \tag{9.50}$$

where V_{DC} and I_{DC} are the inverter input voltage and current and Φ is the phase
angle between the voltage phasor V_{inv} and current phasors I_{inv}. The voltage V_{inv}
is applied to the AC bus or the load. In actuality, inverters have losses associated
with switching, copper losses, and controller losses. Some inverters have fans that
actuate to reduce the internal temperature but consume power in doing so. Including
the inverter losses:

$$P_{\text{inv,out}} = \eta_{\text{inv}} P_{\text{inv,in}}. \tag{9.51}$$

The efficiency curve of an inverter is nonlinear, as shown in Fig. 9.37. Under
no load, inverters consume standby power, and so the efficiency is low. Standby
power is consumed even when there is no load. As the output power increases,
the inverter efficiency increases and then slowly tapers. Peak efficiency is typically
between 92 % and 95%, but we should not assume the inverter always operates at
this efficiency.

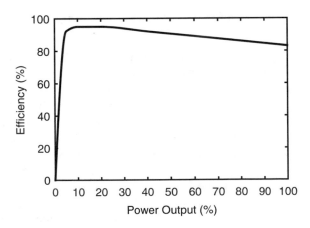

Fig. 9.37 Typical inverter
efficiency curve

The daytime load of a mini-grid that primarily serves households is low, which means the inverter tends to operate far below its peak capacity. An average efficiency over the course of the day could be about 70%. One strategy to improve efficiency is to disconnect the inverter when the load is zero or near zero. This of course reduces the availability of the electricity. Some inverters have sleep modes which automatically reduce the standby consumption when the load is low.

To account for the variation in efficiency, some manufacturers report the "European" efficiency, which is the weighted average efficiency of the inverter at different loading levels, X, $\eta_{inv}(X)$:

$$\eta_{inv} = 0.03\eta_{inv}(0.05) + 0.06\eta_{inv}(0.10) + 0.13\eta_{inv}(0.20)$$
$$+0.10\eta_{inv}(0.30) + 0.48\eta_{inv}(0.50) + 0.20\eta_{inv}(1.00) \quad (9.52)$$

where the loading level is $P_{inv,out}$ divided by the rated power of the inverter.

9.10 Solar Inverters

Solar inverters, also known as "PV inverters," are designed to couple a PV array directly to the AC bus, bypassing or eliminating the DC bus. Solar inverters are commonly used in applications where there is no need for energy storage or a DC bus. However, solar inverters are not able to form the AC bus voltage. Thus, they must be used with another source, for example, a gen set, in an AC-coupled architecture.

The basic premise of a solar inverter is shown in Fig. 9.38. Shown is a two-stage solar inverter. The first stage is a DC–DC converter. Maximum power point tracking is performed in this stage. The converter is typically a buck–boost converter.

A large capacitor is placed either at the input or the output of the DC–DC converter to stabilize the voltage. This conceptually replaces the voltage source (battery) on the DC side of the inverter as previously discussed. Although the two-stage inverter is conceptually simple, greater efficiencies can be achieved using a single-stage inverter [3]. These inverters combine the DC–DC and AC–DC conversion in a single stage.

Solar inverters can be arranged in four architectures as shown in Fig. 9.39 [9]. The centralized architecture uses just one inverter, whereas in the module architecture, there is one inverter per PV module. The primary advantage of using more inverters is that the MPPT tend to be more effective. The disadvantage is primarily the cost of using more inverters.

Fig. 9.38 Two-stage solar inverter

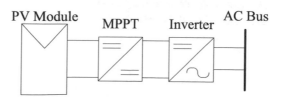

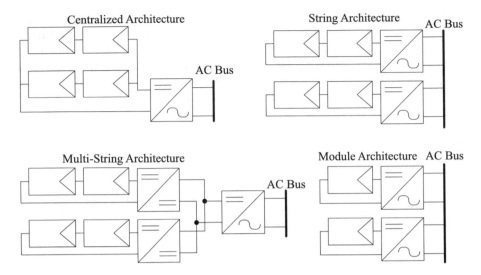

Fig. 9.39 Solar inverters can be arranged in different architectures

Centralized Architecture

In the centralized architecture, one or more PV strings are combined and input into a single solar inverter. This is the lowest cost-per-kilowatt architecture because there is a single inverter. However, there are several drawbacks. The MPPT is performed on the entire array. That is, all strings in the array operate at the same voltage. Most likely, the modules have slightly different irradiance, temperatures, and perhaps shading. The optimal point for the array is likely not the optimal point for each string. There is also a single point of failure. If the inverter fails, no load can be served. In addition, because the power rating is larger than in the other architectures, a larger capacitor must be used in the filter. Electrolytic capacitors are often used because they can offer high values of capacitance. However, they are often the first component to fail in an inverter. Their lifespan, perhaps 10–15 years, quickly reduces as temperature increases [3]. This can be of concern in off-grid systems where the inverter is not located in a temperature-controlled environment. Electrolytic capacitors are not just used in solar inverters, but in any inverter with a larger power rating.

Multi-string Architecture

Multi-string inverters use a single DC–AC stage, but each string is connected to its own DC–DC converter. This allows the MPP to be tracked on a per-string basis, likely improving the power production. There is no need for blocking diodes and so the associated cost and losses are avoided. The multi-string architecture is more expensive than the centralized architecture, and a single point of failure remains.

String Architecture

String solar inverters are similar to the centralized architecture, but each string is connected to a solar inverter. The costs are increased, but the there is no longer a single point of failure, and the power is maximized at the string level. Blocking diodes are also omitted.

Module Architecture

Module inverters (also known as "micro-inverters") are solar inverters that are connected to an individual PV module. DC–DC conversion, MPPT, and DC–AC are performed in a single unit. The units are typically compact and can be integrated into the module's junction box. The maximum power point is tracked at the module level, yielding the highest power output of any of the architectures. Because the PV modules are only connected at the AC bus, they can be added or removed (in case of failure) easily. The ratings are typically less than 300 W, which means that electrolytic capacitors can be avoided. This can substantially increase the lifespan of the inverter. Module inverters are more expensive, ranging from about US$0.40 to 0.50/W, but this is decreasing as the industry matures. Module inverters are almost always single phase. Most are not capable of forming the AC bus voltage.

9.11   Grid-Tied Inverters

Inverters can be classified as being stand-alone or grid-tied. We have discussed stand-alone inverters. The output of a stand-alone inverter is controlled so that it approximates an ideal voltage source, establishing the voltage magnitude and frequency of the AC bus. When an inverter is "grid-tied," it does not necessarily mean it is connected to the national grid. Rather, it means that it is capable of being coupled to an AC bus where another source—a gen set or another inverter—has formed the AC bus voltage. An example is shown in Fig. 9.40.

Grid-tied inverters (GTIs) operate on the same principles as stand-alone inverters. However, they must be controlled so that their output synchronizes with the frequency of the AC bus voltage. If they do not, large, damaging current could be exchanged between the inverter and the AC Bus. They are controlled so that they provide the desired current or power to the AC bus. Not all inverters are capable of synchronizing to the AC bus. Serious damage can result if an inverter capable only of stand-alone operation is connected to an AC bus whose voltage is established by another source.

Fig. 9.40 Grid-tied inverter connected to the AC bus through an inductance. Other sources and loads on the DC and AC bus are not shown

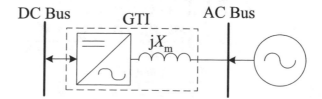

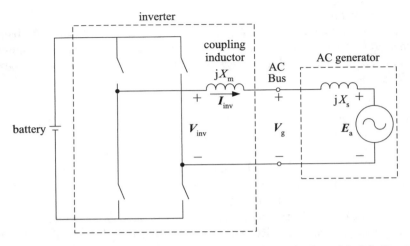

Fig. 9.41 Example of an off-grid system with grid-tied inverter; a simple model of the inverter is shown

Internal to a GTI is an output inductance. A more detailed view is shown in Fig. 9.41. The inductance allows the internal inverter voltage V_{inv} to be somewhat different than the voltage at the AC bus. This is conceptually similar to how the synchronous reactance X_s in a generator decouples the induced voltage from the terminal voltage. The output inductance allows the inverter to control the real and reactive power it supplies to the AC bus.

Here the gen set line-to-neutral output voltage is V_g, and the internal inverter line-to-neutral voltage is V_{inv}. Both are represented as phasors. The inverter has a control system that synchronizes its frequency with that of the gen set.

We set $V_g = |V_g|\angle 0$ so that it is the reference, and let $V_{inv} = |V_{inv}|\angle\delta$ where δ is the angle between the inverter voltage and generator voltage.

The current from the inverter to the AC bus is

$$I_{inv} = \frac{V_{inv} - V_g}{jX_m}. \tag{9.53}$$

The real power output by the inverter through the coupling inductance is:

$$P_{inv} = \text{Re}\left\{V_{inv} I_{inv}^*\right\} = \text{Re}\left\{V_{inv}\left(\frac{V_{inv} - V_g}{jX_m}\right)^*\right\} \tag{9.54}$$

$$= \text{Re}\left\{\frac{|V_{inv}|^2}{-jX_m} - \frac{V_{inv} V_g}{-jX_m}\right\} \tag{9.55}$$

$$= \text{Re}\left\{\frac{|V_{inv}|^2}{-jX_m} - \frac{|V_{inv}||V_g|\cos(\delta) + j|V_{inv}||V_g|\sin(\delta)}{-jX_m}\right\} \tag{9.56}$$

$$= \frac{|V_{inv}||V_g|\sin(\delta)}{X_m}. \tag{9.57}$$

The angle δ and magnitude $|V_{inv}|$ are controlled by the inverter. Adjusting the magnitude of δ controls the magnitude of the real power exchanged between the buses. Adjusting the magnitude of the inverter voltage also affects the real power, but to a lesser extent. Because δ is usually small, the real power is more sensitive to changes in δ than $|V_{inv}|$. The inverter can also supply reactive power. It can be shown that the reactive power produced by the inverter is

$$Q_{inv} = \frac{|V_{inv}|^2}{X_m} - \frac{|V_{inv}||V_g| \cos \delta}{X_m}. \tag{9.58}$$

Note that some of this reactive power is absorbed by the inductance X_m. The reactive power is more sensitive to the inverter voltage than the angle. This conveniently establishes a basic strategy for the control of the inverter: adjust δ to control the real power; adjust $|V_{inv}|$ to control the reactive power. Since there is not a complete de-coupling of these variables in (9.57) and (9.58), the control of the real and reactive power cannot be completely independent. A phase-locked loop is used to determine the firing angle of the MOSFET in relation to the AC bus voltage, and the modulation index (9.48) can be used to adjust the magnitude. The formulations in (9.57) and (9.58) are for single-phase inverters. As usual, they are multiplied by three if the three-phase real and reactive power are of interest.

9.12 Bi-directional Converters

There is nothing about (9.57) that restricts the real or reactive power supplied by the inverter to be positive. For example, controlling the inverter such that δ is negative results causes power to flow from the AC bus to the DC bus. The reactive power can also flow from AC bus to the DC bus. Some AC–DC-coupled mini-grids benefit from the flexibility of bi-directional conversion. For example, a hybrid solar/gen set mini-grid may require power to flow from the DC bus to AC bus when the battery is discharging. But when the gen set charges the battery, power flows from the AC bus to the DC bus. When power can flow in both directions through a converter, it is known as a "bi-directional converter." Most bi-directional converters encountered in mini-grids are inverters whose switches allow bi-directional conduction, for example, an IGBT with antiparallel diode.

Example 9.5 A lossless single-phase bi-directional inverter is connected to an AC bus. A single-phase gen set is also connected to the AC bus. The real portion of the AC load is 13 kW. The gen set supplies 10 kW. Compute the required power from the inverter and the associated angle δ. Compute the

(continued)

reactive power from the inverter. The decoupling reactance is 0.5 Ω; let the generator voltage be 230 V and the inverter voltage be 232 V.

Solution The real power required from the inverter is $P_{inv} = 13 - 10 = 3$kW. From (9.57), the angle δ is computed as

$$\sin(\delta) = P_{inv}\frac{X_m}{|V_{inv}||V_g|} = 3000\frac{0.5}{230 \times 232} = 0.0281$$

$$\sin^{-1}(0.0281) = 0.0281 \text{ rad} = 1.61°.$$

The reactive power of the inverter is found using (9.58)

$$Q_{inv} = \frac{|V_{inv}|^2}{X_m} - \frac{|V_{inv}||V_g|\cos\delta}{X_m} = \frac{232^2}{0.5} - \frac{232 \times 230 \times \cos 1.61°}{0.5} = 970 \text{ VAR.}$$

9.13 Inverter Practical Considerations

Commercially available inverters are typically compatible with nominal DC-side voltages of 12, 24, 48, or 96 V. Some inverters are compatible with several nominal voltages. Inverters with higher-power ratings tend to be compatible with higher DC voltages.

The rating of an inverter is based on the power it can continuously supply without overheating. Many inverters "peak" rating, which is the power the inverter can supply for short intervals. This is an important consideration for loads that draw additional power at start-up or draw power irregularly. Inverter ratings can be specified in watts or volt–amps, with the watt rating always being less than the volt–amps rating. If the units of the rating are overlooked, the inverter might be over- or under-designed. If the power factor of the load is unknown, the designer must make an assumption to convert between watts and volt–amps; a power factor of 0.85 is often reasonable.

The power an inverter can continuously supply is limited by the temperature increase associated with its internal losses as it supplies power. The ambient temperature also affects this—the higher the ambient temperature, the lower the power the inverter can supply. Many manufactures provide temperature-dependent ratings. A 10% de-rating of the rated power is appropriate for an inverter operating at 40°C instead of 25°C. For example, if the continuous load is expected to be 3.0 kW, an inverter rated at 3.3 kW can be selected if the inverter is expected to operate in a high-temperature environment.

There are many types and sizes of commercially available inverters. High-quality inverters feature the following:

- output sinusoidal voltage with little distortion and constant frequency and magnitude;
- good voltage regulation;
- high efficiency at low loading;
- insensitivity to changes in input voltage;
- short-term increased surge capacity;
- low-voltage disconnect capability;
- can be configured in the field;
- include data logging and diagnostic features.

9.14 Summary

This chapter covered the basic principles and functions of the converters found in off-grid systems. A brief description of each follows:

- DC–DC converters: these devices can increase or decrease the input DC voltage by varying their duty cycle. They are also used in maximum power point trackers and battery chargers.
- Maximum power point trackers (MPPT): MPPTs allow PV arrays to operate at or near their point of maximum power by decoupling the array voltage from the battery or load voltage.
- Solar battery charger: these devices are found in solar home systems, solar lanterns, and some PV-based mini-grids. They are designed to prevent the battery from being overcharged, and those controlled using PWM are able to employ three-stage charging.
- AC battery charger: AC battery chargers are common in AC-coupled mini-grids that also have a DC bus. Single-phase and three-phase chargers use diode-based rectifiers that convert AC to DC. Rectifiers are also used to connect variable speed AC generators like WECS to the DC bus for battery charging applications.
- Automatic voltage regulator (AVR): AVRs often use phase-angle-controlled rectifiers to control the field current of synchronous generators, allowing the terminal voltage to be regulated.
- Electronic load controller (ELC): ELCs are used for speed control in MHP systems. The power to a ballast load—a resistor with high-power rating—is controlled to electrically govern the system frequency. There are several types of ELCs, including those that use binary-weighted resistor networks, phase-angle control or impedance control. Impedance control is achieved by a chopper in series with the output of phase-angle-controlled rectifier. This allows greater control over the real and reactive power consumed by the ballast load. Fixed capacitors can also be included, depending on whether or not consumption of reactive power is an issue.

- Inverter: inverters convert DC voltage to AC. They are required when a DC-coupled architecture supplies an AC load. There are several types of inverters.
- Solar inverter: solar inverters convert the DC output by PV modules, strings, or arrays to AC directly, without a dedicated DC bus. Most solar inverters are unable to form the AC bus voltage.
- Grid-tied inverter (GTI): GTIs transfer power from a DC bus to an AC bus whose voltage is formed by another generator. The angle and magnitude of the inverter voltage are controlled to achieve the desired injection of real and reactive power to the AC bus.
- Bi-directional inverter: bi-directional inverters allow power to flow from the DC Bus to the AC bus and vice versa. This capability is useful in AC–DC-coupled mini-grids with AC and DC loads or AC loads and a battery bank.

Problems

9.1 A square wave like that produced by a modified sinewave inverter only has odd harmonics. The magnitude of the first, third, fifth, and seventh harmonics are 1.273, 0.424, 0.255, and 0.182, respectively. Compute the THD associated with these harmonics.

9.2 A DC mini-grid incorporates a battery bank whose nominal voltage is 49.5 V. A boost converter is used to increase the voltage to 230 V to the distribution system. At the user's home, the voltage is reduced to 12.8 V using a buck converter. Compute duty cycle of the boost converter and the buck converter. Assume the voltage drop of the distribution system is 5%.

9.3 An MPPT is used to connect a PV module to a load whose resistance is 10 Ω. The MPP of the module under the present conditions is 350 W corresponding to a voltage of 38.54 V. The MPPT uses a buck–boost converter. Compute the duty cycle of the converter for maximum power supplied to the resistor. Compute the current into the MPPT and to the load.

9.4 An MPPT is used to connect a PV array to a battery whose voltage is 49.5 V. The PV array consists of two PV modules connected in series. Under the present conditions, the maximum power of each modules is 185 W corresponding to a voltage of 36.4 V. Compute the duty cycle of the MPPT if a buck converter is used. Compute the current to the battery. Ignore the resistance of the battery.

9.5 A single-phase AC source is used to charge a nominal 24 V battery bank through a full-bridge rectifier. The battery resistance is 0.3 Ω, and V_{SoC} is 25.6 V. The RMS value of the AC source is 25 V. Compute the power to the battery. Ignore the voltage drop of the diodes. Plot the current into the battery and the battery terminal voltage V_{DC} for one period.

9.6 A three-phase source with line–line RMS voltage of 43 V is used to charge a battery through a three-phase rectifier. The battery resistance is 0.15 Ω and V_{SoC}

is 50.9 V. Compute the power to the battery. What is the average battery terminal voltage and average battery current? Ignore the voltage drop of the diodes.

9.7 A three-phase MHP is used to charge a battery bank through a three-phase rectifier. The synchronous reactance of the MHP's generator X_s is 1.0 Ω. The magnitude of the induced phase voltage $|E_a| = 40V$. The battery resistance is 0.13 Ω and V_{SoC} is 25.0 V. Use the phasor model of the AC charger to determine the power to the battery and the current I_r, voltage V_r, and angle of the induced phase voltage δ.

9.8 Compute the required firing angle α of a three-phased controllable rectifier used in an AVR to achieve an average DC field winding voltage of 63 V. The RMS value of the line–line voltage input to the AVR is 220 V.

9.9 A single-phase inverter supplies 1350 VA to the AC bus. The DC bus voltage is 25.2 V. Compute the current into the inverter, assuming the inverter efficiency at this operating point is 74%.

9.10 A single-phase bi-directional grid-tied inverter is connected to an AC bus. The inverter voltage is $V_{inv} = 121\angle 5°$ V. The voltage of the AC bus is $V_g = 120\angle 0°$ V. The inductance X_m is 2.6 Ω. Compute the inverter current and the real and reactive power supplied by the inverter.

9.11 A three-phase bi-directional grid-tied inverter is connected to an AC bus. The inverter line-to-neutral voltage is $V_{inv} = 234\angle 0°$ V. The line-to-neutral voltage of the AC bus is $V_g = 230\angle 0°$ V. The inductance X_m is 2.6 Ω. Compute the current I_r and the real and reactive power supplied by the inverter.

9.12 A three-phase bi-directional grid-tied inverter is connected to an AC bus. The inverter line-to-neutral voltage is $V_{inv} = 230\angle 5°$ V. The line-to-neutral voltage of the AC bus is $V_g = 230\angle 0°$ V. The inductance X_m is 2.6 Ω. Compute the current I_r and the real and reactive power supplied by the inverter.

References

1. Bonert, R., Hoops, G.: Stand alone induction generator with terminal impedance controller and no turbine controls. IEEE Trans. Energy Convers. **5**(1), 28–31 (1990). DOI 10.1109/60.50808
2. Bonert, R., Rajakaruna, S.: Self-excited induction generator with excellent voltage and frequency control. IEE Proc. Gener. Transm. Distrib. **145**(1), 33–39 (1998). DOI 10.1049/ip-gtd:19981680
3. Çelik, Ö., Teke, A., Tan, A.: Overview of micro-inverters as a challenging technology in photovoltaic applications. Renew. Sustain. Energy Rev. **82**, 3191–3206 (2018). DOI https://doi.org/10.1016/j.rser.2017.10.024. URL http://www.sciencedirect.com/science/article/pii/S1364032117313850
4. Crompton, T.: 48 - taper charging of lead-acid motive power batteries. In: Crompton, T. (ed.) Battery Reference Book (Third Edition), 3rd edn., pp. 1–6. Newnes, Oxford (2000). DOI https://doi.org/10.1016/B978-075064625-3/50049-4. URL https://www.sciencedirect.com/science/article/pii/B9780750646253500494

5. Espinoza, J.R.: 15 - inverters. In: Rashid, M.H. (ed.) Power Electronics Handbook (Third Edition), 3rd edn., pp. 357–408. Butterworth-Heinemann, Boston (2011). DOI https://doi.org/10.1016/B978-0-12-382036-5.00015-X. URL http://www.sciencedirect.com/science/article/pii/B978012382036500015X

6. Gao, D.Z., Sun, K.: 16 - DC–AC inverters. In: Rashid, M.H. (ed.) Electric Renewable Energy Systems, pp. 354–381. Academic Press, Boston (2016). DOI https://doi.org/10.1016/B978-0-12-804448-3.00016-5. URL https://www.sciencedirect.com/science/article/pii/B9780128044483000165

7. Grady, W.M.: Harmonics and how they relate to power factor. In: Proc. of the EPRI Power Quality Issues & Opportunities Conference (1993)

8. Harvey, A., Brown, A., Hettiarachi, P., Inversin, A.: Micro-Hydro Design Manual. Practical Action Publishing (1993)

9. Jana, J., Saha, H., Bhattacharya, K.D.: A review of inverter topologies for single-phase grid-connected photovoltaic systems. Renew. Sustain. Energy Rev. 72, 1256–1270 (2017). DOI https://doi.org/10.1016/j.rser.2016.10.049. URL http://www.sciencedirect.com/science/article/pii/S1364032116306943

10. Krein, P.T.: Elements of Power Electronics. Oxford University Press (1998)

11. Muljadi, E., Drouilhet, S., Holz, R., Gevorgian, V.: Analysis of permanent magnet generator for wind power battery charging. In: Industry Applications Conference, 1996. Thirty-First IAS Annual Meeting, IAS '96., Conference Record of the 1996 IEEE, vol. 1, pp. 541–548 (1996). DOI https://doi.org/10.1109/IAS.1996.557087

12. Nayar, C., Islam, S., Dehbonei, H., Tan, K., Sharma, H.: 27 - power electronics for renewable energy sources. In: Rashid, M.H. (ed.) Power Electronics Handbook (Second Edition), Engineering, 2nd edn., pp. 673–716. Academic Press, Burlington (2007). DOI https://doi.org/10.1016/B978-012088479-7/50045-6. URL http://www.sciencedirect.com/science/article/pii/B9780120884797500456

13. Sari, Z., Amgoud, K., Bouabdallah, M.A.: Microprocessor based switching mode power supply with a standby battery. IFAC Proc. Vol. 25(8), 591–598 (1992). DOI https://doi.org/10.1016/S1474-6670(17)54114-2. URL http://www.sciencedirect.com/science/article/pii/S1474667017541142. IFAC Workshop on Automatic Control for Quality and Productivity (ACQP'92), Istanbul, Turkey, 3–5 June 1992

14. Singh, R.R., Chelliah, T.R., Agarwal, P.: Power electronics in hydro electric energy systems – A review. Renew. Sustain. Energy Rev. 32, 944–959 (2014). DOI https://doi.org/10.1016/j.rser.2014.01.041. URL http://www.sciencedirect.com/science/article/pii/S1364032114000525

15. Pistoia, G.: Chapter 7 - battery safety, management and charging. In: Pistoia, G. (ed.) Batteries for Portable Devices, pp. 163–191. Elsevier Science B.V., Amsterdam (2005). DOI https://doi.org/10.1016/B978-044451672-5/50007-7. URL https://www.sciencedirect.com/science/article/pii/B9780444516725500077

Part IV
Off-Grid Systems

Chapter 10
Operation and Control of Off-Grid Systems

10.1 Introduction

The previous chapters have given us the knowledge to understand how a mini-grid operates as a complete system, which we focus on in this chapter. We are primarily concerned with understanding how power flows through a mini-grid and, in particular, how the action of different controllers together affects the operation. The electrical aspects of the mini-grid take center stage, as we are less concerned with the physical details of the underlying energy conversion and storage technologies. This chapter begins by describing how a mini-grid manages the power to and from its batteries. We next develop a model of the mini-grid, which is useful in determining the flow of power under various conditions. Parallel operation of AC-coupled generators are then covered. The chapter concludes by presenting the algorithms used by maximum power point trackers to improve power production from PV arrays.

10.2 Battery Charging and Discharging

Off-grid systems that incorporate battery storage have special control requirements. The control scheme determines when to charge and discharge the battery and how this can be done in a way that prolongs the lifespan of the battery. This should be done without sacrificing the desired level of reliability and availability.

Some battery charging control schemes are simple. A series- or shunt-type controller can be used as described in the previous chapter. However, in many cases it is more complex, especially as the power output of the system exceeds several kilowatts. Often there is not a single controller that manages the charging and discharging of the battery. Rather, it is a coordinated effort among many charge and diversion load controllers. Charging and discharging also become complex in

© Springer International Publishing AG, part of Springer Nature 2018
H. Louie, *Off-Grid Electrical Systems in Developing Countries*,
https://doi.org/10.1007/978-3-319-91890-7_10

the sense that sophisticated algorithms are used. Remember, the battery bank is a significant capital expense and is often the first major component to fail. Although lithium–ion (LI) batteries are increasing in popularity, especially in solar lanterns and solar home systems, lead–acid batteries are much more common in mini-grids. We therefore focus on lead–acid batteries and make comments regarding lithium–ion batteries when appropriate. The reader may wish to review Chap. 8 before continuing as many of the concepts introduced in that chapter will be used.

10.2.1 Charge Control

Lead–acid batteries must be carefully charged. In particular, they should be charged to avoid evolution of hydrogen and oxygen gas caused by overvoltage conditions and thermal-related degradation caused by excessive current. In addition to the basic function of restoring charge, proper battery charging can break up the sulfate crystals that reduce capacity and can replenish energy lost by self-discharge.

Chapter 9 introduced several circuits that can be used to charge a battery. The control strategy for shunt- and series-type charge controllers is similar. When the battery terminal voltage reaches a predefined threshold, the switch operates— closing in a shunt controller and opening in a series controller. This stops the battery from being charged. When the voltage drops below a predefined threshold, the state of the switch is reversed.

Higher-quality charge controllers use PWM to regulate the current into the battery as shown in Fig. 10.1. The charge controller uses sensors to measure the battery's terminal voltage. The measured voltage is used in a closed-loop control system that automatically adjusts the charging current to prevent the battery from being overcharged. The charging process consists of three-stages [5]:

1. Bulk stage
2. Absorption stage
3. Float stage

The time required to complete the three-stage process depends primarily on the battery's state-of-charge (SoC) when charging begins. Five to 10 h is a reasonable range. On occasion, a fourth "equalizing" stage is used for periodic maintenance. When charging LI batteries, only the bulk and absorption stages are used, and the charging is usually faster. The bulk, absorption, and float stages are identifiable from the plot of the battery terminal voltage, as shown in Fig. 10.2.

Figure 10.3 shows a schematic representation of a charge controller connected to a battery at the DC bus. For now, we will consider a simplistic case in that nothing else is connected to DC bus so that

$$I_{CC} = -I_B, \qquad (10.1)$$

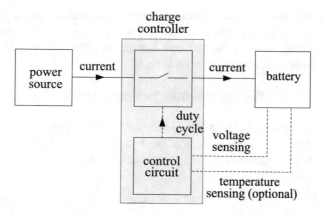

Fig. 10.1 Block diagram of a charge controller

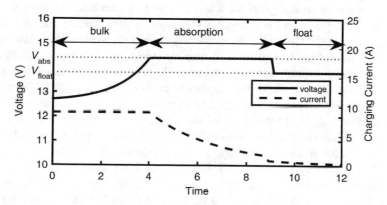

Fig. 10.2 Idealized voltage and current profile of a 12 V lead–acid battery undergoing a three-stage charging process

Fig. 10.3 Circuit model of a charge controller and battery

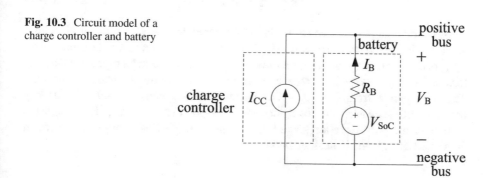

and we note that the charge controller current I_{CC} is positive when charging the battery, whereas the battery current has the opposite polarity. Recall from Sect. 8.5.2 that the relationship between the terminal voltage V_B, the voltage representing the SoC V_{SoC}, current, and battery resistance R_B is

$$V_B = V_{SoC} - I_B R_B = V_{SoC} + I_{CC} R_B. \tag{10.2}$$

We must always remember that the battery resistance depends on many factors, including the SoC and the magnitude of the battery current. We therefore interpret R_B as the battery resistance under the present conditions.

10.2.2   Bulk Stage of Charging

The bulk stage of charging begins when a battery is connected to the charger or when the input power to the charger is sufficient for it to operate. For example, a solar battery charger will begin the bulk charging stage sometime after sunrise.

The purpose of the bulk stage is to quickly charge the battery. Fast charging is desirable mostly for convenience. During the bulk stage, the battery's initial SoC is typically low, and so the battery is able to rapidly accept charge without worry of it being damaged by overvoltage. The current supplied by the controller is set to a predefined set-point value I_{CC}^*, which is typically no more than 13 to 20% of capacity at a C-rate of 0.05C. For example, I_{CC}^* should not exceed $0.20 \times 200 = 40$ A for a 200 Ah battery. The bulk stage is also referred to as the "constant current" stage. This is apparent from Fig. 10.2. Note that the controller is only able to supply its set-point current if there is sufficient power supplied to it. For now, we will assume this is the case. With nothing else connected to the DC bus, the terminal voltage is

$$V_B = V_{SoC} + I_{CC}^* R_B. \tag{10.3}$$

Most of the charge provided to the battery regenerates the active material, with a small portion going toward unwanted side reactions. As the battery charge increases, V_{SoC} rises. The battery resistance increases as well. These effects together cause the terminal voltage to rise more rapidly, even though the charging current is constant. This is apparent from Fig. 10.2. The current remains constant until the battery terminal voltage reaches the absorption stage set-point voltage, V_{abs}. The bulk stage then ends and the absorption stage begins. The battery's SoC is typically between 70 and 80% when the bulk stage ends.

10.2.3 Absorption Stage of Charging

Continuing to charge at a high current risks damage to the battery. During the absorption stage, the charge controller automatically adjusts the battery current so that the terminal voltage remains constant at a predefined set-point V_{abs}. This is to avoid an overvoltage condition. The absorption stage is also known as the "constant voltage" stage. Figure 10.2 shows that unlike the bulk stage, it is the voltage that is constant and the current that is variable. The absorption set-point voltage is selected to be below the voltage at which gassing becomes problematic. For a nominal 12 V lead–acid battery, a set-point of 14.4 V (2.40 V/cell) is common. For a nominal 24 V battery, it doubles to 28.8 V and so on.

With $V_B = V_{abs}$, and with V_{SoC} and R_B continuing to rise as the battery charges, it is inevitable that I_{CC} decreases with time. The current from the charge controller is

$$I_{CC} = \frac{V_{abs} - V_{SoC}}{R_B}. \tag{10.4}$$

Because I_{CC} is decreasing, the rate at which the SoC increases slows. The current gradually decays as the difference between V_{SoC} and V_{abs} reduces, as shown in Fig. 10.2. The efficiency during the absorption stage is reduced as some gassing inevitably occurs at this higher voltage.

The absorption stage can be of fixed duration, for example, 4 h, or variable based upon an estimate of the most recent depth of discharge. Ideally, the absorption stage would end when the battery bank has reached a full SoC, but this cannot be precisely determined from the terminal voltage.

Example 10.1 A primary school is supplied with electricity by a stand-alone PV system. The PV module is used to charge a 12 V, 280 Ah lead–acid battery. At one point during the absorption stage, the battery resistance is 0.015 Ω/cell and V_{SoC} is 2.05 V/cell. Compute the current required to keep the terminal voltage at the absorption set-point of 14.4 V.

Solution A nominal 12V lead–acid battery has six cells in series. The terminal voltage of the battery is found from (10.4) to be:

$$I_{CC} = \frac{V_{abs} - V_{SoC}}{R_B} = \frac{14.4 - 6 \times 2.05}{6 \times 0.015} = 23.33 \text{ A}.$$

Example 10.2 Consider the same battery from the last example. Sometime later during the absorption stage, the battery resistance increases to 0.060 Ω/cell, and V_{SoC} increases to 2.10 V/cell. The absorption set-point is 14.4 V. Compute the current required to keep the terminal voltage at the absorption set-point of 14.4V.

Solution As before, the terminal voltage of the battery is found from (10.4) to be:

$$I_{CC} = \frac{V_{abs} - V_{SoC}}{R_B} = \frac{14.4 - 6 \times 2.10}{6 \times 0.060} = 5.0 \text{ A}.$$

We see that as the absorption stage progresses, the current decreases as V_{SoC} and R_B increase.

10.2.4 Float Stage of Charging

The float stage begins as the absorption stage ends. The two stages are similar in that the charge controller regulates the terminal voltage according to a predefined set-point. The float stage set-point V_{float} is lower than the absorption set-point V_{abs}, for example, 13.4 V (2.23 V/cell) for a nominal 12 V lead–acid battery. The purpose of this stage is to keep the battery fully charged by off-setting any self-discharge. The charge controller current is close to zero during the float stage, again assuming that there are no other components connected to the battery. The charge controller current required to regulate the terminal voltage at the float stage set-point is

$$I_{CC} = \frac{V_{float} - V_{SoC}}{R_B}. \tag{10.5}$$

Lithium–ion battery chargers do not include a float stage as the prolonged higher voltage shortens their lifespan.

10.2.5 Equalizing Stage of Charging

The equalizing stage is done periodically, perhaps once per month, to break up sulfate crystals that may have formed. The equalizing stage occurs immediately after the absorption stage. During this stage, the voltage is regulated at a much higher voltage, for example, 15 to 16 V for a nominal 12V lead–acid battery. The battery current will be low, perhaps 3 to 6% of the battery's 20-h capacity rating. The stage lasts for 3 to 6 h. Some chargers automatically schedule equalizing; in others, the user must initiate it by pushing a button. The equalizing stage prolongs the total charging time. This can be a problem in PV-powered systems where the battery can only be charged during daylight hours. One solution is to reduce the overnight discharge the evening before the equalization is to take place. This allows the bulk and absorption stages to end early in the day, leaving sufficient time for the equalization stage to complete.

10.2.6 Power-Constrained Charging

The power provided by the charge controller is the product of the current it supplies and the DC bus voltage. Since the DC bus voltage is equal to the battery terminal voltage, we can write the charge controller output power $P_{CC,out}$ as

$$P_{CC,out} = V_B I_{CC}. \tag{10.6}$$

The power output from the charge controller is also equal to its input power $P_{CC,in}$ multiplied by the charge controller's efficiency:

$$P_{CC,out} = \eta_{CC} P_{CC,in} \tag{10.7}$$

where $P_{CC,in}$ is computed from the voltage and current at the input side of the charge controller. This can be, for example, a PV array or gen set.

When the power input to the charge controller is insufficient to maintain the bulk stage set-point current or the absorption or float stage voltage set-points, it is said to be "power-constrained." This often occurs in solar- and wind-powered mini-grids. A charge controller connected to a PV array might not be able to supply the bulk stage set-point current I_{CC}^* when it is cloudy, for example.

The maximum current I_{CC}^{max} that can be provided by the controller at a particular moment is the lesser of its set-point current and that allowed by its power supply

$$I_{CC}^{max} = \min\left\{ I_{CC}^*, \frac{\eta_{CC} P_{CC,in}^{max}}{V_B} \right\}. \tag{10.8}$$

where $P_{CC,in}^{max}$ is the maximum power that can be input to the charge controller by its source at a particular moment.

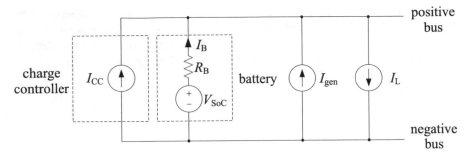

Fig. 10.4 Circuit model of DC bus with battery, charge controller, generator and load

10.2.7 Battery Charger Circuit Analysis

The previous section considered the simplest case of battery charging. In the absence of other generation sources, controllers, or load, the charge controller current is equal to the negative of the battery current. We now consider a more general case. Consider the mini-grid shown in Fig. 10.4. Here we have modeled the current from a generic generator and the load as constant current sources. The generator does not have its own controller, but it is connected to the DC bus through an uncontrolled rectifier, which is not explicitly modeled in the circuit diagram. This is sometimes the case when a WECS is used. The load can be connected to the DC bus directly or through an inverter. In either case, I_L is the load current as seen by the DC bus. Although only one generator and load are shown, the current I_{gen} and I_L can easily be replaced with the sum of the current from several generators and loads, respectively. We should not expect I_{gen} and I_L to ever be negative.

Applying Kirchhoff's Current Law at the positive DC bus:

$$I_B = -I_{CC} - I_{gen} + I_L. \tag{10.9}$$

From this we see that it is possible for the battery current to be positive, indicating that it is discharging, despite being connected to a charger.

The terminal voltage of the battery is

$$V_B = V_{SoC} - I_B R_B = V_{SoC} - \left(-I_{CC} - I_{gen} + I_L\right) R_B. \tag{10.10}$$

The current from the charge controller depends on the charging stage and whether or not the charge controller itself is power-constrained, as discussed next.

10.2.7.1 Bulk Stage

During the bulk stage, the current from the charge controller is equal to its maximum value I_{CC}^{max} (the lesser of the set-point current or power-constrained current). The corresponding battery current and terminal voltage can be computed from (10.9) and (10.10) by replacing I_{CC} with I_{CC}^{max}.

10.2.7.2 Absorption and Float Stages

In the absorption stage, the charge controller acts to balance the changes in net load and generator current to maintain the terminal voltage at V_{abs}. The *required* charge controller current $\hat{I}_{CC}$ to maintain the terminal voltage at the absorption set-point is

$$V_B = V_{abs} \tag{10.11}$$

$$\hat{I}_{CC} = \frac{V_{abs} - V_{SoC}}{R_B} + I_L - I_{gen}. \tag{10.12}$$

However, the actual charge controller current might not be $\hat{I}_{CC}$. The charge controller current cannot be negative; otherwise, it would be a load. The charge controller also cannot supply current above its maximum current as determined by (10.8). Therefore, the actual charge controller current is

$$I_{CC} = \begin{cases} 0 & : \hat{I}_{CC} \leq 0 \quad \text{(Overvoltage)} \\ \hat{I}_{CC} & : 0 < \hat{I}_{CC} < I_{CC}^{max} \\ I_{CC}^{max} & : \hat{I}_{CC} \geq I_{CC}^{max} \quad \text{(Undervoltage)} \end{cases} \tag{10.13}$$

Whenever $I_{CC} \neq \hat{I}_{CC}$, the absorption set-point voltage is not maintained and $V_B \neq V_{abs}$. Instead, the battery's terminal voltage is found using (10.10). The terminal voltage can be higher or lower than the absorption set-point. It is higher when there is excessive current from other generators; it is lower when the load current is too large. Neither condition is desirable. However, overvoltage is worse as it can damage the battery; undervoltage indicates that the battery is not being charged at the desired rate or is even being discharged.

If the load is constant, then an increase or decrease in I_{gen} is countered by an equal and opposite change in $\hat{I}_{CC}$ as seen in Fig. 10.5a. However, near the center of the plot, the generator current becomes so low that $\hat{I}_{CC} > I_{CC}^{max}$. The charge controller produces its maximum current, but this is insufficient to regulate the battery voltage at V_{abs}. This is seen by the dip in Fig. 10.5b. Similarly, when the generator current increases near the end of the plot, $\hat{I}_{CC}$ is negative, and so from (10.13) the actual charge controller current is zero. The battery voltage rises past the absorption set-point voltage. In practice, the charge controller will not be able to perfectly and instantly balance the changes in net load and generation even if $0 < \hat{I}_{CC} < I_{CC}^{max}$. Some slight fluctuation in the terminal voltage is therefore to be expected.

The analysis of the float stage is similar to the absorption stage, but V_{abs} is replaced with V_{float} in (10.12). As in the absorption stage, it is possible for the load current to be such that the float voltage cannot be maintained.

To prevent overvoltage, it is important that any source supplying current to the battery be connected to the DC bus through a charge controller (or a bi-directional converter with battery charging feature) or for there to be a diversion load, as discussed in the following section.

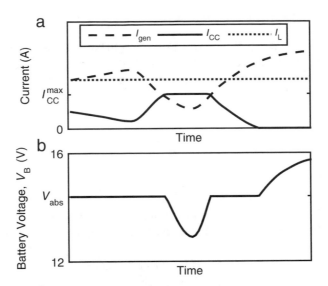

Fig. 10.5 (**a**) The charge controller current balances the change in generator current unless the charge controller's limits are reached (**b**) the battery voltage is constant when the charge controller is able to balance changes in the generator current

Example 10.3 Consider a hybrid DC-coupled mini-grid consisting of a PV array with charge controller, MHP generator with uncontrolled rectifier, and a battery. Let $V_{abs} = 14.4$ V, and $R_B = 0.075$ Ω. Compute the absorption stage charge controller current when $V_{SoC} = 12.5$ V. The load is 8 A. The current from the generator is 36 A. The charge controller is rated at 30 A, has an efficiency of 98%, and is supplied by a PV array whose input power is 300 W.

Solution We first calculate the charge controller current required to maintain the terminal voltage at 14.4 V using (10.12)

$$\hat{I}_{CC} = \frac{V_{abs} - V_{SoC}}{R_B} + I_L - I_{gen} = \frac{14.4 - 12.5}{0.075} + 8 - 36 = -2.67 \text{ A}.$$

Immediately we see that the charge controller cannot provide this current as it is negative. Applying (10.13), we set $I_{CC} = 0$ A. The charge controller is unable to maintain the absorption stage set-point. We can determine the battery voltage using (10.10) to be:

$$V_B = V_{SoC} - \left(-I_{CC} - I_{gen} + I_L + \right) R_B = 12.5 - (0 - 36 + 8)\,0.075 = 14.6 \text{ V}.$$

The battery is being overvoltaged by 0.2 V. To prevent this overvoltage, either the MHP generator current must be decreased or the load increased.

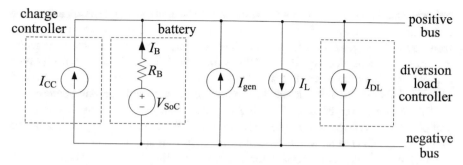

Fig. 10.6 Circuit model of a mini-grid with a diversion load

10.2.8   Diversion Load Control

Diversion loads are connected to the DC bus of the system through a diversion load controller. The diversion load controller is usually a PWM-controlled chopper circuit. It controls the current to the diversion load I_{DL}. The hardware and control circuitry of a diversion load controller is nearly identical to a charge controller. Instead of regulating the battery voltage by controlling current injected into the DC bus, the current withdrawn from the DC bus into the diversion load I_{DL} is controlled. In fact, diversion load controllers and charge controllers are so similar that some manufacturers produce a single unit that can function as either one (not simultaneously). The user selects the mode while configuring the set-points. The diversion load controller set-points for the absorption and float stages are designated $V_{abs,DL}$ and $V_{float,DL}$, respectively.

A mini-grid with diversion load is shown in Fig. 10.6. Applying Kirchhoff's Current Law at the positive bus:

$$I_B = -I_{CC} - I_{gen} + I_L + I_{DL}. \tag{10.14}$$

The terminal voltage of the battery is

$$V_B = V_{SoC} - I_B R_B = V_{SoC} - \left(-I_{CC} - I_{gen} + I_L + I_{DL}\right) R_B. \tag{10.15}$$

Similar to a charge controller, the diversion load controller current I_{DL} is limited between zero and its maximum value. The maximum value is based on the resistance of the diversion load R_{DL} and can be computed as

$$I_{DL}^{max} = \frac{V_B}{R_{DL}}. \tag{10.16}$$

Although the maximum current will vary somewhat based on the battery terminal voltage, we will make the assumption that it is a constant value. Like all resistors, diversion load resistors have a resistance value and a power rating. The power rating is based on the power that the resistor can dissipate without overheating. The diversion load's power rating should be such that it can dissipate the total power from the DC bus generators that do not have a charge controller of their own. For example, a system with two 3 kW WECS and one 5 kW PV array (with its own charge controller) requires a diversion load rated at $2 \times 3 = 6$ kW. It is often prudent to somewhat oversize the diversion load in case of unexpected operating conditions.

The mini-grid modeled in Fig. 10.6 has two controllers: a charge controller and diversion load controller. The absorption and float voltage set-points must be carefully coordinated between the controllers. Current should never flow through the charge controller and diversion load controller at the same time. The additional current from the charge controller can cause the diversion load to overheat, potentially causing a fire. In theory, the set-points can be exactly the same, but in practice, even with identical set-points, small measurement error can lead to the diversion load overheating. Instead, the diversion load controller set-points are made slightly higher than the charge controller set-points

$$V_{abs,DL} = V_{abs} + \Delta \tag{10.17}$$

$$V_{float,DL} = V_{float} + \Delta \tag{10.18}$$

where Δ is perhaps 0.1 V.

10.2.8.1 Bulk Stage Charging with Diversion Load

If the absorption set-points are programmed according to (10.17), then the diversion load current during the bulk stage is $I_{DL} = 0$. As usual, $I_{CC} = I_{CC}^{max}$, and the terminal voltage is found using (10.15). From (10.17), the charge controller absorption set-point will be reached first, and the charge controller will enter the bulk stage before the diversion load controller does.

10.2.8.2 Absorption and Float Stage Charging with Diversion Load

During the absorption and float stages, the charge controller current is controlled as before. The diversion load will not draw current unless the battery voltage rises to the diversion load controller's absorption set-point $V_{abs,DL}$. If the charge and diversion load controllers are properly coordinated, then the battery voltage will only reach $V_{abs,DL}$ when $\hat{I}_{CC} < 0$. This occurs when the current from the generators is excessive. If the diversion load controller absorption voltage set-point is reached, then the diversion load current will increase to regulate the battery voltage at $V_{abs,DL}$. This is shown in Fig. 10.7. Compare this result to Fig. 10.5 where there is no diversion load.

The current required by the diversion load controller $\hat{I}_{DL}$ to regulate the battery voltage at $V_{abs,DL}$ is

Fig. 10.7 (a) The current from the diversion load increases near the end of the plot to regulate the voltage as the generator current increases. (b) The voltage is regulated at $V_{abs,DL}$ when current is supplied to the diversion load

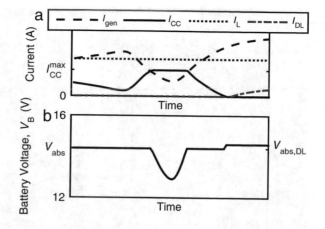

$$\hat{I}_{DL} = -\frac{V_{abs,DL} - V_{SoC}}{R_B} - I_L + I_{gen} + I_{CC}. \tag{10.19}$$

The actual diversion load current when $V_B > v_{abs}$ is

$$I_{DL} = \begin{cases} 0 & : \hat{I}_{DL} \leq 0 \quad \text{(Undervoltage)} \\ \hat{I}_{DL} & : 0 < \hat{I}_{DL} < I_{DL}^{max} \\ I_{DL}^{max} & : I_{DL}^{max} \leq \hat{I}_{DL} \quad \text{(Overvoltage)} \end{cases} \tag{10.20}$$

In a well-designed system, $\hat{I}_{DL}$ will never exceed I_{DL}^{max}. However, should it do so, the terminal voltage is found using (10.15). If the set-points of the controllers are set properly, then I_{DL} and I_{CC} will never simultaneously be positive. The diversion load current during the float stage is similarly computed using (10.19) and (10.20) but using the diversion load controller's float voltage set-point instead.

10.2.9 Coordinating Controller Set-Points

In systems with higher capacities, multiple charge controllers are often needed. See, for example, Fig. 10.8. A mini-grid might have several PV strings, each capable of producing high enough current or voltage that they require their own charge controller. In these systems, the controllers should be of the same model and have identical set-points. The charge controllers can then operate autonomously without communication to each other. As detailed previously, the set-points for diversion load controllers should be slightly higher than that of the charge controllers. If multiple diversion load controllers are needed, they all should have the same set-points.

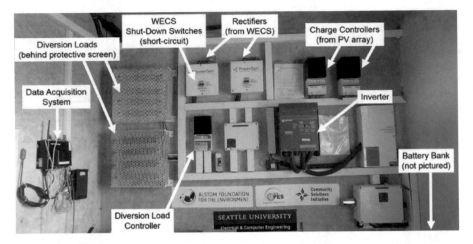

Fig. 10.8 The control room of a 5 kW hybrid wind/solar mini-grid in Kenya. Note the dual charge controllers and rectifiers (courtesy of author)

Fig. 10.9 LVD incorporated into a charge controller

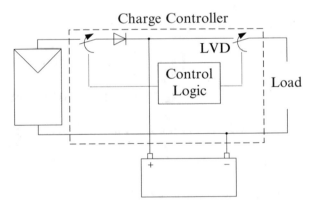

10.3   Battery Discharge Control

Batteries should be prevented from excessive discharge to avoid permanent damage. This is usually achieved by disconnecting the load when the terminal voltage of a battery falls below a certain threshold. This feature is realized through a *low-voltage disconnect* (LVD). A solid-state switch with high current rating, typically a power MOSFET, is placed in series between the load and battery. A voltage sensor measures the battery terminal voltage. A signal is sent to the switch to disconnect the load if the measured value falls below a predefined disconnect set-point voltage $V_{dis,LVD}$. A LVD can be incorporated into a charge controller, as shown in Fig. 10.9, and/or the inverter and individual loads.

Care must be taken in determining the set-point value of V_{LVD}. The battery's internal voltage drop during discharge must be accounted for. A simple procedure is to consult the battery's specification sheet to determine the steady-state open circuit voltage V_{SoC} corresponding to the SoC the LVD is desired to operate at. The anticipated internal voltage drop $I_B R_B$ is subtracted from V_{SoC} to determine the set-point of $V_{dis,LVD}$. Failure to correct for the internal voltage drop will result in the battery disconnecting prematurely. That the LVD does not directly measure the battery's SoC directly is an important and often misunderstood limitation of a LVD. It can be exploited to discharge the battery more deeply than expected, as explored further in the following example.

Example 10.4 A solar home system with a nominal 12 V AGM lead–acid battery powers two lights, each drawing 0.5 A. To prolong the life of the battery, the LVD should disconnect both lights when the battery reaches a SoC of 50%, corresponding to an open-circuit voltage of 12.24 V (see Table 8.1). Determine the LVD disconnect set-point if the battery resistance during discharge at a 50% SoC is 0.4 Ω. If the user only powers one light instead of two, will the LVD prevent the battery from being discharged to less than 50% SoC?

Solution The LVD disconnect set-point should be

$$V_{dis,LVD} = V_{SoC} - I_B R_B = 12.24 - 2 \times 0.5 \times 0.4 = 11.84 \text{ V}$$

If the user only uses one light, then the battery terminal voltage when 50% discharged is

$$V_B = V_{SoC} - I_B R_B = 12.24 - 1 \times 0.5 \times 0.4 = 12.04 \text{ V}$$

This is above the LVD voltage, and so power will continue to be supplied to the light, draining the battery further. The LVD will actuate when $V_B = V_{dis,LVD} = 11.84$ V. The SoC voltage will be:

$$V_{SoC} = V_B + I_B R_B = 11.84 + 1 \times 0.5 \times 0.4 = 12.04 \text{ V}$$

corresponding to a SoC of 25%, assuming R_B curve does not change. This is much lower than the target of 50% and will result in a shorter lifespan than the designer intended.

Solar home system users have been known to exploit the load dependence of the battery voltage drop to continue supplying some of the load past the designed minimum SoC target.

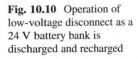

Fig. 10.10 Operation of low-voltage disconnect as a 24 V battery bank is discharged and recharged

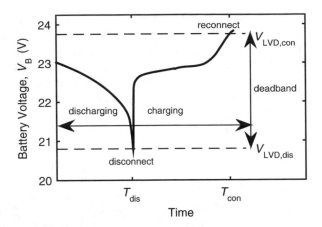

As soon as the LVD disconnects the load, the battery terminal voltage will quickly and sharply rise above $V_{LVD,dis}$. This is shown in Fig. 10.10 at time T_{dis}. This occurs even if the battery is not immediately recharged, as the voltage drop internal to the battery $I_B R_B$ becomes zero. Additional voltage rise may occur due to mixing of the battery electrolyte.

The battery should be reconnected to the load after its SoC has increased. Reconnection occurs only after V_B exceeds a second set-point value $V_{LVD,con}$ where $V_{LVD,con} > V_{LVD,dis}$. This occurs in Fig. 10.10 at time T_{con}. The difference between the reconnect and disconnect voltages is known as the "deadband." The reconnect voltage does not need to be the fully charged open-circuit voltage. However, it should be high enough to prevent nuisance disconnect—reconnect oscillations and attempts to defeat the LVD described in Example 10.4. For example, if the disconnect set-point is 11.5 V, then $V_{LVD,con}$ could be set to 12.6 V.

High-quality LVD systems will use dedicated voltage sensor wires that carry very low current to reduce the voltage drop between the LVD circuit and the battery terminals. They will also compensate the set-points based on battery temperature—increasing the disconnect voltage as temperature rises. Other improvements are possible, for example, adjusting the LVD set-point based upon the measured discharge current, so that under higher discharge current, the set-point voltage is reduced.

Solar home systems and solar lanterns usually feature a LVD. Some are remotely configurable so that the LVD set-points can be adjusted based upon the user's behavior. In certain business models, this feature can be used to allow customers to "upgrade" their solar home system after purchase by paying to reduce the LVD set-point.

10.4   Charge and Diversion Load Controller Practical Considerations

Charge controllers are commercially available with a wide variety of features and ratings. They are typically rated based on their bulk-stage current, compatibility with battery nominal voltages, and input power or voltage and current ratings. Manufacturers often describe their solar charge controllers as "MPPT" or "PWM" (pulse width modulation). The former have integrated maximum power point trackers, as discussed in the previous chapter, whereas the PWM do not. This is somewhat confusing because most MMPT trackers also use pulse width modulation in controlling the built-in DC–DC converter. Higher-end charge controllers include a display of battery voltage and charging stage, and an integrated data acquisition and remote monitoring system. Some might include a SoC indicator, but these are often inaccurate and are best used as a general indicator rather than an exact measure.

Diversion load controllers are sometimes referred to as "wind turbine" charge controllers. The reason for this is that diversion loads are needed with WECS— WECS cannot be controlled by a series-regulated controller without risking an overspeed or overvoltage. However, a diversion load controller does not control the wind turbine—it controls the current to the diversion load.

It is very important to properly set and coordinate the absorption and float set-points of all controllers. Most charge controller manufacturers provide recommended settings based on the type of battery being charged; see, for example, Table 10.1. Battery manufacturers will usually provide values for the set-points, which should be followed.

Table 10.1 Typical charging set-points for a 24 V system

Battery	Absorption	Float	Equalization
AGM	28.2	27.6	31.8
Flooded lead–acid	28.8	27.6	32.4
LiFePO$_4$	28.4	27.0	–

In Chap. 8, we saw that the open-circuit voltage and battery resistance of lead–acid batteries are affected by temperature. At all but the lowest electrolyte concentration levels, the terminal voltage increases with temperature, whereas the battery resistance decreases with temperature. Gassing also occurs at a lower voltage as temperature increases. The set-points of the charge controller should be adjusted to account for these changes. Despite the increase in the voltage V_{SoC} with increasing temperature, the controller set-points should be *decreased* as temperature increases. For example, for a 24 V nominal lead–acid battery, the absorption and float stage voltage set-points are decreased by 0.032 V for every degree Celsius increase in temperature above the rated value. Higher-quality charge controllers with temperature sensors will automatically make this adjustment.

10.5   Interpreting Battery Voltage Profiles

A lot can be inferred from the daily or weekly profile of the battery bank voltage. In fact, the battery terminal voltage is perhaps the single most important and useful quantity to measure in an off-grid system. The interpretation of the profile is part art and part engineering. The profile provides insight into whether or not the controllers are functioning properly. It further provides information on whether the energy supply is appropriately sized to meet the load. It can also be used in troubleshooting technical problems. We will consider examples of daily operation from an actual 1.8 kW PV-powered mini-grid next. Each example considers a different day. In each example, three plots will be presented: the power from the PV array, the power to the inverter (load) and the battery bank's terminal voltage over a 24-h period.

To begin, consider the three plots in Fig. 10.11. The battery voltage profile is indicative of a healthy, properly functioning mini-grid. The three charging stages are distinct. The controller is able to rapidly charge the battery during the bulk stage. The controller tightly regulates the terminal voltage during the absorption and float stages, indicating that the controller is appropriately sized and not power-constrained. Notice how the power from the PV array begins to decay as soon as the absorption stage beings. This is not caused by a decrease in irradiance, but rather there is no need for additional power to be produced as the battery is nearly full and the load is low. The bulk charging is completed by 8:00, and the absorption stage is finished a few hours later. The charging is complete several hours before sunset; and we conclude the system can supply additional load during the evening and still be fully charged during the day.

In mini-grids powered by variable sources such as PV or WECS, the voltage profile will often not look like Fig. 10.11. Figure 10.12 is from the same mini-grid on a different date. The load is much higher than for the day depicted in Fig. 10.11. We see the battery voltage fluctuates during the absorption stage, indicating that the charge controller was power-constrained and unable to regulate the terminal voltage. This is most likely due to partial cloud coverage. The float stage is not reached before sunset. We should not be overly worried. The battery still spent considerable time in the absorption stage and is likely nearly full.

Now consider the voltage profile in Fig. 10.13. This is again for the same system as in Figs. 10.11 and 10.12. During this day, the absorption voltage is never reached, and so the charge controller never progresses past the bulk charging stage. As a result, the voltage lacks the distinct plateaus shown in the previous figures. This is a sign that during this day, the energy supplied by the battery is more than was input to it. This is obviously not sustainable in the long-term. Should the load not be reduced or the PV power increase, perhaps from increased irradiance or by cleaning the PV array, then the voltage will continue to decrease until the LVD actuates. This variability in supply is one reason that the reliability of gen sets are appealing.

To an untrained eye, it might appear that the battery is being properly charged. For example, for a 24 V nominal system, the open-circuit steady-state voltage is 25.2 V. The voltage profile in Fig. 10.13 exceeds this for most of the afternoon. This could lead to the erroneous conclusion that the battery is fully charged or even

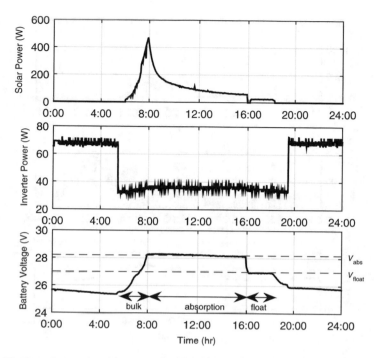

Fig. 10.11 The solar power, inverter power, and voltage profile of a mini-grid exhibiting typical behavior

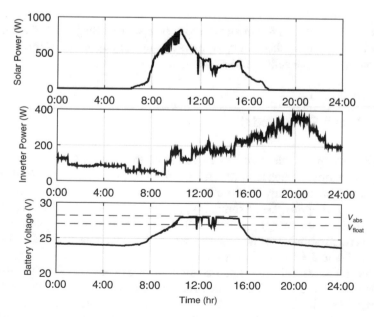

Fig. 10.12 The solar power, inverter power, and voltage profile of a mini-grid showing the operation of power-constrained charge controller during the absorption stage

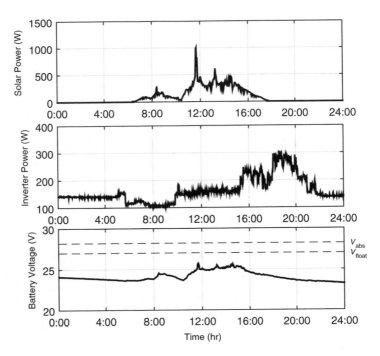

Fig. 10.13 The solar power, inverter power and voltage profile of a mini-grid showing the operation on a cloudy day

overcharged. It is easy to forget that the battery terminal voltage should not be used to infer the SoC unless the battery is open-circuit and has rested. Neither of those conditions apply to an active mini-grid.

10.6 Power Flow Model

We next develop a model that lets us calculate the flow of power from, to, and through the various components of a mini-grid. It is a steady-state algebraic model. Power flow models of this type are used in computer programs that simulate the operation of mini-grids. These simulation programs are described in Chap. 12. The model pertains to the energy production system of the mini-grid, not the distribution system. A more complicated nonlinear power flow model is needed to determine the flow of power in the distribution system [6].

The reader may recall from a course on basic circuit analysis that the sum of the real power produced by all the sources in a circuit equals the sum of the real power consumed by the loads. This is an extension of the more general law of conservation of energy and is the basis of the power flow model. When applied to a mini-grid with J components, the conservation of power is written as

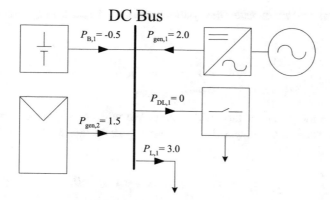

Fig. 10.14 Example of power flow in a mini-grid

$$\sum_{j=1}^{J} P_j = 0 \qquad (10.21)$$

where P_j is the power associated with a component j, which could be a generator, load, or battery. This equation is the basis of the power flow model. Readers well-versed in circuit analysis should keep in mind that (10.21) and those that follow are in reference to the average real power, not the instantaneous power.

We adopt the convention that the power associated with a generator is positive when the generator is producing power and the power associated with a load is positive when it is consuming power. The power associated with these components will never be negative. Batteries follow the convention that the power is positive when discharging (supplying power) and negative when charging (consuming power).

Rewriting (10.21) using these conventions

$$\sum_{g=1}^{G} P_{\text{gen},g} + \sum_{b=1}^{B} P_{\text{B},b} - \sum_{l=1}^{L} P_{\text{L},l} - \sum_{m=1}^{M} P_{\text{DL},m} - \sum_{n=1}^{N} P_{\text{BL},n} = 0 \qquad (10.22)$$

where $P_{\text{gen},g}$, $P_{\text{B},b}$, $P_{\text{L},l}$, $P_{\text{DL},m}$, and $P_{\text{BL},n}$ are the power associated with the G generators, B batteries, L loads, M diversion loads, and N ballast loads of the mini-grid, respectively. Recall from Chap. 9 that a ballast load acts like a diversion load but on the AC bus. Since we are focused on the electrical aspects of the system, the term "generator" refers to any energy conversion technology that generates electricity. This equation shows that the power associated with any component can be algebraically determined if the power of all other components is known. Figure 10.14 shows an example of the power flow in a DC-coupled mini-grid. The reader should verify that (10.21) is obeyed.

In general, we will consider the system batteries collectively as one battery bank so that

$$\sum_{b=1}^{B} P_{B,b} = P_B \qquad (10.23)$$

where P_B is the power of the battery bank. We will use the same convention to collectively refer to the power from the other components so that (10.22) can be concisely written as:

$$P_{gen} + P_B - P_L - P_{DL} - P_{BL} = 0. \qquad (10.24)$$

We use the term "total load" to refer to the sum of the power to the load, the diversion load and the ballast load.

10.6.1   Power Balance at DC and AC Buses

If we let $\sum P_{DC}$ and $\sum P_{AC}$ be the summation of the power associated with the components connected to the DC and AC buses, respectively, then the power balance for the whole system is

$$\sum P_{DC} + \sum P_{AC} = 0. \qquad (10.25)$$

As suggested by (10.25), the sum of the power at each bus separately does not necessarily equal zero. An imbalance of generation, total load, and battery power at an individual bus is possible. When this occurs, power necessarily flows from one bus to the other through a rectifier, inverter, or bi-directional converter that links the buses. For simplicity, we will assume that a bi-directional converter links the buses. We use the convention that the power through the converter P_{con} is positive when the power flows from the DC bus to the AC bus. If the converter is a rectifier, then $P_{con} \leq 0$; and if it is an inverter then, $P_{con} \geq 0$.

The power through the converter is equal to the imbalance on the DC bus

$$P_{con} = \sum P_{DC} = P_{DC,gen} + P_B - P_{DC,L} - P_{DL} \qquad (10.26)$$

which must be equal to the imbalance on the AC bus

$$P_{con} = -\sum P_{AC} = -P_{AC,gen} + P_{AC,L} + P_{BL} \qquad (10.27)$$

where the subscripts DC and AC are used to denote which bus the generation and load are connected to. Recall that batteries and diversion loads can only be coupled to the DC bus and the ballast load can only be coupled to the AC bus.

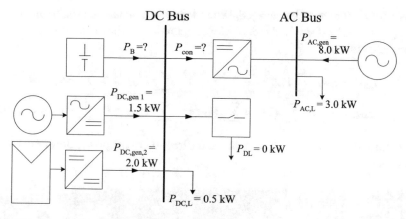

Fig. 10.15 Mini-grid for Example 10.5

Example 10.5 Consider the AC–DC-coupled mini-grid in Fig. 10.15. Compute the battery power and converter power.

Solution The battery power is computed by solving (10.24) for P_B

$$P_B = -\sum_{g=1}^{G} P_{\text{gen},g} + P_L + P_{DL}$$

$$P_B = -1.5 - 2 - 8 + 3 + 0.5 + 0 = -8 \text{ kW}$$

The battery is therefore being charged. The converter power can be computed using (10.26) or (10.27). The AC bus has fewer components, so we use (10.27) to see $P_{\text{con}} = -8 + 3 = -5$ kW (rectifier mode).

10.6.2   Including Losses

The losses in a mini-grid occur within the components themselves and in the wiring between the components. In a well-designed system, the losses in wiring and connections are minimal, perhaps a few percent. The assumption hereafter is that all losses are included in the load. For example, a load of 100 W becomes 102 W when 2% losses are assumed. Generator losses are already included in P_{gen}, so their power does not need to be adjusted. When losses are included, (10.26) and (10.27) are replaced with either (10.28) or (10.29), depending on the direction of power flow through the inverter:

$$P_{con} = \eta_{con} \sum P_{DC} = -\sum P_{AC} : \quad \sum P_{AC} < 0, \sum P_{DC} > 0 \quad (10.28)$$

$$P_{con} = -\eta_{con} \sum P_{AC} = \sum P_{DC} : \quad \sum P_{AC} \geq 0, \sum P_{DC} \leq 0 \quad (10.29)$$

Be mindful that η_{con} varies with power and direction of the power flow as discussed in Chap. 9.

Example 10.6 Repeat the previous example, but assume the inverter efficiency is 90%.

Solution We begin by computing the inverter power. We know $\sum P_{AC} = 8.0 - 3.0 = 5.0\,$kW. This is positive, and so we use (10.29)

$$P_{con} = -\eta_{con} \sum P_{AC} = -0.90 \times (8 - 3) = -4.5\,\text{kW}.$$

so that 4.5 kW are supplied to the DC bus through the converter. Next, we apply (10.29) and again solve for the power to the battery bank:

$$P_{con} = \sum P_{DC} = P_{DC,gen} + P_B - P_{DC,L} - P_{DL}$$
$$-4.5 = (2.0 + 1.5) + P_B - 0.5 - 0$$
$$P_B = -4.5 - 3.0 = -7.5\,\text{kW}.$$

The losses in the inverter reduce the power flowing from the AC bus to DC bus. Subsequently, less power is supplied to the battery when compared to the lossless case.

10.6.3   *Including Constraints*

In actual mini-grids, the power associated with any component cannot exceed the component's rating. More generally, there are minimum and maximum limits that must be enforced in the power flow model, as discussed next.

10.6.3.1 Load Constraints

Under steady-state conditions, a load cannot consume more than its rated power, $P_{L,l}^{max}$, nor can it supply power. Therefore the following constraints are imposed:

$$0 \leq P_{L,l} \leq P_{L,l}^{max} \quad \forall l. \tag{10.30}$$

Some components such as motors temporarily consume power in excess of their rated value when starting up, in which case $P_{L,l}^{max}$ is variable.

Diversion and ballast loads cannot supply power or consume more than their rating:

$$0 \leq P_{DL,m} \leq P_{DL,m}^{max} \quad \forall m \tag{10.31}$$

$$0 \leq P_{BL,n} \leq P_{BL,n}^{max} \quad \forall n. \tag{10.32}$$

10.6.3.2 Generator Constraints

Generators are limited in the power they can provide:

$$P_{gen,g}^{min} \leq P_{gen,g} \leq P_{gen,g}^{max} \quad \forall g \tag{10.33}$$

where $P_{gen,g}^{max}$ and $P_{gen,g}^{min}$ are the maximum and minimum power that the generator can output. These limits might vary over time. There are three ways that the maximum power can be limited. The first is the generator cannot output more power than its rated value in steady state. The second is that the generator cannot output more power than input to it by the energy conversion device, minus the generator losses. This constraint often applies to PV arrays and WECS. The third is that the power can be limited by a controller. The constraint $P_{gen,g}^{max}$ is the lowest of these three limits. The minimum power cannot be negative, but it might be limited to a positive value. For example, a MHP system without a governor is unable to decrease the power it produces in steady state.

10.6.3.3 Battery Limits

The power into and out of a battery is limited by its SoC. A battery that is fully discharged cannot supply power; a battery that is fully charged cannot absorb power. More generally, the constraints on the battery power at any moment are:

$$P_B \leq P_B^{max} \quad \text{(discharge limit)} \tag{10.34}$$

$$P_B \geq P_B^{min} \quad \text{(charge limit).} \tag{10.35}$$

Note that per convention, P_B^{min} will be less than or equal to zero. The values of P_B^{max} and P_B^{min} battery depend on the SoC of the battery. The charging limit is enforced by the charge controller. It decreases in magnitude as the SoC increases, for example, as the controller transitions from the bulk to absorption stage. The discharge limit is enforced by the LVD of the charge controller, inverter or load. In the case where a LVD is not used, then the lower limit is dictated by the energy remaining in the battery.

10.6.3.4   Converter Limits

The power that can flow through a converter is limited to its rated power. This is expressed as the following constraints:

$$P_{con} \le P_{con}^{max} \quad \text{(inverter mode : DC to AC)} \tag{10.36}$$

$$P_{con} \ge P_{con}^{min} \quad \text{(rectifier mode : AC to DC)} \tag{10.37}$$

If a converter is not bi-directional, then either P_{con}^{min} or P_{con}^{max} is set to zero.

10.7   Operation Priority Schemes

In mini-grids with multiple controllable components, there are often several ways that the power balance (10.22) can be maintained. For example, a decrease in load can be balanced by increasing the power to the diversion load or decreasing the gen set power output. While both actions are viable, the latter is preferred because it reduces fuel costs. The actions that a mini-grid can take in response to increases or decreases in load or uncontrollable generation should be prioritized by the designer and implemented in the mini-grid's control scheme.

A mini-grid should be operated in a way that minimizes fuel costs while achieving the desired reliability and availability targets. To do this, certain actions are given higher priority than others. An example of an operation priority scheme for an increase in load, or decrease in uncontrollable generation, is shown in Table 10.2. Uncontrollable generation refers to those whose power output cannot be changed on demand. The actions proceed in order, from one to the next, as applicable. The last and therefore least desirable action is shedding load, meaning part or all of the load is disconnected.

A priority scheme for a decrease in load (or an increase in uncontrollable generation) is shown in Table 10.3. It is similar to Table 10.2 but in reverse order. There are other considerations that might affect this scheme. For example, if the battery bank is at a low SoC, then charging it (priority 2) might be more desirable than saving fuel by decreasing power from the gen sets (priority 1). Priority 3, intentionally reducing the power from a source with no fuel cost—solar, wind, and hydro—is known as "throttling." Rather than throttling, it is usually better to

Table 10.2 Operational priority for increasing load (decreasing uncontrollable generation)

Priority	Action
1	Reduce power to the diversion load and ballast load
2	Increase power from zero-energy-cost resources (WECS, MHP, PV)
3	Increase power from the battery (discharge)
4	Increase power from fossil-fuel or biomass gen sets, starting with the least expensive
5	Reduce user load, starting with least critical

Table 10.3 Operation priority for decreasing load (increasing uncontrollable generation)

Priority	Action
1	Decrease power from fossil-fuel or biomass gen sets, starting with the most expensive
2	Increase power to battery bank (charge)
3	Decrease power from zero-energy-cost resources (throttle)
4	Increase power from diversion load or ballast load

increase the user load if possible so that some benefit is derived from the irradiance, wind speed, or water flow. For example, heating water and reducing the temperature in a freezer add meaningful load to the mini-grid. Discounted electricity rates might be offered to encourage use.

Priority schemes are implemented by the controllers in a mini-grid. As previously discussed, charge and diversion load controller set-points can be programmed so that, for example, the power from a PV array is reduced before the diversion load draws power. Controlling gen sets to follow the priority scheme is more complicated and often requires specialized controllers and the ability for the gen sets to automatically start-up and shut down. These aspects are discussed in the following sections.

There are other factors that affect the priorities in an operation scheme. For example, if the only source capable of forming the AC bus is the gen set, then the gen set must be continuously operated. Switching, for example, by LVD, can change the topology of the mini-grid. This can restrict the ability of AC-coupled sources to serve DC loads and DC-coupled sources to serve AC loads. Gen sets might also be operated to maximize their efficiency and to reduce their run times—which prolongs their lifespan—as discussed next.

10.8 Gen Set Control Schemes

The operation of gen sets described in Tables 10.2 and 10.3 is known as the "load-following" scheme. The gen set only produces as much power as is needed at any given time. It can be thought of as the default scheme. The gen set is only used when the power from the battery and renewable resources is insufficient to supply the load. This is basically how a hybrid electric vehicle operates.

Fig. 10.16 Schematic of the
PV system with DC load

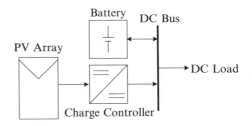

Alternatively, the gen set can be controlled according to a "cycle charging" scheme. In this approach, when the gen set is running it is loaded as high as possible, serving the load and charging the battery. The gen set shuts off when the battery reaches a predefined SoC as estimated by its terminal voltage. There are two benefits to this. The first is that the efficiency of gen sets increases with load (see Sect. 5.3.4); second, it limits the number of hours each day the gen set operates, prolonging its lifespan and reducing its noise pollution.

Automated cycle charging requires a sophisticated control system including a bi-directional converter (or a separate rectifier and inverter) with synchronization capability, a gen set with an auto-start feature, and a controller that coordinates the operation of the converter and gen set. There are commercially available products that can be used. Cycle charging can also be approximated manually. The operator switches from the inverter to the gen set when the load is expected to exceed the inverter rating (during the peak hours) and uses a battery charger to recharge the battery.

10.9 Examples of Mini-Grid Operation

To highlight how constraints, operation priorities and gen set control schemes affect the power flow in a mini-grid, we consider the three following illustrative cases. The cases are idealized in that they are based on simulation and that losses are ignored.

10.9.1 PV System with Battery-Constrained

We begin with a simple DC-coupled mini-grid consisting of a PV array, charge controller, battery, and DC load, as shown in Fig. 10.16. A simple system like this could be found in a stand-alone system providing LED lighting to a home. The power and SoC plots are shown in Fig. 10.17. The load is low overnight and during most of the day but has a sharp peak in the early evening, as is typical in most households. The power from the PV array follows the typical pattern of production under clear sky conditions.

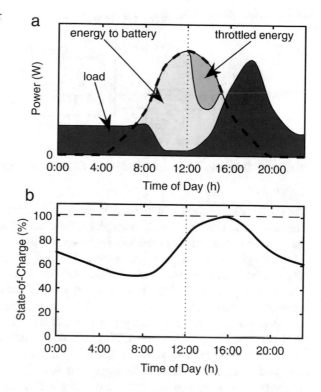

Fig. 10.17 (a) Flow of power in an off-grid PV system. (b) The corresponding battery state-of-charge

The vertical line at 12:00 shows when the absorption stage begins as the battery approaches a full SoC. The charge controller begins to throttle the PV array. Although this can be viewed as "wasted" power, in fact with the battery approaching a full SoC, it is an example of the power balance equation at work. There is simply nowhere for the power to go and so it is not produced. The power actually produced by the PV array and the power it is capable of producing diverge. The throttled energy is shown as the dark gray area in Fig. 10.17. This is the energy the PV array was capable of producing but did not. Throttling energy from PV arrays is a common side effect of the three-stage battery charging process.

10.9.2 Hybrid System—Load-Following

We next consider a hybrid AC–DC-coupled mini-grid consisting of a gen set, PV array, charge controller, battery, inverter with LVD and an AC-coupled load as shown in Fig. 10.18. The gen set is controlled using a load-following scheme and the priority list of actions are shown in Tables 10.2 and 10.3.

The mini-grid serves a somewhat unusual load that peaks in the morning as shown in Fig. 10.19a. The dashed white line is the gen set power output, and the gray

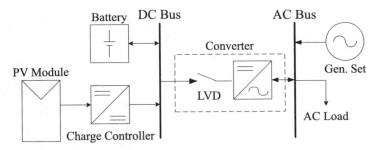

Fig. 10.18 Schematic of mini-grid used in load-following and cycle-charging examples

area is the energy produced by the PV array. The middle plot is the power through the bi-direction inverter. The bottom plot is the battery SoC. The SoC corresponding to the voltage that the LVD actuates is shown as E_{min}.

We begin by conceptually assuming that just prior to time 0:00 (midnight), there is no load or generation. At 0:00, the load increases to the value shown in Fig. 10.19a. From Table 10.2, the first priority in balancing this increase in load is to reduce power to the diversion and ballast loads. This mini-grid has neither of these components, and so the second priority is considered: increase power from zero-fuel-cost sources (the PV array). This cannot be done when there is no sunlight, and so the third priority action is considered: discharge the battery. For the first few hours, the battery is able to supply the load. However, around 2:00 the load exceeds the inverter's power rating, and so the fourth priority is considered: increase power from the gen set. For this to happen, the gen set would need to be capable of automatically starting and synchronizing to the AC bus, perhaps initiated by a control signal sent by the inverter.

The battery continues to supply the maximum power allowed by the converter because it is the higher priority (lower cost) than the gen set. However, around 6:00, its LVD operates when the battery SoC becomes too low. The DC bus and AC bus are now isolated from each other, and the gen set supplies the entire load. The power from the PV array has a higher priority than the gen set. After sunrise its power output should increase while the gen set power decreases. However, this architecture does not allow the PV array to supply power to the AC bus unless the LVD has reconnected the DC bus to the AC bus. Therefore, the PV array power goes toward recharging the battery. Clearly the architecture of the mini-grid affects how the priority lists are implemented. On this day, the battery SoC does not increase enough for its terminal voltage to initiate a reconnection by the LVD. However, the load is fully served by the gen set.

Fig. 10.19 (a) Power from the gen set, PV array, and to the load when the gen set is controlled using a load-following scheme. (b) Power through the converter (positive is from DC bus to AC bus). (c) Battery state-of-charge

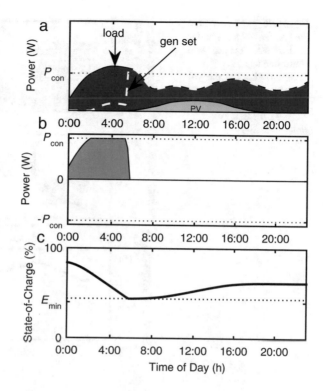

10.9.3 Hybrid System—Cycle Charging

We next consider the same mini-grid architecture as the previous case but with the gen set controlled using cycle-charging scheme and the converter being bidirectional. The operation is plotted in Fig. 10.20.

At the start of the day (0:00), the load is served by the battery through the converter acting in inverter mode. At 2:00, the load has surpassed the converter's rating, as seen in Fig. 10.20b. The gen set begins producing power so that the load can be served. Because it is following a cycle-charging scheme, the gen set will produce as much power as possible, thereby operating at a higher efficiency. The gen set's power output is traced by the dashed line in Fig. 10.20a. The shaded area between the dashed line and the load is the energy used to recharge the battery. Note that the gen set produces rated power for several hours before tapering as the battery enters the absorption stage. At around 7:00, the battery is fully charged, and the load can be entirely supplied by the converter. The gen set therefore shuts down. Whenever practical, the PV array also begins producing power around this time. The PV array has the highest priority because it has zero fuel costs. However, it is not able to supply the entire load. From Table 10.2, the battery rather than the gen set is used to supplement the PV production.

Fig. 10.20 (a) Power from the gen set, PV array and to the load when the gen set is controlled using a cycle-charging scheme. (b) Power through the converter (positive is from DC bus to AC bus). (c) Battery state-of-charge

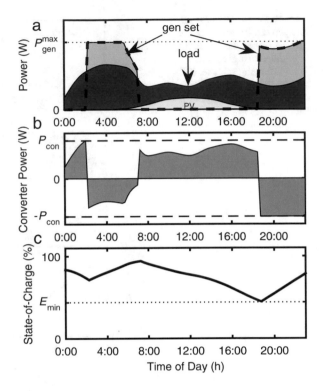

Around 19:00, the battery reaches its minimum SoC and cannot continue supplying power. The gen set restarts and supplies the load while recharging the battery. It is unable to produce its rated power due to the converter's power limit (now acting as a rectifier).

Comparing Figs. 10.19 and 10.20, it is clear that the gen set's control scheme greatly affects the operation of a mini-grid. Both load-following and cycle-charging schemes are able to accomplish the basic function of supplying the load. However, cycle-charging has the advantage of reducing the number of hours the gen set is operated and it is operated at a more efficient loading point.

10.10   Parallel Generator Operation

We next consider the operation of an AC-coupled system with multiple synchronous generators. This is a common architecture in mini-grids whose capacities exceed 100 kW and are powered by gen sets. A mini-grid of this size could likely serve several hundred if not thousands households. Simultaneous operation of gen sets connected to the same bus is known as "paralleling." Paralleling can also be done with inverters and in general any group of AC generators, but we shall only consider the case of gen sets. The same basic concepts broadly apply. A paralleled AC-coupled architecture offers the following benefits:

- Reliability—paralleling adds redundancy to the system so that if one gen set fails, a portion of or even the entire load can be served;
- Scalability—gen sets can be added or removed from the mini-grid as needed;
- Serviceability—maintenance can be done on one gen set at a time while the other(s) continue to serve the load.

Paralleled architectures are more complex to control and install, and are more expensive. Additional space is also required, but this is usually not a constraint in mini-grids in developing countries.

For gen sets to be operated in parallel, the generators must:

1. have the same number of phases;
2. have the same phase rotation (e.g., a-phase leads b-phase by 120°which leads c-phase by 120°);
3. have the same open-circuit terminal voltage at a given speed;
4. each have voltage and speed control, for example, through an AVR and a governor.

The voltage output by generators will not be perfectly sinusoidal. One factor affecting the distortion is the generator's "pitch factor." The pitch factor has to do with how the windings are physically arranged in the stator. We want the voltage waveform output by paralleled generators to as similar as possible, and so generators with the same pitch factor should be used—although this is not strictly required. The complexity of paralleling generators is greatly simplified if the gen sets are of the same model by the same manufacturer.

When paralleling generators, we are concerned with how they are synchronized, switched, and share load. Extra precaution is needed to protect the generators, for example, if the synchronism is done improperly or the control malfunctions. This can lead to conditions such as motoring, in which a generator operates as a motor, consuming real power.

10.10.1   Synchronizing and Switching

Generators should not be connected together until they are synchronized. Generators are synchronized when their terminal voltages have the same—or very nearly the same—frequency, phase, and magnitude. This can be done by an operator, with care, provided the speed and excitation of the generators can be manually adjusted.

Some gen sets can be synchronized through an external controller, as shown in Fig. 10.21. The control of the governor and AVR of each gen set is connected to this controller. It adjusts the excitation and speed until synchronization is achieved. After the generators are synchronized, they can be connected together. Manual or automatic transfer switches can be used to connect generators to the AC bus.

Fig. 10.21 Parallel operation
of two gen sets using an
external controller

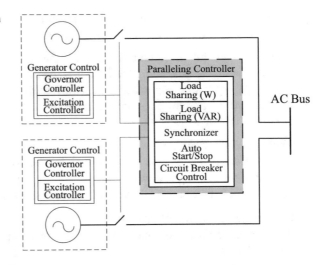

10.10.2 Load-Sharing

From the conservation of power, the sum of the real power from G-paralleled gen sets must equal the total load P_L (inclusive of losses). It can be shown that this must also hold true for the reactive power:

$$P_L = P_{gen,1} + \cdots + P_{gen,g} + \cdots + P_{gen,G} \qquad (10.38)$$

$$Q_L = Q_{gen,1} + \cdots + Q_{gen,g} + \cdots + Q_{gen,G} \qquad (10.39)$$

Load-sharing refers to how each gen set contributes to the real and reactive power required by the load [3]. In some situations each gen set contributes evenly, in which case

$$P_{gen,1} = \cdots = P_{gen,G} = \frac{P_L}{G} \qquad (10.40)$$

and a similar expression holds true for reactive power. In some situations, equal sharing of the load is not desirable. In particular, one gen set might be larger or more fuel efficient than the rest. A common practice is to share the load in proportion to the gen sets' size. A larger-capacity gen set provides a larger share so that

$$P_{gen,g} = P_L \times \frac{P_{rated,g}}{\sum_{k=1}^{G} P_{rated,k}} \qquad (10.41)$$

where $P_{rated,g}$ is the rated power of gen set g. Keep in mind that the efficiency of a gen set is very low at low loading. It is more fuel efficient when one generator operates at full capacity versus two at half. A good practice is to only turn on another gen set if the load is expected to exceed the capacity of those already supplying power. Regardless of how the load is to be shared, a control scheme is

Fig. 10.22 Droop curve for
two gen sets; an isochronous
curve is also shown

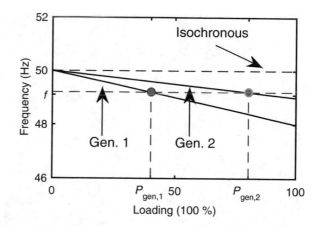

needed to implement the sharing. There are two general control approaches: droop
and isochronous. The difference is how the gen set's governor responds to a change
in load.

10.10.2.1   Droop Control

A simple approach to load-sharing is droop control. In this approach, each gen set
is automatically and autonomously controlled by its governor so that as the power it
supplies increases, its frequency *slightly* decreases (droops). For a gen set *g*, this is
expressed as

$$f = f_{g,0} - d_g \frac{P_g}{P_{\text{rated},g}} \tag{10.42}$$

were f is the operating frequency when the gen set outputs real power P_g, and $f_{g,0}$
is the frequency of the gen set under no-load conditions—usually 50 or 60 Hz—
and d_g is the droop slope [2]. When expressed graphically, (10.42) is known as the
"droop curve." A droop curve is shown in Fig. 10.22 for two different gen sets each
with a no-load frequency of 50 Hz but with different droop slopes.

We must keep in mind that the sum of the gen sets' power must equal that
consumed by the load and that because the gen sets are in parallel, they operate at
the same frequency. Under no load, the gen sets operate at their no-load frequency.
In order for the gen sets to be paralleled, their no-load frequency must be the
same. As the load increases, the frequency begins to decrease. See Sect. 5.2.5 for an
explanation of why this happens. Each governor, sensing the reduction in frequency,
responds by increasing the power produced by each gen set according to (10.42).
The frequency will stabilize when the sum of the power from the generators matches
the load. The system is at a new steady-state operating point, whose frequency is
somewhat less than the no-load frequency.

Example 10.7 Consider a mini-grid with two gen sets operated in parallel. Gen set 1 is rated at 75 kW with droop slope of 0.6, and Gen set 2 is rated at 37.5 kW with a droop slope of 0.3. Both have a no-load frequency of 50 Hz. Determine the operating frequency and the power output by each gen set if the load increases to 60 kW.

Solution From (10.42), the droop for each gen set can be written as

$$f_1 = f_{1,0} - d_1 \frac{P_1}{P_{rated,1}} = 50 - 0.6 \frac{P_1}{75}$$

$$f_2 = f_{2,0} - d_2 \frac{P_2}{P_{rated,2}} = 50 - 0.3 \frac{P_2}{37.5}$$

According to the power balance, the sum of the generation must equal the load:

$$60 = P_1 + P_2.$$

Further, because the gen sets are operating in parallel, their frequencies must be the same in steady-state. Setting f_1 and f_2 equal to each other yields:

$$50 - 0.6 \frac{P_1}{75} = 50 - 0.3 \frac{P_2}{37.5}$$

$$P_1 = \frac{75}{0.6} \times 0.3 \frac{P_2}{37.5}$$

$$P_1 = P_2.$$

Applying the power balance equation shows that each gen set produces 30 kW. The operating frequency is therefore:

$$f_1 = f_2 = 50 - 0.6 \frac{30}{75} = 49.76 \, \text{Hz}$$

In this case, the droops of the gen sets have been set so that the load is evenly shared. By selecting different relative values of d_1 and d_2, the load can be shared differently. For example, if each gen set had the same droop slope, then the power is shared in proportion to the rated capacity of each gen set.

A gen set's droop is often expressed as a percentage, calculated as:

$$Droop = 100 \times \frac{f_0 - f_{FL}}{f_{FL}} \qquad (10.43)$$

where f_{FL} is the full-load frequency—the frequency when the generator is producing its rated power—and f_0 is the no-load frequency.

The gen sets share reactive power by adjusting their excitation. Recall from Sect. 9.11 that reactive power is sensitive to voltage magnitude. Therefore a voltage-based droop scheme can be used:

$$|V| = V_{g,0} - d_{Q,g} \frac{Q_g}{Q_{rated,g}} \qquad (10.44)$$

where $|V|$ is the magnitude of the AC bus voltage, $V_{g,0}$ is the gen set's no-load (open-circuit) voltage, $d_{Q,g}$ is the reactive power droop slope, and Q_g is the reactive power output of gen set g with rated reactive power $Q_{rated,g}$.

An advantage of using droop control is that the generator's governors do not need to communicate with one another. They operate entirely autonomously. In fact, some gen sets will use droop control regardless of whether or not it is paralleled. The disadvantage of droop control is that the frequency and voltage magnitude varies somewhat with load.

10.10.2.2   Isochronous Control

Isochronous control is used in most newer or higher-end gen sets. The governor is automatically controlled so that the frequency is constant. The droop curve appears as a horizontal line, as in Fig. 10.22. Isochronous control of paralleled generators requires a separate controller to coordinate the load-sharing, as shown in Fig. 10.21. Should the communication fail, the generators default to droop-based control. If the generators tried to achieve isochronous operation without communicating to each other, the uncoordinated interaction of the separate controllers in each generator would likely cause the frequency and voltage to be unstable.

10.11   Maximum Power Point Tracking Algorithms

As detailed in Sect. 9.4, a maximum power point tracker (MPPT) is a DC–DC converter that is controlled so that a PV array operates at its maximum power point (MPP). This is an idealization. In practice, the MPP is constantly changing as irradiance and temperature change from one moment to the next. The MPPT then must constantly search for the MPP. There are several approaches for doing this [1, 4]:

- open-circuit voltage method
- short-circuit current method
- perturb and observe method
- incremental conductance method

Some of these approaches are simple and can be realized with an analog circuit. Others require a digital controller and sensors. The more sophisticated approaches are usually more expensive to implement as they require voltage and current sensors. In general they offer improved MPP tracking.

10.11.1 Open-Circuit Voltage Method

In this method, the open-circuit voltage of the PV array is periodically measured by the MPPT. The maximum power point voltage is assumed to be related to the open-circuit voltage V_{OC} by some constant k_v. The PV array voltage V_{PV} is controlled so that

$$V_{PV} = k_v V_{OC}. \tag{10.45}$$

Typical values of k_v range from 0.73 to 0.80. To be effective, k_v is selected based on the parameters of the PV array the MPPT is connected to. The MPPT is therefore not "plug-and-play," as it must be custom-programmed for a given array. Since the relationship between the open-circuit and MPP voltage is not constant, the PV array does not necessarily operate at the true MPP. In addition, the PV array must be periodically open-circuited to measure the voltage, which somewhat reduces the average power output.

10.11.2 Short-Circuit Current Method

This method is similar to the open-circuit method, but the short-circuit current is measured instead of the open-circuit voltage. The PV array current I_{PV} is controlled such that

$$I_{PV} = k_i I_{SC}. \tag{10.46}$$

Values of k_i typically range from 0.78 to 0.92, depending on the parameters of the PV array. Like the open-circuit method, the power output is periodically interrupted, and the array does not necessarily operate at the true MPP. It is generally more accurate and efficient than the open-circuit method. However, it requires a current sensor which generally is more expensive and less accurate then a voltage sensor.

Fig. 10.23 Example of the
PO algorithm at various steps

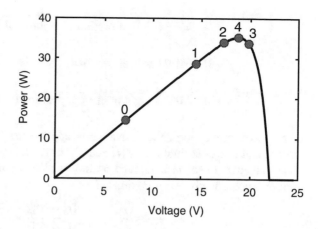

10.11.3   Advanced Methods

Most MPPTs use control algorithms to continuously adjust V_{PV} as environmental
and load conditions change. The most common are the perturb and observe (PO)
and the incremental conductance (IC) methods. These methods are widely used
in commercially available MPPTs. Both methods require voltage and current
measurements but are not dependent on the parameters of the PV array. We will
discuss the PO method next. The reader can consult references such as [1, 4] for
additional details on the IC and other methods

The PO method uses a "hill-climbing" approach. Hill-climbing approaches use
an iterative method for finding the maximum power point voltage V_{PV}^*. Consider
the I–V and power characteristic shown in Fig. 10.23. Notice that the power curve
resembles a hill, with the maximum power point at the top of the hill. Assume the
present operating point of the PV module is at point (0). Although from the figure
we can readily identify the location of the maximum power point, the MPPT must
only rely on current and voltage sensors at its present or past positions to determine
if the voltage should be increased or decreased to increase the power output. In
other words, it cannot "see" the entire hill. Different algorithms employ a variety
of methods to determine which voltage direction is "up-hill", how large of a step
to take in that direction, and then adjust the duty cycle to the DC–DC converter
accordingly. The result is a sequence of operating points that lead to the top of the
hill as shown by the numbered operating points (0)–(4) in Fig. 10.23. It is unlikely
that the algorithm will arrive at the top of the power curve without overshooting,
shown as point (3), in which case the algorithm identifies that decreasing the voltage
will increase the power, eventually leading to point (4), the maximum power point.

The PO method is based on the idea that the sign of the derivative of the power
with respect to voltage can be used to tell if the voltage should be increased or
decreased to reach the MPP:

$$\frac{\mathrm{d}P}{\mathrm{d}V} < 0 \quad \text{MPP is left of present position (voltage too high)} \qquad (10.47)$$

$$\frac{\mathrm{d}P}{\mathrm{d}V} = 0 \quad \text{MPP has been reached} \qquad (10.48)$$

$$\frac{\mathrm{d}P}{\mathrm{d}V} > 0 \quad \text{MPP is right of present position (voltage too low).} \qquad (10.49)$$

In practical applications, the derivative is approximated as a difference. Let the voltage and power at time t be $V[t]$ and $P[t]$, respectively. The voltage is directly measured; the power is computed as the product of the measured voltage and current. The derivative is approximated as:

$$\frac{\mathrm{d}P}{\mathrm{d}V} \approx \frac{\Delta P}{\Delta V} = \frac{P[t] - P[t-1]}{V[t] - V[t-1]} \qquad (10.50)$$

If the result is negative, then the voltage $V[t]$ is too low. At the next time step, the voltage is perturbed in the positive direction by some step size α:

$$V[t+1] = \alpha + V[t]. \qquad (10.51)$$

The step size can be fixed, for example, 0.1 V, or it can be based on the magnitude of (10.50). The voltage change from one step to the next is achieved by adjusting the duty ratio of the DC–DC converter.

Despite the conceptual simplicity, PO methods tend to oscillate around the MPP, resulting in submaximal power output. PO methods have slower convergence than some other methods, in particular when irradiance rapidly changes. For example, the PO method does not know if an increase in power after a perturbation is because the MPP is being approached or if the irradiance increased. This can lead the PO to perturb in the incorrect direction. The operating point can also be "trapped" at a local maximum in the power-voltage curve that occurs when part of the PV array is shaded. PO methods can be implemented using analog circuits or a simple digital controller.

Example 10.8 In this example, the perturb and observe method with variable step size is demonstrated. The I–V curve of a PV module in Fig. 10.24 is considered. The important values for each of the first several steps are provided in Table 10.4. The starting point (step 0) is arbitrarily selected as 12 V with a positive initial perturbation of 0.1 V in the positive direction. For simplicity, we assume that the I–V curve does not change. The step-size magnitude is set to

$$|\alpha| = \frac{1.5}{I[k]} \times \left| \frac{\Delta P[k]}{\Delta V[k]} \right|$$

(continued)

Fig. 10.24 The I–V and power characteristics for the module in Example 10.8

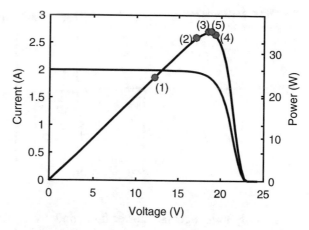

Table 10.4 Example of perturb and observe method

Step	V	I	P	ΔP	ΔV	α
0	12.00	2.00	24.00	–	–	0.10
1	12.10	2.00	24.20	24.20	0.10	5.00
2	17.10	1.97	33.64	9.44	5.00	1.44
3	18.54	1.89	35.11	1.47	1.44	0.81
4	19.35	1.79	34.61	−0.51	0.81	−0.53
5	18.82	1.87	35.11	0.50	−0.53	0.77

with a maximum value of 5.0. The reader should confirm the values in the table. The actual MPP is 35.13 W. Note that the algorithm nearly obtains this value in step 3 and then overshoots it, before returning.

10.12 Summary

This chapter covered how off-grid systems, mini-grids in particular, are controlled to achieve the desired balance of power among and between its components. Off-grid systems are generally controlled to minimize operating costs, maximize reliability, and prolong the life of their components. The charging and discharging of batteries require special consideration as their lifespan and performance can be threatened if the system is not properly operated. Most batteries are charged using a three-stage approach consisting of bulk, absorption, and float stages. The terminal voltage and battery current associated with each stage have a distinct profile. Inspection of the voltage profile in particular is useful in qualitatively assessing the health of the system and understanding if it is functioning properly.

Fig. 10.25 Mini-grid for
Problem 3 and 4

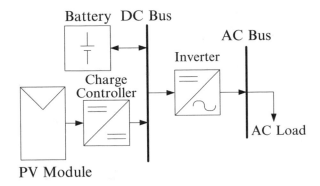

A steady-state algebraic power flow model was developed for a system with
multiple sources. The model is based on conservation of power. The model can
be combined with the operation scheme of the mini-grid to analyze and simulate
it. Operation schemes show the priority of actions that should be taken as load and
generation from uncontrolled sources change. They are usually based on minimizing
the operational cost of the mini-grid. When multiple generators or inverters are
coupled to the AC bus, they must be synchronized. Droop or isochronous control
can be used to share the real and reactive power between the generators. The chapter
also covered different maximum power point tracking methods for PV arrays.

As we proceed in the next chapters to discuss the design of off-grid systems,
we should not forget that the operation and control of the mini-grid require careful
consideration prior to implementation.

Problems

10.1 Describe, in your own words, the purpose of each stage in a three-stage battery
charging process.

10.2 Sketch a charging plot (voltage) of a 24 V system with the following set-
points: $V_{abs} = 28.4$ V and $V_{float} = 27$ V. Assume the battery's initial voltage is 26 V,
and the absorption stage lasts for 4 h.

10.3 Consider the mini-grid in Fig. 10.25. The AC load current is 17.5 A. The
battery resistance is 0.09 Ω. The battery is being charged in the absorption stage
with set-point 14.2 V. The state-of-charge voltage V_{SoC} is 12.1 V. What is the current
from the charge controller? What is the required power from the PV array assuming
the charge controller's efficiency is 95%?

10.4 Consider the mini-grid in Fig. 10.25. The AC load current is 9.1 A. The battery
resistance is 0.09 Ω. The battery is being charged in the absorption stage with set-
point 14.2 V. The state-of-charge voltage V_{SoC} is 12.1 V. The charge controller is

Fig. 10.26 Mini-grid for
Problem 5, 6 and 7

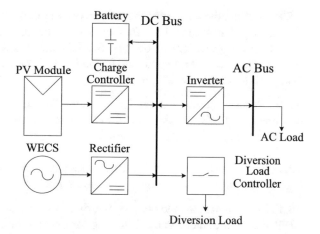

power-constrained due to a cloud reducing the irradiance. The maximum charge controller current is 11 A. Determine the battery's terminal voltage and comment on whether or not the terminal voltage is able to be regulated at the absorption stage set-point.

10.5 Consider the mini-grid in Fig. 10.26. The power from the PV array is 10 kW and the load is 16 kW. The battery is in the bulk charging stage, and so the diversion load is not consuming power. Compute the power required from the WECS to charge the battery with 2 kW if the inverter efficiency is 85%.

10.6 Consider the mini-grid in Fig. 10.26. The charge controller absorption set-point is 57.6 V; the diversion load set-point is 58.4 V. The battery is being charged in the absorption stage. The state-of-charge voltage V_{SoC} is 52 V. The current from the wind turbine is 56 A. The load current is 19 A. The battery resistance is 0.10 Ω. Compute the battery terminal voltage and current from the charge controller and to the diversion load, if any.

10.7 Consider the mini-grid in Fig. 10.26. The charge controller absorption set-point is 57.6 V; the diversion load set-point is 58.4 V. The battery is charged in the absorption stage. The state-of-charge voltage V_{SoC} is 52 V. The current from the WECS is 74 A. The load current is 8 A. The battery resistance is 0.10 Ω. Compute the battery terminal voltage and current from the charge controller and to the diversion load, if any.

10.8 Consider a mini-grid with two gen sets connected in parallel to the AC bus. Gen Set 1 is rated at 100 kW with droop slope of 0.15 and no-load frequency of 60 Hz. Gen Set 2 is rated at 80 kW with droop slope of 0.13 and no-load frequency of 60 Hz. Compute the power output and frequency of each gen set when the load is 110 kW.

10.9 Consider the gen sets in the previous example. The reactive power droop slope for Gen Set 1 is 10.0 and 8.0 for Gen Set 2. The no-load voltages are 230 V. The rated reactive power for Gen Set 1 is 65 kVAR and 50 kVAR for Gen Set 2. Compute the reactive power output and voltage magnitude of each gen set when the reactive power of the load be 60 kVAR.

10.10 What must the relationship between the droops, in percent, be for three gen sets rated at 50 kW, 75 kW, and 100 kW be for the power to be shared (1) equally among the gen sets or (2) in proportion to the rating of the gen sets?

10.11 Consider the plots in Fig. 10.27. Identify if the gen set is being operated in a load-following or cycle-charging scheme. Explain your reasoning.

10.12 Write a perturb and observe algorithm using a computer language to find the MPP of a PV module with the following parameters: $I_0 = 8 \times 10^{-10}$ A, illumination current of 4 A, and $V_T = 0.0258$ V. The module has 60 series-connected cells. Ignore the shunt and series resistance of the PV module.

Fig. 10.27 Mini-grid for
Problem 11

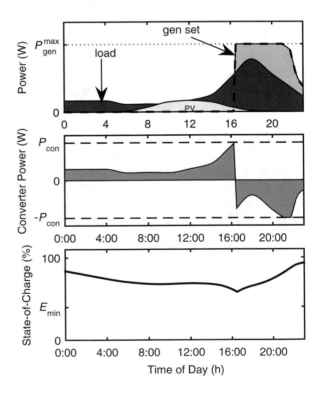

References

1. Joshi, P., Arora, S.: Maximum power point tracking methodologies for solar PV systems – A review. Renew. Sustain. Energy Rev. **70**, 1154–1177 (2017). doi:https://doi.org/10.1016/j.rser.2016.12.019. URL http://www.sciencedirect.com/science/article/pii/S1364032116310826
2. Lopes, J.A.P., Moreira, C.L., Madureira, A.G.: Defining control strategies for MicroGrids islanded operation. IEEE Trans. Power Syst. **21**(2), 916–924. https://doi.org/10.1109/TPWRS.2006.873018
3. Olson, G.: Paralleling dissimilar generators: Part 3 – load sharing compatibility. Tech. Rep. 9017, Cummins Power Generation (2010)
4. Ram, J., Rajasekar, N., Miyatake, M.: Design and overview of maximum power point tracking techniques in wind and solar photovoltaic systems: A review. Renew. Sustain. Energy Rev. **73**, 1138–1159 (2017). doi:https://doi.org/10.1016/j.rser.2017.02.009. URL http://www.sciencedirect.com/science/article/pii/S1364032117302137
5. Vader, R. (ed.): Energy Unlimited. Victron Energy (2011)
6. Zimmerman, R.D., Chiang, H.D.: Fast decoupled power flow for unbalanced radial distribution systems. IEEE Trans. Power Syst. **10**(4), 2045–2052 (1995). doi:10.1109/59.476074

Chapter 11
Load and Resource Estimation of Off-Grid Systems

11.1 Introduction

From a technical viewpoint, an appropriately designed off-grid system is one that strikes a reasonable balance between the cost of implementing and operating the system and the ability of the system to reliably and safely meet the needs of its users. To do this, we need to know the basic characteristics of the load the system is expected to serve. We must also know qualities of the energy resources that are locally available. For example, the effective head and flow rate of a water resource are needed to determine if micro hydro power is a suitable choice. This chapter discusses the load and resource characteristics that are most relevant from a design perspective and introduces approaches to their modeling and estimation.

11.2 Load Characterization

The load characteristics that are of most consequence when designing off-grid systems are:

1. Average daily load
2. Load profile
3. Peak load
4. Load variation and growth

Depending on the off-grid system, we might be interested in these characteristics as they pertain to a single user or a group of users.

© Springer International Publishing AG, part of Springer Nature 2018
H. Louie, *Off-Grid Electrical Systems in Developing Countries*,
https://doi.org/10.1007/978-3-319-91890-7_11

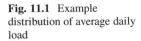
Fig. 11.1 Example distribution of average daily load

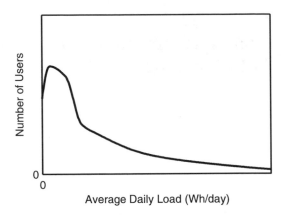

11.2.1 Average Daily Load

The average daily load $\bar{E}$ is the total energy consumed by a user or group of users over a period of time, typically 1 year, divided by the number of days in the period

$$\bar{E} = \frac{\text{total energy consumed over } D \text{ days}}{D}. \tag{11.1}$$

An off-grid system is designed so that its energy sources are capable of supplying at least the total average daily load plus losses.

A user's average daily load is influenced by several factors. Most important is the type of user. Households tend to have the lowest consumption, followed by social institutions and businesses. Agricultural and industrial facilities consume the most energy. There is often, however, wide variation even within each type of user.

The distribution of average daily load often resembles that in Fig. 11.1 [1]. Some households will consume little or no electricity, even though they are connected to a mini-grid. It could be that the property is abandoned or that the user is unable to afford electricity or appliances. Most households will use some electricity, but their average consumption tends to be very low, perhaps 30 to 50 Wh per day. This is far below the 1 kWh per day target discussed in Sect. 2.5. At these levels of consumption, it may take several years to recoup the financial investment made in connecting the user to the mini-grid. The distribution in Fig. 11.1 has a "long tail," meaning that a small number of users will consume much more than the others. This inflates the overall average daily load of the mini-grid to perhaps 50 to 200 Wh per user per day [1]. This of course will vary from one system to the next.

The long tail is evident in Table 11.1, which shows the breakdown of consumption by user type for a mini-grid in Haiti. Approximately three quarters of the users consume on average just 30 Wh/day. The largest commercial users consume about 75% of the total energy, despite constituting just 0.4% of all users. The economic viability of a mini-grid often depends on serving these large consumers. These consumers are referred to as "anchor" loads.

Table 11.1 Average consumption of mini-grid users (courtesy of Sigora Haiti)

User class	Consumption (kWh/yr)	No. of users	Percent of users (%)	Average consumption (Wh/day/user)
Residential (low)	0–299	844	76.1	30
Residential (high)	300–999	224	20.2	240
Commercial (low)	1000–4999	37	3.3	1479
Commercial (high)	>4999	4	0.4	5282

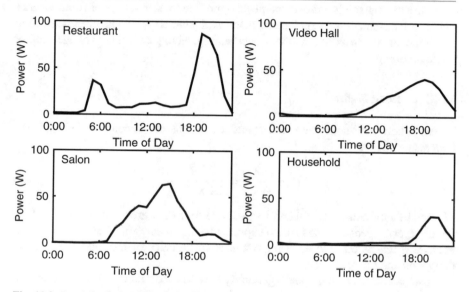

Fig. 11.2 Load profiles of several types of mini-grid users

11.2.2 Load Profile

The average daily load, while important, does not tell us when the energy is consumed. The timing of the load can be important, especially in systems using solar power. A larger battery is needed if the load primarily occurs in the evening than if the load occurs during the day. A load profile graphically shows how the average consumption varies throughout the day. Examples are shown in Fig. 11.2 for four different types of users [4]. The average daily load, in terms of energy consumed, is equal to the integral of the load profile (the area under the load profile).

The load profile for most off-grid households is similar to that shown in Fig. 11.2. Very little electricity is consumed during the day. The load steadily increases after sunset, when lights and perhaps a television or radio are turned on. Security lighting might be kept on throughout the night.

11.2.3   Peak Load

The peak load is defined as the maximum power consumed by the user (or group of users) within some time period, for example, 1 year. The peak load does not generally correspond to the peak of the load profile (the load profile is based on the *average* load computed over many days). The actual peak could have occurred on a holiday or day with extreme weather conditions. The peak load is important because it sets the minimum acceptable power rating for several system components. The inverters, generators, conductors, fuses, and other components must be rated to accommodate the peak load without overheating, having excessive voltage drop or malfunctioning.

11.2.3.1   Load Factor

The ratio between the average load $\bar{P}$ and the instantaneous peak load P_{peak} is the *load factor*[1] (*LF*):

$$LF = \frac{\bar{P}}{P_{\text{peak}}} = \frac{\bar{E}}{24 \times P_{\text{peak}}}. \tag{11.2}$$

The load factor can be calculated for a single user or a group of users. The load factor never exceeds 1.0 and is often expressed as a percent. The load factor is useful because it relates the two most important characteristics of the load: the average daily load and the peak load.

Load profiles with sharp peaks generally have low load factors. Low load factors are undesirable. To understand why, consider two different users each served by a stand-alone PV system. Each user has the same average AC load $\bar{P}$ but with different load factors. The load on the day with the peak power is shown in Fig. 11.3 for both users. The inverter serving each user must be rated to supply at least the user's peak power. Consequently, the user with the higher load factor can be supplied with a lower-rated (and less expensive) inverter.

11.2.3.2   Demand Factor

A related, but different, way of characterizing the peak load is the demand factor. The demand factor (*DF*) is the ratio of the peak power to the power that would occur if a user simultaneously turned on all of their appliances:

$$DF = \frac{P_{\text{peak}}}{\sum_{a=1}^{A} P_{a,\text{max}}} \tag{11.3}$$

where $P_{a,\text{max}}$ is the maximum power draw of the ath appliance and A is the number of appliances. The DF never exceeds 1.0 and is often expressed as a percent. For

---

[1] The Load Factor should not be confused with the power factor.

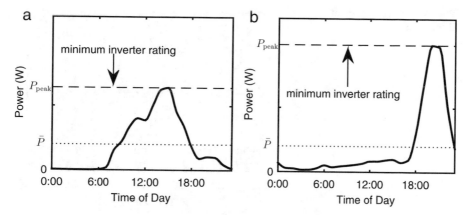

Fig. 11.3 (a) Power consumption of User A. (b) Power consumption of User B. Both have the same average load, but User A has a higher load factor and can be served by a smaller inverter than User B

the purposes of design, it is common practice, even in countries with mature and robust national grids, to assume that the demand factor of most users is significantly less than 1.0. In this way, smaller and less expensive equipment can be used. As an example, a user whose total appliance load is 5000 W with a demand factor of 0.6 can be served by a transformer rated at 3000 W instead of a more expensive 5000 W transformer.

Example 11.1 Consider a mini-grid serving Household A. For simplicity, assume that Household A's load is the same each day. The load profile is shown in Fig. 11.4. The power rating of the equipment is provided in Table 11.2. Compute the average daily load, the peak load, the load factor, and the demand factor. Make the simplifying assumption that each appliance consumes exactly the rated power when turned on.

Solution The average daily load is found by integrating the load profile. The reader should verify that this results in $\bar{E} = 258$ Wh. The peak load is 46 W, which occurs when the TV and the indoor CFL are both on. The load factor is found from (11.2):

$$LF = \frac{\bar{E}}{24 \times P_{\text{peak}}} = \frac{258}{24 \times 46} = 0.234.$$

The demand factor is found from (11.3):

$$DF = \frac{P_{\text{peak}}}{\sum_{a=1}^{A} P_{a,\text{max}}} = \frac{46}{11 + 11 + 15 + 35} = 0.639.$$

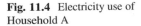

Fig. 11.4 Electricity use of Household A

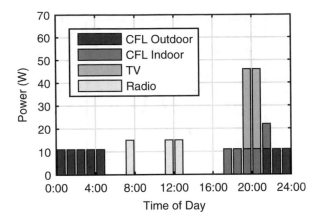

Table 11.2 Example appliance ratings

Appliance	Rating (W)
CFL	11
TV	35
Radio	15

11.2.4 Coincidence Factor

Larger-capacity systems such as mini-grids usually serve many users. We will refer to the total load supplied by the system as the "aggregate load." The ratio of the peak aggregate load $P_{\text{peak,agg}}$ to the sum of the individual peak loads is the coincidence factor (CF):

$$CF = \frac{P_{\text{peak,agg}}}{\sum_{n=1}^{N} P_{\text{peak},n}} \tag{11.4}$$

where $P_{\text{peak},n}$ is the peak power of the nth user. The coincidence factor is sometimes expressed as a percent. The coincidence factor can be similarly computed for a subset of users, rather than all of them. The inverse of the coincidence factor, known as the *diversity factor*, is sometimes also used to characterize the load.

The CF for a single user is 1.0. The CF tends to decrease as the number of users increases because the chance of the individual peaks overlapping in time is usually low. The CF is often expressed as a function of the number of users N as $CF(N)$. When $CF(N)$ is plotted, it is known as a coincidence factor curve (CF curve). An example CF curve for a mini-grid is shown in Fig. 11.5. Off-grid users, particularly households, tend to have much higher coincidence factors than those in developed countries. Residents in a village might all have access to the same limited set of appliances which are used around the same time (usually around sunset).

Fig. 11.5 Coincidence Factor of mini-grid users

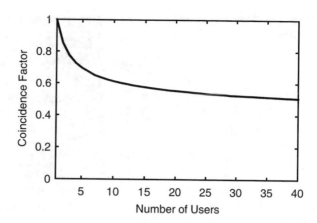

In general, it is more economical to serve users that as a group have a low coincidence factor. The reasoning is similar to why high load factors are desirable—lower-rated components can be used to supply the same total energy.

One strategy to reduce the CF is to seek users whose load profiles are anticipated to be different. For example, the peak load of many businesses is during the daytime, whereas for households it is in the evening. The CF when both user types are served will be lower than if one type is exclusively served by the system.

Example 11.2 Consider Household B whose daily consumption is shown in Fig. 11.6. Determine the coincidence factor of Household A and Household B.

Solution The aggregate load profile for the households is shown in Fig. 11.7. The coincidence factor is found from (11.4) to be

$$CF = \frac{P_{\text{peak,grp}}}{\sum_{n=1}^{N} P_{\text{peak},n}} = \frac{68}{46 + 46} = 0.739.$$

11.2.5 Load Variation and Growth

The consumption of most users will vary day-to-day and over time. Day-to-day variation in load poses a challenge for off-grid system design. To understand why, consider two users with similar average daily load, but with different daily load

Fig. 11.6 Electricity use of Household B

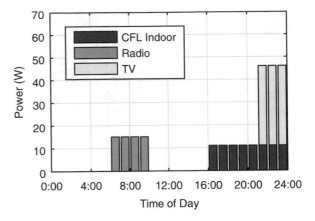

Fig. 11.7 Combined electricity use of Household A and B

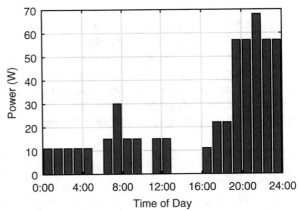

variability. The variability is evident in the histogram of the daily load, as in Fig. 11.8. A stand-alone system designed to serve User A with high reliability must be capable of supplying the maximum load, about 330 Wh/day. The system for User B, on the other hand, must be capable of supplying its maximum load, about 630 Wh/day. This system will likely be much more expensive, even though the average daily load is the same as User A. In practical terms, reliably serving users whose daily load variation—particularly their maximum load—is high is more expensive than a user with low variability. One way to mitigate the variation is to serve many, diverse, users from the same system. The overall variation will likely decrease because on days when one user has a higher-than-average load, another user might have a lower-than-average load.

The load of off-grid users tends to increase over time. Various load growth rates have been reported, but between 5 and 10% per year for the first several years seems reasonable. The growth rate does vary over time, however.

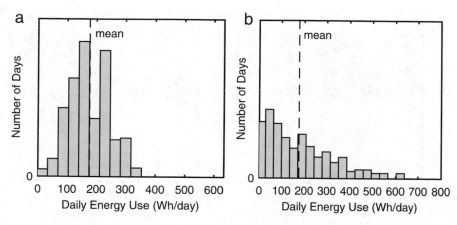

Fig. 11.8 Histogram of two users with similar average daily consumption but with different day-to-day variability

As strange as it may seem, for-profit mini-grid developers often must stimulate consumption. Although electricity is a high-quality fuel, most households will continue to "fuel stack" by using other fuels (see Sect. 2.3.4). If electricity is deemed to be too expensive, it will not be used. For most renewable energy-supplied mini-grids, the variable costs are low but the fixed costs are high. Therefore, the grid is more profitable when consumption increases (so long as reliability does not suffer). Ways of stimulating consumption include starting leasing or financing programs that reduce the barriers to high-power appliance ownership and setting tariffs that offer discounts as more energy is consumed. This is the opposite of what many large utilities do, where electricity becomes more expensive as consumption increases.

There are two general approaches to accommodating load growth. The first is to design the system based on the projected load at some number of years in the future. This is shown in Fig. 11.9a for an exponential load growth. The dashed line is the energy production capability of the system. The surplus capability is represented as the shaded area. The system will be oversized initially but will be able to serve the increased load in the future without modification.

The second approach is to use a modular design, adding energy production capability incrementally as the load increases. This happens at time T_{add} in Fig. 11.9b. Some upfront considerations are needed, such as reserving enough physical space for the future components to be added, as well as making sure certain components like inverters are capable of synchronizing with other sources. It is generally not recommended to add strings of new batteries to an existing battery bank, as the mixing of old and new batteries can cause uneven charging and discharging, resulting in failure. There are also additional logistic and installation costs when the modular approach is used.

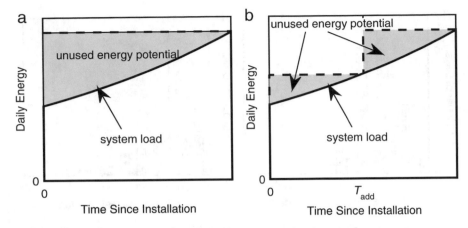

Fig. 11.9 Load growth can be accommodated by initially installing an oversized system as in (**a**) or by incrementally adding capacity as needed over time as in (**b**)

11.3 Load Estimation

We have so far discussed the characteristics of the load that affect the design and reliability of an off-grid system. In most cases, these characteristics cannot be known in advance, and so estimates are needed. Load estimation is considered by many to be the greatest challenge in appropriate off-grid system design. Errors of over 300% have been reported [1, 10].

If the estimated load is larger than the actual load, then the system will be over-designed. That is, the components will be larger than they need to be. This is an inefficient use of capital. Conversely, if the estimated load is less than the actual, then the system is under-designed. It will operate less reliably than needed. Neither of these outcomes are desirable.

There are four basic approaches to estimating the load:

1. Bottom-up
2. Survey
3. Regression
4. Data-driven

These are discussed in the following sections.

11.3.1 Bottom-Up Approach to Load Estimation

The bottom-up approach to load estimation is straightforward but can only practically be applied under certain circumstances. It requires knowing exactly what appliances are used, how much power each consumes, and when or how long they will be used each day. This information is virtually impossible to know ahead of time for most users, especially households. However, in some situations, for example, a school whose load is only lighting, it is a viable approach.

The load estimate for a single user is formed by assigning a power rating and loading percentage to each appliance and estimating the average daily load as

$$\hat{E}_n = \sum_{a=1}^{A} P_{\text{adj},a} T_{\text{app},a} \tag{11.5}$$

where $\hat{E}_n$ is the estimated average daily energy load in watthours of user n, A is the total number of appliances to be used, and $P_{\text{adj},a}$ and $T_{\text{app},a}$ are the assumed adjusted power rating and hours of use of appliance a, respectively. The adjusted power rating can be decomposed into $p_a \times K_a/100$ where p_a is the power rating of the appliance and K_a is the loading percent of the appliance. The loading percent varies between 0 and 100. It is less than 100 for appliances like refrigerators that do not continuously consume their rated power when turned on. Typical power ratings for various appliances are found in Table 11.3. Note that the rated power and loading percent can vary widely depending on the exact model of the appliance and the operating conditions. If the time of day that each appliance is used is known, then the load profile can be constructed by summing $P_{\text{adj},a}$ for all appliances in use at a given point in time from the start to the end of the day. If the usage is expected to change based on the day of the week—for example, if the weekend consumption is higher or lower than weekday—then a weighted average is used to estimate the average daily load.

The aggregate load is estimated by summing the estimated daily energy for all N users:

$$\hat{E} = \sum_{n=1}^{N} \hat{E}_n. \tag{11.6}$$

11.3.2 Survey Approach to Load Estimation

A survey approach can be used when the appliances and their rating and usage are not known in advance. This is a popular approach to load estimation which has been in use since at least the 1990s [5, 7, 12]. In it, surveys are administered to each customer prior to the installation of the mini-grid. A typical survey might ask about the types and quantities of appliances the customer believes they will own, as well as the anticipated hours of use on average each day. Assumptions are made about the power rating and loading percent for each appliance. The load estimation is then done as in the bottom-up approach, using (11.5).

Table 11.3 Example
appliance ratings

Appliance	Rating (W)	Load Per. (%)
Blow dryer	1369	100
CFL bulb	11	100
Cooker	5000	Varies
Desktop computer	225	33
DVD player	21	100
Fan	20	100
Freezer	19	100
Laptop computer	60	40
LED light	7	100
Phone charger	3.7	100
Television	35	100

Example 11.3 In a survey, a mini-grid customer states they will use four light bulbs for 8 h each night, a TV set for 3 h each night, and a refrigerator that will be continuously plugged in. Estimate their daily consumption using the values in Table 11.3.

Solution The average daily load is estimated from (11.5) to be

$$\hat{E}_c = \sum_{a=1}^{A} \frac{p_a K_a T_{\text{app},a}}{100} = \frac{4\,(8 \times 100 \times 11)}{100} + \frac{(35 \times 100 \times 3)}{100}$$

$$+ \frac{(90 \times 50 \times 24)}{100} = 1537 \text{ Wh}.$$

The refrigerator is responsible for about two-thirds of the total load.

Although the survey approach seems straightforward and is widely adopted, it is generally inaccurate and difficult to implement [1, 9]. Respondents may have trouble accurately predicting the appliances they might acquire and when they will be used. Remember, it will likely be the users' first time having access to electricity. They might be unfamiliar with how much energy each appliance consumes and whether or not they can afford to use it as often as they would like. It is also possible that the survey or interviewer poses questions in biased, confusing, or ambiguous way or that the respondent is untruthful. The survey method can be tedious—interviewing dozens or more customers per mini-grid takes considerable time and resources.

Experienced practitioners are wary of using surveys. Many view them as a useful starting point in estimating the load, but expect the actual load will not reconcile with that estimated by a survey. There is a tendency for the survey approach to overestimate consumption. One study found that 70% of the users overpredicted their consumption when compared to their actual usage [1].

If the survey asks about what time of day each appliance is used, then a load profile can be created, and the load factor and demand factor can be computed. The responses from several users can be combined to estimate the coincidence factor.

Example 11.4 A mini-grid with a single gen set is to be designed to serve 18 users. A survey approach was used to estimate the load. It is estimated that half of the users will each have appliances whose combined power rating is 89 W and the other half will each have appliances whose combined power rating is 119 W. The demand factor for each group is estimated to be 0.7. The coincidence factor is estimated to be 0.60. Compute the minimum rated power of the gen set.

Solution The peak load for each user in the first group is found from (11.3) to be

$$P_{\text{peak}} = DF \sum_{a=1}^{A} P_{a,\text{max}} = 0.7 \times 89 = 62.3 \text{ W}.$$

The peak load for the second group is found using the same equation to be $0.7 \times 119 = 83.3$ W. The minimum rated power of the gen set is equal to the aggregate peak load, which is found from (11.4) to be

$$P_{\text{peak,grp}} = 0.6 \times (9 \times 62.3 + 9 \times 83.3) = 786.24 \text{ W}.$$

11.3.3   Regression Approach to Load Estimation

In some cases, the average daily load can be estimated from basic demographic or census data using regression. Regression is a statistical technique that models an output variable—in this case average daily load—from explanatory variables. Several explanatory variables have been proposed: the number of people living in the house, education level, income, distance to the nearest grid, and even whether or not the household has a flushing toilet. Many others have been considered [6, 11, 16].

Since the explanatory variables are often available in existing databases, a regression-based load estimation can be done quickly and cheaply. The biggest disadvantage to regression methods is that the underlying models have not been shown to be widely applicable. It is possible that a model that is accurate in estimating consumption in one community is not valid in another community or country.

11.3.4 Data-Driven Approach to Load Estimation

Perhaps the most promising method for load estimation is the data-driven approach [1]. In this approach, the load is estimated from the measured load of existing and similar mini-grid users. It is important that the mini-grids have similar technical and economic characteristics—especially the tariff—and the demographics of the users be comparable. As a simple example, the average aggregate daily load for a proposed mini-grid with 30 households in Tanzania would be estimated to be twice that of an existing mini-grid in a nearby community in Tanzania with 15 households. The data-driven approach requires robust data sets of historical consumption, which presently are not readily accessible.

A novel, but perhaps expensive alternative to the load estimation approaches discussed, is to use temporary gen sets to serve the load for a period of time. Load data are collected during this period, which are used to inform the design of the permanent system. Once the system is installed, the gen sets are removed.

11.4 Energy Resource Characterization

The suitability of a specific energy conversion technology for powering an off-grid system largely depends on the characteristics of the underlying energy resource. Table 11.4 shows the information needed to begin the electrical design of an off-grid system for a given conversion technology. Other characteristics might be needed to refine the design. For example, the ambient temperature is needed to model the effect of temperature on the power produced by a PV array.

The quality of the data required differs for each resource. Wind tends to be the most erratic resource, and so it is important to directly measure it at regular intervals to create a time series data set; on the other hand, the average solar insolation for almost any location on earth is accessible from online databases. In most cases, monthly or seasonal averages are needed to model the resource's variation over the course of the year.

Once the relevant resource data are gathered, the energy production potentials are estimated. The reader may wish to consult Chaps. 5 to 7 to review the principles of operation of the different energy conversion technologies. The energy production potential is often expressed as the *capacity factor*. The capacity factor is useful in evaluating the suitability of a conversion technology to the local resource and is discussed next. The rest of the chapter covers methods for measuring and evaluating each resource and how their characteristics affect the potential energy production.

Table 11.4 Energy resource characterization data requirements

Conversion technology	Physical characteristic	Data quality	Typical data source
PV array	Insolation	Average per day by month	Solar database
Wind Energy Conversion System (WECS)	Wind speed	1–2 year time series (preferred) or average by month	Direct measurement
Biomass	Feedstock type, yield	Average yield by month	Agricultural database, local observation
Micro hydro power (MHP)	Head, flow rate	Average flow by month or season	Direct measurement
Conventional Gen Set	Fuel availability	Availability by season	Local observation

11.4.1   Capacity Factor

In the context of resource estimation, the capacity factor is an indicator of how suitable a resource is for electricity generation. We modify the classic definition of the capacity factor to be appropriate for our discussion on off-grid systems. We define the capacity factor as the ratio of the *estimated* energy production over a period of time to the energy it is capable of producing if operated continuously at its rated capacity. We assume that downtime for maintenance and repair is negligible and the load is large enough so that the energy conversion technology is never throttled (see Sect. 10.7). We consider the efficiency of the energy conversion technology itself, but ignore other losses such as those associated with wiring, aging, dust and shading. That is, its power production is only limited by its rating and the availability of the energy resource. With this understanding, the capacity factor is generically expressed as

$$Capacity\ Factor = \frac{\hat{E}}{T \times P_{\text{rated}}} \tag{11.7}$$

where $\hat{E}$ is the estimated production. The period of time considered T matches that of the resource data—typically monthly, seasonally, or yearly. The capacity factor is also often expressed as a percent. The capacity factor is useful because it incorporates the technical aspects of the energy conversion technology as well as the characteristics of the resource. It is unitless and so energy conversion technologies of different ratings can be compared.

Table 11.5 shows the typical ranges of capacity factor for different energy conversion technologies. The last column of Table 11.5 shows the rated power capacity required to equal the energy produced by a hypothetical generator supplying 1 kW of continuous power (24 kWh/day). Because most components are priced based on their rated power capacity, the capacity factor can be used to determine which technology is most economical.

Table 11.5 Energy
conversion technology
capacity factors

Type	Approx. capacity factor	Capacity needed to produce 24 kWh/day (kW)
Gen set	0.90–1.00	1.00–1.11
PV	0.15–0.24	4.1–6.67
MHP	0.90–0.95	1.05–1.11
WEC	0.20–0.30	3.33–5.00

Example 11.5 A mini-grid will have an estimated load of 48 kWh per day.
If served by a two-nozzle Pelton turbine, the water resource is such that the
capacity factor will be 0.91. If the load is served by a PV array, the capacity
factor will be 0.18. Compute the required capacity of the Pelton turbine and
the PV array to supply the load.

Solution The required capacity of the Pelton turbine is found from by
rearranging (11.7) to be

$$P_{rated,MHP} = \frac{\hat{E}}{T \times Capacity\ Factor} = \frac{48.0}{24 \times 0.91} = 2.20\ \text{kW}.$$

Following the same approach, the PV array must be rated at

$$P_{rated,PV} = \frac{48.0}{24 \times 0.18} = 11.11\ \text{kW}$$

which is much larger than the MHP system. Even if the PV array was five
times less expensive per kilowatt of capacity, the MHP solution would be
less expensive. This of course ignores the cost of other components, such as
batteries, inverters, and the conveyance system.

11.4.2   Solar Resource

A solar resource is characterized by its average daily insolation. The insolation I_d
on day d is the integral of the irradiance it receives:

$$I_d = \int_{t=0}^{24} G_d(t)\mathrm{d}t \tag{11.8}$$

where $G_d(t)$ is the irradiance at time t on day d in kilowatts per meter squared and t is the time in hours. The average daily insolation $\bar{I}_{\mathscr{D}}$ is simply

$$\bar{I}_{\mathscr{D}} = \frac{\sum_{d \in \mathscr{D}} I_d}{D} \tag{11.9}$$

where $\mathscr{D}$ is the set of days the average is calculated for—usually a particular month or year—and D is the number of days in the set. The units of average daily insolation are kilowatthours per meter squared per day. The average daily insolation is sometimes referred to as "sun hours" or "full sun hours." For example, a location with insolation of 5.1 kWh/m^2/day is said to receive 5.1 sun hours per day. Do not confuse this to mean that there are only 5.1 h of daylight. Most locations suitable for PV-based mini-grids have at least four sun hours of insolation.

The average daily insolation for many parts of the world is available from online solar resource databases. Some databases include the average insolation by month and also consider the effects of tilting the PV array. If a database is unavailable, then a device known as a pyranometer can be used to measure the insolation.

11.4.2.1 Capacity Factor

Recall from Sect. 7.9 that the maximum power produced by a PV module or array, ignoring temperature effects, with the irradiance expressed in kilowatts can be approximated as

$$P_{\text{PV}} = P_{\text{STC}}^* \times G \tag{11.10}$$

where P_{STC}^* is the rating of the PV array. The estimated energy production of the PV array on day d is the integral of P_{PV}, so that

$$\hat{E}_{\text{PV},d} = P_{\text{STC}}^* \int_{t=0}^{24} G_d(t) \mathrm{d}t = P_{\text{STC}}^* I_d \tag{11.11}$$

and the estimated total production is

$$\hat{E}_{\text{PV},\mathscr{D}} = P_{\text{STC}}^* \sum_{d \in \mathscr{D}} I_d = P_{\text{STC}}^* \times \bar{I} \times D. \tag{11.12}$$

From (11.7), the capacity factor for the days considered is

$$Capacity\ Factor = \frac{\hat{E}_{\text{PV},\mathscr{D}}}{24 \times D \times P_{\text{STC}}^*} = \frac{P_{\text{STC}}^* \bar{I} D}{24 \times D \times P_{\text{STC}}^*} = \frac{\bar{I}}{24}. \tag{11.13}$$

This simple result shows that the capacity factor of a PV array can be estimated solely from the average insolation and is independent of the rating of the PV array.

Example 11.6 A large mini-grid located in Cameroon has a PV array rated at 30 kW. The average daily insolation is 4.9 kWh/m^2/day. Compute the capacity factor.

Solution Using (11.11), the average daily energy is

$$E_{PV} = P^*_{STC} I = 30 \times 4.9 = 147 \text{ kWh}.$$

The capacity factor is computed using (11.7)

$$\text{Capacity Factor} = \frac{E}{T \times P_{\text{rated}}} = \frac{147}{24 \times 30} = 0.204.$$

This can also be computed from (11.13) to be 4.9/24 = 0.204.

11.4.3   Other Solar Resource Considerations

Aside from the atmospheric conditions, the average daily insolation received by a PV array depends on its location, in particular the latitude, and on its orientation. We next briefly discuss these factors.

11.4.3.1   Latitude

Locations near the equator tend to have the most consistent insolation throughout the year, whereas near the north and south pole the seasonal variation is most pronounced. The insolation by month for Kisumu, Kenya, located on the equator and Cape Town, at the southern tip of South Africa, are shown in Fig. 11.10. We see that Cape Town's insolation, and therefore capacity factor, is sharply reduced in the winter months (June and July in the southern hemisphere) and is at its highest in the summer. On the other hand, the insolation is fairly consistent year round in Kisumu.

11.4.3.2   Orientation of PV Array

The orientation of a PV array refers to its tilt with respect to the horizontal plane and its azimuth. Figure 11.11 shows a tilted PV array for a mini-grid in Nigeria. The azimuth is the skewness from true north or south. The orientation affects the angle that the sun's rays strike the array. The angle at any moment can be calculated

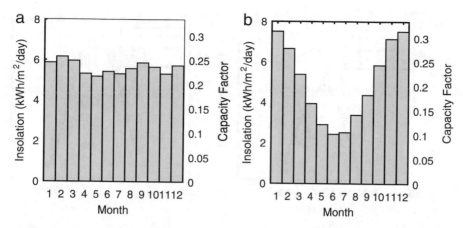

Fig. 11.10 Average daily insolation and capacity factor by month of a horizontal PV array for (**a**) Kisumu, Kenya (0°), and (**b**) Cape Town, South Africa (34°S)

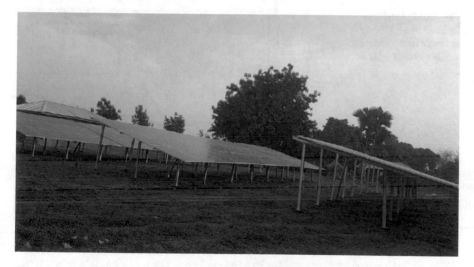

Fig. 11.11 PV modules like these for a mini-grid in Angwan Rina, Nigeria, are often mounted on a rack to obtain the desired tilt (courtesy of GVE Projects)

from the location, orientation, time of day, and day of the year [14]. The irradiance is maximized when the sun's rays are normal to the surface of the PV array.

Figure 11.12a shows how the tilt affects irradiance in the winter and summer months on a north-facing PV array located in the southern hemisphere. The area under an irradiance curve is the insolation, a measure of the total solar energy over a day. Three different tilt angles are shown: not tilted (horizontal, 0°), tilted at the latitude (ϕ), and tilted at twice the latitude (2ϕ). Clearly the tilt affects the irradiance received and therefore the total energy produced by a PV array.

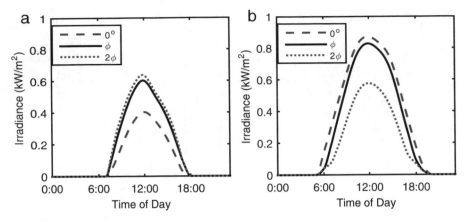

Fig. 11.12 Irradiance on a PV array at various tilts during (**a**) winter and (**b**) summer months for Cape Town, South Africa

Fig. 11.13 Irradiance incident to a PV array in the southern hemisphere for various azimuth angles

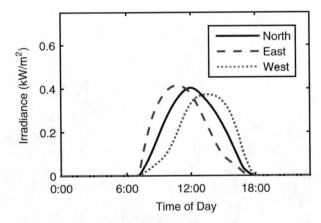

An example of how the azimuth affects the irradiance is shown in Fig. 11.13. A PV array facing east receives greater irradiance during the morning, whereas a west-facing array captures more irradiance in the afternoon. Generally, the insolation in either case is less than if the array was not skewed away from true north (or south).

11.4.3.3   Optimal Orientation

As a rule of thumb, the tilt of a PV array should be set to the latitude of its location and its azimuth should be so that the array faces toward the equator. Figure 11.14 shows the insolation at various tilts and azimuth angles and the corresponding capacity factors when compared to a north-facing PV array tilted at latitude (the black line). Tilting at latitude yields the greatest average yearly insolation and capacity factor. This tends to be the case in most locations and so has become

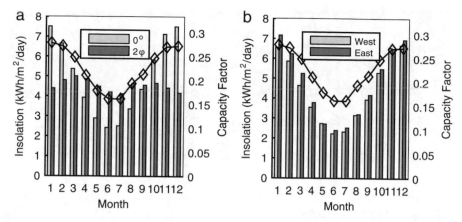

Fig. 11.14 Average daily insolation and capacity factor for a PV array at various (**a**) tilts, and (**b**) azimuths for a location near Cape Town, South Africa

the default practice. However, as was shown in Fig. 11.12, the insolation can be increased by adjusting the tilt throughout the year, particularly in the summer and winter. In most situations the effort and expense involved in adjusting the tilt of the array throughout the year are not worth the additional energy yield.

While the tilt-at-latitude rule of thumb will generally achieve optimal or near-optimal power production potential, there are a few caveats. The minimum tilt is recommended to be no less than 5 to 10°. This prevents debris from collecting on the surface of the PV module. If the load has a pronounced seasonal component—for example, it is much larger in January than June—then the tilt can be set to maximize the capacity factor during that month. Sometimes the roof line or site conditions only allow a PV array to face east or west. In this case, a west-facing array often yields higher production—evenings tend to be less cloudy than mornings, and the load tends to be greater near sunset than sunrise.

Example 11.7 Estimate the average daily energy production and capacity factor for a north-facing 2.4 kW PV array tilted at twice its latitude (2ϕ) in Cape Town, South Africa, in the months of January and March, based on the insolation in Fig. 11.14.

Solution From Fig. 11.14a, the insolation in January is 4.4 kWh/m²/day. The energy the PV array is able to produce on average in January from (11.11) is therefore

$$E_{PV} = P^*_{STC} \times \bar{I} = 2.4 \times 4.4 = 10.56 \, \text{kWh/day}.$$

(continued)

The capacity factor can be found from (11.7) or equivalently by dividing the
sun hours by 24: 4.4/24 = 0.183. Repeating the calculation for March where
the insolation is 5.0 kWh/m^2/day shows

$$E_{PV} = 2.4 \times 5.0 = 12.0 \, \text{kWh/day}$$

and a capacity factor of 5.0/24 = 0.208.

11.4.4 Wind Resource

The energy a wind energy conversion system (WECS) is capable of producing
primarily depends on its power curve and the wind speed at the hub's height. It
is good practice to measure the wind speed at the location prior to committing to
implementing the system. Hourly sampling is usually sufficient. If this not possible,
then a rough model can be developed based only on the estimated average wind
speed. This model is discussed later.

General rules of thumb or localized wind speed maps can be used to prospect
for windy areas. Inspection of trees and shrubs for evidence of consistent high
wind can be useful for prospecting but should not replace direct measurement for
final decision-making. Talking with local residents to identify windy areas can be
useful in prospecting, but the reliability of this method can be questionable as most
firsthand experiences are at ground level and few people have firsthand knowledge
of what a suitable wind resource feels like.

It is useful to create a histogram of the data. An example is shown in Fig. 11.15a.
The histogram shows how often the wind speed will be below the cut-in or
above the cut-out speed of a particular WECS. At these wind speeds no power is
produced. Some locations or wind turbines can be quickly eliminated from further
consideration by inspecting the histogram.

11.4.4.1 Capacity Factor

The capacity factor of a particular WECS for a given location is found by first
applying the WECS's power curve to each measured wind speed. Recall from
Sect. 6.2.7 that the power curve relates a wind speed $v(t)$ at time t to the power
output by the WECS. The power curve can be generically expressed as the function
$g(v)$ so that the power produced by the WECS is

$$P_{WECS}(t) = g(v(t)). \tag{11.14}$$

For example, applying the power curve of the 10 kW WECS from Fig. 6.13 in
Chap. 6 to the wind speed data in Fig 11.15a results in the histogram of WECS

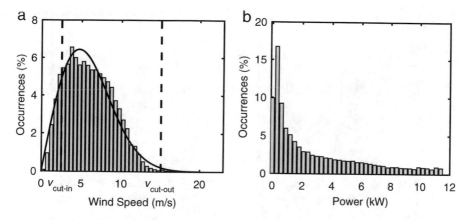

Fig. 11.15 (a) Histogram of wind speed with Rayleigh model superimposed with the black line and (b) histogram of WECS power production

power shown in Fig 11.15b. Although specific to this particular wind resource and WECS, the shape of the histogram is fairly typical. Most WECS do not produce power about 10 to 30% of the time, as the wind is below the cut-in wind speed or, occasionally, above the cut-out speed.

The total energy production for each month is estimated as

$$\hat{E}_{\text{WECS},\mathscr{D}} = 24 \times D \times \frac{\sum_{t \in \mathscr{D}} P_{\text{WECS}}(t)}{s} \qquad (11.15)$$

where $\mathscr{D}$ is the set of measurements that are in the month under consideration with D days and s is the number of wind speed measurements. The capacity factor is found by applying (11.7)

$$Capacity\ Factor = \frac{\hat{E}_{\text{WECS},\mathscr{D}}}{24 \times D \times P_{\text{WECS,rated}}} \qquad (11.16)$$

where $P_{\text{WECS,rated}}$ is the rated capacity of WECS. This is same as the average power divided by the rated power of the WECS. The result is a capacity factor for each month. An example is shown in Fig. 11.16. Note that the capacity factor is not proportional to the average wind speed due to the nonlinear nature of the power curve. It is for this reason that the average wind speed alone is insufficient to characterize the wind resource without other assumptions, as discussed next.

11.4.4.2   Wind Speed Model

When measured wind speed data are not available or the data set is incomplete, a probabilistic model can be used. The only required information is the average wind

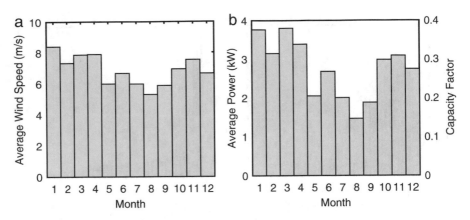

Fig. 11.16 (**a**) Wind speed and (**b**) WECS capacity factor by month based on the power curve in Fig. 6.13

speed (or an estimation of it). The wind speed at any moment is assumed to be a random variable $\tilde{v}$. Associated with any random variable is a probability density function (pdf) that describes the likelihood that the random variable will assume any value. You might be familiar with the Gaussian or uniform pdfs. A Rayleigh pdf $f(\tilde{v})$ is often used to model the wind resource:

$$f(\tilde{v}) = \begin{cases} \frac{\tilde{v}}{c^2} \, e^{(-\tilde{v}^2/2c^2)} : v > 0 \\ 0 \quad : v \leq 0 \end{cases} \tag{11.17}$$

The parameter c is related to the average wind speed:

$$c = \frac{\bar{v}\sqrt{2}}{\sqrt{\pi}}. \tag{11.18}$$

Random samples drawn from this distribution can be used as synthetic wind speed data. The capacity factor is calculated from the synthetic wind speed as if it were measured. However, we should expect the estimated capacity factor to be less accurate than if measured data were used. If additional but still incomplete data are available, a Weibull pdf can be fit to the data, often yielding better results [2].

11.4.4.3   Considerations

An important characteristic of the wind resource is that it varies throughout the day. Figure 11.17a shows the wind speed for two different days. When averaged over the course of a month or year, a daily pattern might emerge. An example is shown in Fig. 11.17b. Although this pattern is useful in knowing if the power from the WECS coincides with the load, the wind speed during each day can deviate considerably from the average.

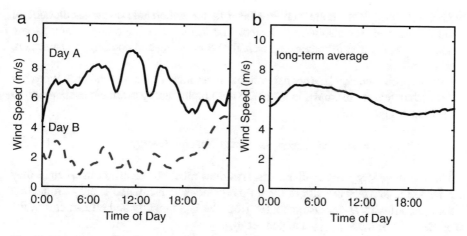

Fig. 11.17 (a) The wind speed on individual days (b) the average wind speed

11.4.5 Biomass

The first step in a biomass resource assessment is to identify the availability of the possible feedstock: what it is, where it is, and what it costs to procure, transport, store, and process. In some countries, high-level regional maps can be consulted for preliminary information on biomass availability, including crop yield per hectare. The United Nations Food and Agriculture Organization maintains a database with basic data for approximately 200 countries [15]. The availability of animal waste usually requires local assessment.

11.4.5.1 Capacity Factor

We will consider the case of a biomass gasification system using crop residue. The calculation using anaerobic digestion or direct combustion using other feedstock is similar.

A rough approach for estimating the biomass resource from crop residue in an area is

$$Potential\ crop\ residues = crop\ yield \times r \quad (11.19)$$

where r is the residue-to-crop ratio. It is the mass of the residue per mass of crop produced [3]. This value can exceed 1.0. For example, the ratio is about 1.8 for cotton. The potential crop residue is often expressed in tonnes per hectare per year. The mass of the residue available for energy production is estimated from

$$Available\ crop\ residues = residue\ production \times \eta_{collection} - alternative\ uses \quad (11.20)$$

where $\eta_{\text{collection}}$ is the collection efficiency factor, which acknowledges that not all the residue can reasonably be collected. The last term is important because some residue is often used for other purposes: fuel for heating and cooking, construction, and fertilizer.

The specific energy (energy per unit mass) associated with several crop types has been tabulated, for example, in [13]. The total input energy available each month to the biomass gasifier is

$$\hat{E}_{\text{in}} = \textit{Available crop residues} \times \textit{Specific Energy}. \tag{11.21}$$

The specific energy and available crop residues must be consistent in whether they refer to the dry or wet mass of the crop residue. The estimated energy production for a month with D days is computed from the input energy and efficiencies of the gasification process η_{reactor} and gen set η_{genset}:

$$\hat{E}_{\text{B}} = \hat{E}_{\text{in}} \times \eta_{\text{reactor}} \times \eta_{\text{genset}} \tag{11.22}$$

The capacity factor is found by applying (11.7):

$$\textit{Capacity Factor} = \frac{\hat{E}_{\text{B}}}{24 \times D \times P_{\text{genset,rated}}} \tag{11.23}$$

where $P_{\text{genset,rated}}$ is the rated power of the gen set powered by the syngas produced by the gasification process. Note that many biomass systems use gen sets fueled by the syngas as well as diesel. It may be necessary to assess the availability of the diesel fuel.

11.4.5.2 Considerations

A biomass project must be mindful of the existing or emerging uses for the feedstock and the pressure that the project might have on the availability and cost of the biomass. Understanding the seasonal availability is extremely important, and so the available feedstock should be computed by month or season.

11.4.6 Hydro

A hydro resource is characterized by two quantities: the available flow rate and the elevation head. The elevation head is the vertical distance from where the conveyance intake will be to the turbine. Recall that this head is different from the effective head. There are several methods of measuring head and flow. A brief description of select methods is provided next. The reader should consult references such as [8] for specific details.

The elevation head can be measured using a portable altimeter or using a survey rod and clinometer level to measure the incline. This can be tedious if the terrain is steep and rough. In some cases the measurements are made across long distances, which can make accurate measurement challenging. A long hose with a pressure gauge at one end can also be used. The hose is allowed to fill up with water, and the pressure at the bottom of the hose is measured. Since the density of water is known, the pressure can be converted to head:

$$elevation\ head(m) = \frac{pressure\ (Pa)}{g(m/s^2)} = pressure\ (Pa) \times 0.000102 \qquad (11.24)$$

where the pressure shown in pascals (Pa) and g is the gravitational constant (9.81 m/s^2). It is a good idea to repeat the measurement a few times using different methods and then average the results.

Measuring flow is more difficult than head, but there are several approaches that can be used. For small streams, the entire stream is diverted into a container with a known volume. How long it takes for the container to fill is documented. The flow rate is the volume divided by the time. Another approach estimates the flow by multiplying the velocity of the water flow through a known cross-sectional area. The cross-sectional area of the stream at a certain location is estimated, usually by measuring and averaging the depth at several areas and multiplying it by the width of the stream. The velocity can be measured using a flow meter or by estimating the average speed by placing a floating object on the surface, measuring its velocity, and then noting that average stream velocity is often about 0.6 to 0.8 of the surface velocity. Another approach involves constructing a weir across the stream. In the weir a notch with known area is cut out, and the height of the water passing over it is measured. The flow rate can be computed from the height of the water as described in [8]. Other methods involve pouring a known mass of salt upstream and plotting the conductivity downstream against time. The flow rate can be deduced from this plot as described in [8].

Although the head will not change with the season, the flow most likely will. Multiple flow measurements will be needed to be taken at different times of the year or at least during the months of the lowest and highest flow. The MHP should only use a portion of the total flow in order to reduce its impact on the streams ecology and on competing uses of the water, especially irrigation and drinking.

The maximum flow available for use in the MHP system $Q_{MHP,max}$ is the difference in the total flow rate minus the minimum flow rate needed in the stream for other purposes:

$$Q_{MHP,max} = Total\ Flow\ Rate - Minimum\ Stream\ Flow\ Rate \qquad (11.25)$$

The flow used by the MHP is designated Q and should not exceed $Q_{MHP,max}$.

11.4.6.1 Capacity Factor

The power produced by a MHP can be determined from the effective head and
flow rate as detailed in Sect. 6.3. The effective head H can be determined from
a preliminary estimate of the conveyance system efficiency, for example 90%.
Although the head is not adjustable, the flow rate is. For example, a spear valve
can adjust the nozzle flow of an impulse turbine. When operating at a lower flow
rate, the jet velocity will remain fairly constant, but the power in the water P_{wa} is
reduced according to (6.24):

$$P_{wa} = \rho_{wa} \times g \times H \times Q \tag{11.26}$$

where ρ_{wa} is the density of the water. The estimated energy production for a month
with D days is computed from the average flow that can be used by the MHP that
month $\bar{Q}$ according to

$$\hat{E}_{MHP} = 24 \times D \times P_{MHP}(\bar{Q}) \tag{11.27}$$

where $P_{MHP}(\bar{Q})$ is the electrical power produced by a specific MHP system when
the flow is $\bar{Q}$ (see Sect. 6.3). This value will change if a different hydroturbine or
MHP configuration is considered. The capacity factor for the month is found by
applying (11.7)

$$Capacity\ Factor = \frac{\hat{E}_{MHP}}{24 \times D \times P_{MHP,rated}} = \frac{P_{MHP}(\bar{Q})}{P_{MHP,rated}} \tag{11.28}$$

where $P_{MHP,rated}$ is the rated power of the MHP generator. If the MHP system is
controlled so that it operates at constant speed, then the efficiency of the turbine will
not vary with the flow rate. In this case,

$$Capacity\ Factor = \min\left\{\frac{\bar{Q}}{Q_{rated}}, 1.0\right\} \tag{11.29}$$

where Q_{rated} is the flow rate at the rated power. The capacity factor cannot
exceed 1.0.

Example 11.8 A small mini-grid is being designed for a community. The
elevation head was measured to be 29 m, and the flow rate was measured
four times during the year. The average measured total flow rates are shown
in the first row of Table 11.6. At least 15 l/s of flow must be maintained in the

(continued)

stream for other purposes. Estimate the capacity factor of a Pelton turbine for each month.

Solution The maximum flow $Q_{MHP,max}$ that can be used by the MHP system is found from (11.25). It is apparent that the hydro resources have a strong seasonal variation. We have a choice of what flow Q_{rated} to design the system around. The total annual energy will be greatest if we design the system around the month with the maximum average flow $Q_{rated} = 185$ l/s. The capacity factor for each month is found from (11.29) with results shown in the third row of Table 11.6. The average capacity factor will be low as much of the capacity will not be used most of the year.

It might be better to design the MHP around a lower flow rate. However, in doing so, the total energy produced over the course of the year will be reduced as some of the available flow is not used by the MHP. If, for example, a flow rate of 8 l/s is sufficient to provide the energy needs of the community, then the resulting capacity factor will be as shown in the fourth row of Table 11.6. The average capacity factor is much higher. Even still, there will be times of the year when the power produced by the system will be lower, and so the load might not be fully served.

Table 11.6   Total flow rate for Example 11.8

	Jan.	Mar.	Jul.	Sept.	Avg.
Total flow (l/s)	30	200	20	28	69.5
Max. MHP flow $Q_{MHP,max}$ (l/s)	15	185	5	13	54.5
Capacity Factor, $Q_{rated} = 185$ l/s	0.081	1.000	0.027	0.07	0.295
Capacity Factor, $Q_{rated} = 8$ l/s	1.000	1.000	0.625	1.000	0.906

11.4.7   Conventional Gen Set

The resource for a conventional gen set is typically diesel, petrol, or natural gas. Although no specific physical measurements are required, the availability of these fuels should be investigated. Transporting fuel to some locations can be problematic during certain times of the year—for example, the rainy season. These disruptions should be planned for, often by increasing the on-site storage. The capacity factor of the gen set is nominally 1.0.

11.5 Summary

The design of an off-grid system requires accurate and detailed information about the load it is expected to serve and the energy resource available to supply the system. There are several characteristics of the load that are relevant: the average daily load, the load profile, the peak load, and the variation and growth of the load. It is important to account for the coincidence in load when designing a system. Designing around the worst case scenario of all users using all their appliances at the same time is not usually warranted and would lead to an unnecessarily costly design.

There are several approaches to estimating the load. The survey approach is common, but in many cases it has been shown to be prone to large errors. A data-driven approach which uses historical data from similar existing systems is promising, but the availability of these data is limited.

Different data are needed to characterize the different energy resources. The type, quality, and source of data needed depend on the resource considered. The solar resource, described by the average daily insolation, is usually the easiest to obtain as an online solar database can be consulted. Characterizing the wind resource requires the highest quality data because it is very site specific and varies throughout the day and year. Direct measurement is recommended. The biomass resource is characterized by the type of feedstock and its yield by month or season. Many countries and international organizations track agricultural production; local conditions should be surveyed to improve the estimate of yield and availability of crop residue or animal waste. A water resource is characterized by its head and flow rate. The latter usually varies seasonally, and several measurements should be made throughout the year. The type of fuel locally available is needed to determine if a conventional gen set is suitable, as well as if shortages are common.

After a resource has been characterized, its suitability for energy production is estimated. Here the capacity factor is useful. The capacity factor is the ratio of the estimated energy produced by a specific energy conversion technology by the energy that would be produced if it were able to operate continuously with certain simplifying assumptions made. A high capacity factor indicates a better use of the conversion technology. Although the capacity factor alone should not be used in selecting which energy conversion technology, it provides in a single unitless number an indication of the quality of the resource and the fit of the conversion technology to that resource.

Problems

11.1 Qualitatively explain the load profiles in Fig. 11.2 based on the appliances and time of use that could be expected for each user type.

Table 11.7 Load profiles

| | Avg. load (W) | | | | | | | | | | | |
	1–2	2–4	4–6	6–8	8–10	10–12	12–14	14–16	16–18	18–20	20–22	22–24
User A	0	0	0	0	0	0	0	0	0	30	30	10
User B	9	9	9	0	0	9	0	0	0	28	9	9
User C	45	45	45	26	20	18	18	22	80	82	50	47
User D	18	18	18	18	80	100	200	180	35	18	18	18
User E	16	16	0	60	80	40	0	0	140	40	16	16

Table 11.8 Average daily insolation by month

| | Month (kWh/m^2/day) | | | | | | | | | | | |
	1	2	3	4	5	6	7	8	9	10	11	12
Tilt												
Latitude	5.51	5.61	5.69	5.76	5.36	4.90	5.15	5.99	6.51	6.78	6.19	5.54

11.2 Develop and administer a survey to estimate the average daily load of classmate or colleague.

11.3 The load profiles for five users is estimated to be shown in Table 11.7. Shown is the estimated average power for each 2-h interval. For each user compute the average daily load and load factor.

11.4 Consider the load estimates in Table 11.7. Compute the average daily load and load factor of the aggregate load.

11.5 The insolation by month for a potential location for a mini-grid is shown in Table 11.8. Determine the capacity factor for each month, and calculate the total energy production each month and for the entire year if a 2.1 kW PV array is used.

11.6 The average wind speed at hub height at a potential location for a mini-grid is 4.7 m/s. Plot the Rayleigh pdf estimate of the wind speed.

11.7 The WECS for the wind resource in the previous problem has a cut-in wind speed of 3.0 m/s. Use the cumulative distribution of the Rayleigh pdf to estimate the percent of time the wind speed will be below the cut-in speed.

11.8 The hydro resource at a potential location for a mini-grid has an effective head of 29 m. The flow rate in the summer is 56 l/s and 80 l/s in the winter. A minimum of 35 l/s must be in the stream at all times. Estimate the capacity factor of the system if a MHP system is designed to use a flow rate of 20 l/s. Repeat the problem assuming a designed flow rate of 30 l/s.

11.9 Consider the hydro resource described in Table 11.6. Estimate the energy produced each month using rated flows of 185 l/s and 8 l/s. Assume that a different Pelton turbine will be used in each design. The efficiency of each turbine is 75%. Let the effective head be 22 m.

References

1. Blodgett, C., Dauenhauer, P., Louie, H., Kickham, L.: Accuracy of energy-use surveys in predicting rural mini-grid user consumption. Energy Sustain. Dev. **41**, 88–105 (2017). doi:https://doi.org/10.1016/j.esd.2017.08.002. URL http://www.sciencedirect.com/science/article/pii/S0973082617304350
2. Burton, T., Sharpe, D., Jenkins, N., Bossanyi, E.: Wind Energy Handbook. Wiley (2001)
3. Ciria, P., Barro, R.: 3 - biomass resource assessment. In: Holm-Nielsen, J., Ehimen, E.A. (eds.) Biomass Supply Chains for Bioenergy and Biorefining, pp. 53–83. Woodhead Publishing (2016). doi:https://doi.org/10.1016/B978-1-78242-366-9.00003-4. URL https://www.sciencedirect.com/science/article/pii/B9781782423669000034
4. Dauenhauer, P., Louie, H.: System usage trends for off-grid renewable energy users in developing communities. In: Proceedings of 4th Symposium on Small PV Applications, pp. 175–180. OTTI (2015)
5. ESMAP: Global Lighting Services for the Poor Phase II: Test Marketing of Small "Solar" Batteries for Rural Electrification Purposes. World Bank (ESM 220 / 99) (1999). URL "http://documents.worldbank.org/curated/en/571301468766818957/Global-lighting-services-for-the-poor-phase-II-test-marketing-of-small-solar-batteries-for-rural-electrification-purposes"
6. Fabini, D., Baridó, D.P.L., Omu, A., Taneja, J.: Mapping induced residential demand for electricity in Kenya. In: Proceedings of the Fifth ACM Symposium on Computing for Development, pp. 43–52. ACM (2014). doi:https://doi.org/10.1145/2674377.2674390. URL http://doi.acm.org/10.1145/2674377.2674390
7. GIZ: What size shall it be? A guide to mini-grid sizing and demand forecasting (2016)
8. Harvey, A., Brown, A., Hettiarachi, P., Inversin, A.: Micro-Hydro Design Manual. Practical Action Publishing (1993)
9. Inversin, A.R.: Mini-grid design manual. World Bank (ESMAP technical paper no. 007.) (2000). URL "http://documents.worldbank.org/curated/en/730361468739284428/Mini-grid-design-manual"
10. Louie, H., Dauenhauer, P.: Effects of load estimation error on small-scale off-grid photovoltaic system design, cost and reliability. Energy Sustain. Dev. **34**, 30–43 (2016). doi:http://dx.doi.org/10.1016/j.esd.2016.08.002. URL http://www.sciencedirect.com/science/article/pii/S097308261630374X
11. Louw, K., Conradie, B., Howells, M., Dekenah, M.: Determinants of electricity demand for newly electrified low-income African households. Energy Policy **36**(8), 2812–2818 (2008). doi:http://dx.doi.org/10.1016/j.enpol.2008.02.032. URL http://www.sciencedirect.com/science/article/pii/S0301421508001201
12. Meier, P., Voravate, T., F., B.D., Bogach, S.V., Daniel, F.: Peru - national survey of rural household energy use. World Bank (Energy Sector Management Assistance Program (ESMAP) energy and poverty special report ; no. 007/10.) (2010)
13. Sørensen, B.: Renewable Energy, 3rd edn. Elsevier Academic Press (2004)
14. Twidell, J., Weir, A.: Renewable Energy Resources, 2nd edn. Taylor & Francis (2006)
15. UNFAO: Food and Agriculture Organization of the United Nations (2017). URL http://www.fao.org/faostat/
16. Zeyringer, M., Pachauri, S., Schmid, E., Schmidt, J., Worrell, E., Morawetz, U.B.: Analyzing grid extension and stand-alone photovoltaic systems for the cost-effective electrification of Kenya. Energy Sustain. Dev. **25**, 75–86 (2015). doi:http://dx.doi.org/10.1016/j.esd.2015.01.003. URL http://www.sciencedirect.com/science/article/pii/S0973082615000071

Chapter 12
Design and Implementation of Off-Grid Systems

12.1 Introduction

This chapter considers the design and implementation of off-grid systems. Mini-grid design is an exercise in design under constraints. The constraints most obviously include economic and technical issues, but often times social, political, legal, and environmental considerations influence the final design. Off-grid system design is almost always multi-objective. The systems are often deployed not only to provide electricity access but also to support development goals such as community resiliency and empowerment. Other off-grid systems are implemented to generate profit.

This chapter is primarily concerned with mini-grids. However, the same general concepts apply to off-grid systems serving single users. Stand-alone systems, such as those serving a single school or hospital, will have a more simplified distribution system than covered here.

Mini-grid design is not only design under constraints but also under considerable uncertainty in both the energy resources and load, as covered in Chap. 11. Compounding this challenge is that there is no universal approach to mini-grid design. Over the last several years, different organizations have developed guidelines or standards that can be used by mini-grid designers. However, the industry and technologies are evolving so swiftly that these documents are either presently being revised or are becoming outdated. This chapter is not intended to be a comprehensive off-grid design manual. Rather, it introduces the important concepts, considerations, and technical theory of mini-grid design.

There are several sources dedicated to the technical and nontechnical best practices of off-grid electrification. Many of these contain insightful case studies. The reader is directed in particular to references [1, 8, 9, 14].

12.2   Mini-grid Life Cycle

We start by considering the life cycle of a mini-grid project. We take the perspective of an organization implementing mini-grids. This could be a for-profit company, a nonprofit organization, a governmental organization, or even an individual. Each will have their own priorities and definition of success, and so the following must be adjusted to fit a particular implementing organization.

From the initial decision to develop a mini-grid to the final decommissioning or expansion, the life cycle is composed of eight steps:

1. Prospecting and screening
2. Site assessment
3. Decision
4. Technical and commercial design
5. Pre-implementation
6. Implementation
7. Ongoing operation
8. Expansion or retirement

Before discussing each step in detail, we offer a comment on the general mindset one should have when implementing an off-grid project: never forget that off-grid communities are among the most underserved and at-risk in the world. These communities deserve better than to be treated as test grounds for electricity access projects. There are probably far more failed or unsustainable off-grid systems than there are those that are successful. Just as access to electricity can dramatically improve a community's outlook, a poorly planned project can ultimately cause more harm than good. The good news is that careful planning and diligence and guard against dramatic failure can provide the best opportunity for success.

12.2.1   Site Selection

The selection of the site for an off-grid system is the first and most important phase in the life cycle of a mini-grid. It includes steps 1 through 3 of the mini-grid life cycle. Each organization has its own site selection process, depending on its priorities and resources. Smaller organizations often use an informal approach. The location is selected based on a personal or organizational connection with a particular community. This approach does not scale well—a larger organization wanting to implement dozens or hundreds of mini-grids will use a more systematic approach.

In general, and from a techno-economic perspective, communities that are better candidates for sustainable mini-grids are those that have some or all of these characteristics:

- the national grid is located far away (20 km or more) with no credible plans for it to be extended to the community in the next several years;
- there are no other off-grid systems installed or planned to be installed;
- the estimated demand for electricity is high, and the planned uses are productive or improve the quality of life;
- one or more energy resources are suitable for electricity generation (see Chap. 11);
- there is sufficient ability and willingness to pay for electricity at a cost-reflective rate or a rate that aligns with the implementing organization's ability to operate the mini-grid;
- the community is densely populated, which reduces distribution infrastructure costs;
- there is basic infrastructure, including roads and cellular network coverage;
- the community is politically stable;
- there is a low risk of theft and vandalism.

Other characteristics can be important, especially if the primary goal of the mini-grid is not commercial:

- access to electricity is a high-priority developmental goal for the community;
- access to electricity aligns with and supports other developmental goals, for example, health care and education;
- the community is able to organize itself to manage the mini-grid (if applicable).

12.2.2 Step 1: Prospecting and Screening

With so many communities without electricity access, a valid question is "Where to begin?" Resources are limited, and the success of a mini-grid largely depends on its location. The basic goal of this step is to develop a "short list" of communities where the mini-grid is believed to have best opportunity to be successful.

There may be thousands of off-grid communities in each country or region. They must be screened quickly and inexpensively. Secondary data from census or other data set that provides basic demographic information about communities can be used. These data are even more powerful when combined with the location of existing distribution lines, energy resources, and other geospatial data. Online tools are being developed to improve the screening process. An example is shown in Fig. 12.1 [11]. Screening is often done based on population density, distance to the national grid, and perhaps income. Other resources that can be consulted include Rural Electrification Authorities (see Chap. 3.6) and government electrification plans. The list of communities can be further reduced based on organizational convenience and priorities. These might include the distance from the implementing organization's offices and cultural or language considerations. Additionally, grant-funded projects might be limited to specific regions or communities.

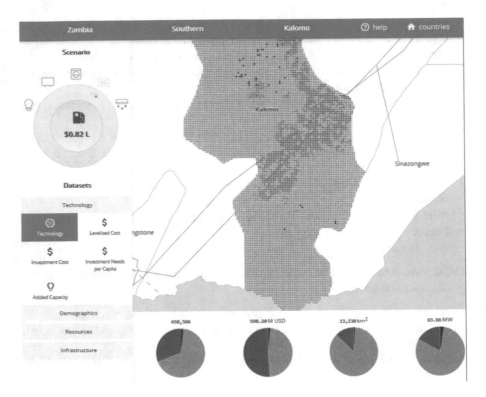

Fig. 12.1 Online screening tools combine demographic and geospatial data to visualize communities where off-grid solutions might be economically viable. Shown is a screenshot of the Electrification Pathways geospatial tool, which is part of the ENERGYDATA.INFO and World Bank Group Energy and Extractives Open Data and Analytics Initiative and created by the KTH University (courtesy of World Bank Group, KTH University, and the Energy Sector Management Assistance Program (ESMAP))

12.2.3   Step 2: Site Assessment

Once the list of communities is reduced to a manageable level, site visits are conducted. At this stage, the organization is investing considerable time and resources evaluating each site. The basic goal of the assessment is to gather primary data to support preliminary design and business modeling of the mini-grid. Typical activities include:

- observe the community firsthand;
- meet with local officials—this might involve governmental officers such as the District Commissioner, as well as traditional leaders such as chiefs, elders, or community groups;
- confirm that the secondary data used in the prospecting and screening step is reasonable;

- conduct surveys or focus groups to estimate present fuel consumption and predict future electric load and consumption patterns;
- check for cellular network coverage. This is important if the mini-grid will use mobile payments and/or remote data monitoring and control;
- identify existing and future local resources and economic activities;
- collect preliminary data for resource assessment;
- identify locations for mini-grid assets—generation equipment and facilities and distribution pathways—including discussion with landowners if land must be leased or purchased.

12.2.4 Step 3: Decision

The findings from the assessment trip can further reduce the list of candidate communities, and allow them to be prioritized. This might be done from a purely commercial perspective. Other tools such as PESTLE (Political, Economic, Social, Technology, Legal and Environmental) analysis and Risk Matrices can be useful in ranking the communities [17]. Depending on the implementing organization's resources, perhaps just one or two communities advance to the next step.

12.2.5 Step 4: Technical Design and Business Plan

Next, the system is designed and, if applicable, the business plan is developed. The technical and business plan should be guided by the targeted electricity access to tier (see Sect. 2.7). The technical design details are discussed later in this chapter. Associated with each design is an economic evaluation including capital and financing requirements, along with operation and maintenance costs. Systems intended to produce a profit or at least break even require a business plan. Most for-profit mini-grids use a metric known as average revenue per user (ARPU), which is the estimated or actual revenue associated with each connection over a period of time—typically 1 month. There might be a threshold ARPU below which the mini-grid is not deemed economically viable. Other financial targets can be similarly considered. The technical and commercial designs are interrelated and iterative. The process may involve further data collection from the site, for example, determining the routing for the distribution system or penstocks. If the mini-grid is found to be economically or technically impractical, then the next community on the priority list is considered.

12.2.6 Step 5: Pre-installation

After viable technical design and business plans have been completed or are nearly complete, pre-installation activities begin. This includes securing the required permits and approvals. This process may need to begin months in advance of a targeted installation date [13]. Many countries exempt mini-grids under a certain capacity, for example, 50 kW, from having to obtain a license or permit. Contracts with vendors, installers, and land owners are finalized. Be aware that importing materials can be a lengthy process.

Most mini-grids do not serve all households or potential customers in an area due to the high upfront connection costs. Therefore, the users need to be selected. Several factors are usually considered. For commercial mini-grids, the customer's expected ARPU and the cost to connect them are especially important. The mini-grid operator assumes a financial risk when connecting a customer. The connection costs are usually too high for the customer to pay upfront. Most operators build these costs into their tariff, essentially financing the connection with slightly higher rates over time. However, if the customer consumes very little energy, which is not uncommon, then the investment in the connection cost is never recuperated. It is therefore very important for the commercial success of a mini-grid to identify customers with high ARPU.

Community meetings and/or customer orientation like that shown in Fig. 12.2 occur during the pre-installation stage. The goal is to inform the community about the pending implementation. The electricity tariff is discussed, questions are answered, and feedback from the community is solicited. This often includes basic safety and technical topics including, for example, the typical cost of running various appliances.

12.2.7 Step 6: Installation

With proper planning, the installation of a mini-grid can occur quickly—often in just a matter of days for PV-powered mini-grids. Installation includes erecting poles or digging trenches for the distribution system and construction and outfitting the power house where the generation, controls, and battery storage (if needed) are located. The mini-grid should be commissioned before final payment is made to installers. Commissioning involves a series of tests and inspections to ensure the mini-grid has been installed properly. Standards such as IEC 62257 [5] offer guidance on commissioning. There is an unfortunate history of poorly installed or counterfeit components being used in many countries, as discussed in Chap. 14.

Fig. 12.2 Community meetings, like this one near Siavonga, Zambia, are important in introducing a project during the pre-installation stage (courtesy of A. Stewart)

12.2.8   Step 7: Ongoing Operation

After installation, the mini-grid begins serving its users. Some mini-grids, such as those powered by a PV array, require very little human intervention; others require daily or continuous oversight. A mini-grid using biomass gasification is one such example. Some mini-grids require administrative and customer support staff, for example, to collect payments. Maintenance should be performed routinely and as required. In some cases, local people can be trained to perform these tasks.

12.2.8.1   Step 8: Expansion or Retirement

Successful mini-grids can be expanded to supply more users or to generally improve the electricity access tier, for example, by increasing its availability. But when mini-grids are not successful, the operator might decide to relocate the major assets—the battery bank and PV array, for example—to another mini-grid, turn the asset over to the community or other organization, or fully retire and decommission the mini-grid. Retiring a mini-grid involves removing its assets, disposing them in an environmentally responsible way, and returning the land to its original condition. Retirement is not always a negative outcome. For example, a mini-grid might be retired because the national grid is finally extended to the community.

12.2.8.2   Modular Design

The desire to rapidly scale and reduce design and implementation costs has pushed some organizations toward using standardized or modular designs. For example, a robust portable structure like a shipping container is outfitted with the energy production components—PV array, batteries, and controllers—prior to being moved to the community.

12.3   Energy Conversion Technology Selection

The most commonly used energy conversion technologies in mini-grids are conventional or biomass gen sets, PV arrays, micro hydro power (MHP), and wind energy conversion systems (WECS). Each has its own advantages and disadvantages. When determining which to use, the designer should consider the characteristics that are important to the particular project. These typically include:

- Quality and availability of the energy resource;
- Fuel cost;
- Capital cost, including balance of system components such as batteries, inverters, conveyance systems, towers, and fuel storage;
- Lifespan of major components;
- Equipment availability, including replacement parts;
- Maintenance requirements;
- Human capital requirements—including technical and business skills;
- Environmental and land-use requirements;
- Design difficulty.

The first consideration should be the quality and availability of the energy resources. This involves estimating the capacity factor of the different technologies, as described in the last chapter. Economic characteristics are the next considered. For many organizations, especially those funded by grants, the upfront capital cost is critically important. However, the cost of fuel and maintenance over the life of the project can easily exceed the upfront cost. A metric such as the levelized cost of energy, which will be discussed in Sect. 12.8, is useful in making economic comparisons between different energy conversion technologies.

Next, characteristics affecting the ongoing operation of the energy conversion technology are considered. What is the expected lifespan of the components, and how long are they covered under the manufacturer's warranty? For example, a gen set might require a major overhaul after 3 years of operation, but a PV array might last 15 to 20 years. Are replacement parts readily available? Is it possible for routine maintenance to be performed without specialized tools and training?

The technical, organizational, and business skills needed to operate the system should be considered. For example, biomass gasification requires constant monitor-

Table 12.1 Energy conversion technology characteristics

	Gen Set	MHP	WECS	PV Array	Biogas	Gasification
Energy source availability	Available	Limited	Limited	Available	Available	Available
Cost of fuel	High	None	None	None	None–Medium	None–Medium
Capital cost	Low	Medium–High	Medium–High	Medium–High	Low	Medium
Cost of operation	High	Low	Low	Low	Low	Medium
Major component lifespan	Short	Long	Medium	PV-long, battery-short	Long	Medium
Availability of components	High	Medium	Low	High	High	Medium
Balance of system requirements	Low	High	High	High	Medium	Medium
Operation skill	Low–Medium	Low	Low	Low	Low	High
Environmental impact	High	Low/Medium	Low	Low	Low/Medium	Low/Medium
Footprint	Low	Medium	Medium	Low/Medium	Low	Low/Medium
Design difficulty	Low	High	Medium	Low	Low	Medium

ing, but a MHP system with an electronic load controller (ELC) requires minimal human intervention. Often the workforce in a rural area will require some training to operate the system.

The environmental impact—including disposal of batteries and equipment—as well as land use should be considered. Conventional gen sets require little land, but noise, fuel spills, and exhaust emissions are a concern. WECSs and PV arrays require considerable land. MHP and biomass might disrupt the traditional use of resources.

Lastly, the time, cost, and difficulty to design the system should be considered. Some systems like MHP must be custom-designed. On the other hand, PV array-powered mini-grids can be modularly constructed and mass manufactured.

Table 12.1 provides a rough ranking of the technologies in the various categories. Of course, these are highly site-specific. The decision of which technology to use can be obvious from this assessment. In other cases, it is worthwhile to consider several different energy conversion technologies separately or together in a hybrid system.

12.4 Design Approaches

We next discuss the design of the energy production components of off-grid systems. We focus on mini-grids in particular. The basic concepts also apply to stand-alone systems. Our primary focus is the electrical design. There are two basic approaches to designing a mini-grid: intuitive and numerical. There are several variations within each approach. The output of the design are the specifications of the energy conversion technologies, energy storage, controllers, and converters. The design of the distribution system is discussed separately in Sect. 12.7. The design of a mini-grid is iterative. Several designs are considered before a final design is selected. Each approach requires an estimation of the load and energy resources as described in the previous chapter.

12.5 Intuitive Approach

Intuitive design approaches have low input data requirements and allow the mini-grid to be designed using relatively simple calculations. The calculations can be based on "rules of thumb" or formalized standards, such as IEEE 1526 [6], IEEE 1013 [7], and IEC TS 62257 [5]. Intuitive approaches, however, provide limited insight and feedback as to how design choices affect the performance of the system, particularly its reliability. The final design may have lower reliability than needed, or higher reliability than necessary, wasting resources.

Through an example, we will produce a basic design which illustrates the fundamentals of an intuitive design approach. A complete design of the mini-grid would include specification of wire sizes, switches, protection equipment and grounding, physical layout, mechanical and civil design, and the distribution system. We stress that example is one of many possible intuitive approaches and do not suggest that it is optimal in any sense.

12.5.1 Project Overview

Assume that after screening and site assessment, the community of Mwase in the southern region of Zambia has been selected for a mini-grid implementation. Mwase is a fictional community. Its characteristics, however, are drawn from real off-grid communities.

The population of Mwase is approximately 3000 people. The nearest electrified town, Bona, is 23 km away. The road to Mwase is not well maintained and becomes impassable during the rainy season (roughly November to February). The people of Mwase primarily support themselves by fishing, raising livestock, and farming. It is anticipated that in addition to serving households, the mini-grid will supply a small number of businesses and a school located on the far side of town. One entrepreneur will start a barber shop/hair salon which will use a high-power blow dryer; another will sell refrigerated drinks from their shop. The households have indicated they will primarily use the electricity for lighting and watching television in the evening. In total, a mix of 24 households, businesses, and the school have been identified as users. These users are located along the main pathway through the town. The households to be connected to the mini-grid are arranged in two dense clusters along the pathway.

None of the users require a three-phase connection. Each will be supplied with a single-phase 230 V, 50 Hz connection, which matches that of the national grid. This means that the appliances readily available in the country can be used. A secure power house will be constructed.

The energy resource estimation showed that WECS, MHP, and biomass are not technically viable due to low availability of their respective resources. This leaves a gen set or a PV array as possible sources for the mini-grid. We will consider the preliminary design of the PV array system, assuming that the logistics of maintaining the fuel supply for a gen set are not practical.

12.5.2 Load Characterization

We begin by characterizing the aggregate load. A survey approach was used to estimate the load. At a minimum, an estimate of the average aggregate daily load and the aggregate peak load is needed. We will assume that the surveys were thorough enough so that a load profile can be produced, as shown in Fig. 12.3. This is inclusive of an assumed 6.5% loss in the distribution system. The daily aggregate average load is computed to be 7.875 kWh/day. The sum of the estimated peak load of the individual users is 4.11 kW. However, with 24 users, some of them businesses, we expect the aggregate peak to be lower because the coincidence factor will not be 1. We will assume a coincidence factor of 0.37, which is reasonable for this situation. The aggregate peak load is therefore $4.11 \times 0.37 = 1.54$ kW. The survey did not indicate if there will be seasonal variation to the load, so we shall assume it to be constant each month. The load growth is assumed to be 5% per year. We will design the system based on the projected load in 5 years. After 5 years the system may need to be upgraded. We will also plan on replacing the battery bank at this time. This brings the average and peak loads to 10.05 kWh/day and 1.94 kW, respectively. The pertinent load values are summarized in Table 12.2.

12.5.3 Architecture Selection

The next step is to determine the high-level architecture of the system. We are primarily concerned with identifying the components needed and how they are

Fig. 12.3 Estimated load
profile of Mwase

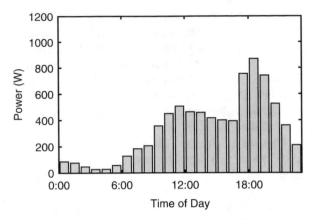

Table 12.2 Estimated load
data

Parameter	Initial	After five years
Avg. daily load (kWh/Day)	7.875	10.05
Peak individual load (kW)	4.11	5.24
Coincidence factor	0.37	0.37
Peak aggregate load (kW)	1.52	1.94
Power factor	0.85	0.85

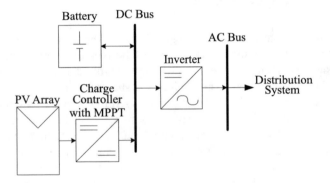

Fig. 12.4 Architecture of the Mwase mini-grid

interconnected. The reader may wish to review Chap. 4 for background information
on mini-grid architecture. The architecture of the mini-grid is dictated by the energy
conversion technology and the load. The only generation source is the PV array,
and so the system is DC-coupled. The users require AC service, and so an inverter
is needed. A battery bank will supply the load during evening hours. Because the
system will be larger than a few hundred watts, a charge controller with maximum
power point tracking (MPPT) will be included. A schematic of the mini-grid is
shown in Fig. 12.4. The PV array will be installed on top of the power house, and
the battery bank and all the DC components will be securely enclosed within.

12.5.4   Voltage Level Selection

The voltage at each bus is selected next. The AC bus voltage is already determined: 230 V at 50 Hz, single-phase. The DC bus voltage is set by the battery bank. Here we consider several factors:

- Nominal voltage of selected battery chemistry (see Chap. 8)
- Electrical distance between DC bus components
- Power capacity of the system
- Safety
- Compatibility with DC bus components

We will assume that the only large-capacity batteries easily available in the country are lead–acid. Therefore, the nominal voltage level will be an integer multiple of 2 V. The battery bank is in a secure area, and so both flooded and valve-regulated lead–acid are viable options. We will use AGM valve-regulated batteries because of the low maintenance requirements. The battery bank will be located in the power house near the PV array and inverter, and so we do not need to be concerned about excessive voltage drop or losses. However, the capacity of the PV array will be several kilowatts. To keep the wires from being too large, a higher voltage is preferred. To be compatible with most charge controllers and inverters, we should select either 12, 24, or 48 V. A common rule of thumb is to use 12 V for systems whose load is less than 1 kWh/day, 24 V for load between 1 and 4 kWh/day, and 48V when the load exceeds for 4 kWh/day. Exceptionally large systems might use 96 V or above. We therefore select a nominal 48 V DC bus voltage. With the DC bus voltage set, the current ratings of the charge controller and inverter can be calculated.

12.5.5   Inverter Selection

Recall from Chap. 9 that an inverter facilitates the flow of power from the DC bus to the AC bus. Inverter parameters that require specification by the designer are shown in Table 12.3. In general, the sum of the power ratings of the inverters and generators connected to the AC bus must be at least as large as the peak load. The minimum power rating of the inverter serving Mwase must therefore be 1.94 kW.

However, this value requires adjustment. We will apply a design margin to mitigate the possible underestimation of the peak load. The inverter will not be in a temperature-controlled environment. Most inverters are rated based on a certain ambient temperature, usually 25°C. If it is expected that the ambient temperature will consistently be higher than the rated temperature, then an inverter with a larger capacity is needed. We will select a high design margin of 0.20 to also account for the effect of temperature:

$$Inverter\ Power\ Requirement = Peak\ Load \times (1 + Design\ Margin). \quad (12.1)$$

Table 12.3 Inverter specifications

Parameter	Requirement	Actual
Max. continuous output power	> 2.33 kW	2.4 kW
Nominal input voltage	48 V	48 V
Nominal output voltage	230 V	230 V ± 2%
Output frequency	50 Hz	250 Hz ± 0.1%
Standby energy consumption	–	25 W
Max. DC current	>57.14 A	125 A
Efficiency	–	85% (assumed)
THD	≤5%	5%
Phase	Single	Single
Bi-directional	No	No
LVD	Yes	Yes

The minimum required rating becomes $1.94 \times (1 + 0.20) = 2.33$ kW. Some inverters are rated in voltamps, not watts. In this case, the assumed power factor is used to convert the requirement to voltamps. For the Mwase system, we will assume a power factor of 0.85 so that the inverter rating in voltamps is $2.33/0.85 = 3.74$ kVA.

We next compute the maximum input current on the DC side of the inverter:

$$Max.\ Inverter\ DC\ Current = \frac{Required\ Inverter\ Power}{Nom.\ Battery\ Voltage \times Inverter\ Efficiency}$$

(12.2)

Inspection of Mwase's estimated load profile in Fig. 12.3 shows that for most hours the load is between 400 and 600 W. This is roughly 20 to 25% of the required inverter rating. Given the general shape of inverter efficiency curves shown in Sect. 9.9.5, we will assume the average efficiency' of the inverter is 85%. This estimate could be improved by calculating the efficiency for each hour of the load profile using the selected inverter's efficiency curve and averaging the result. From (12.2), the maximum current on the DC side of the inverter is

$$Max.\ Inverter\ DC\ Current = \frac{2330}{48 \times 0.85} = 57.14\ A.$$

For convenience, we will round the maximum current up to 58 A.

There are a few other inverter parameters that we should specify. We are not sure if the appliances will tolerate high levels of distortion, so a modified sinewave inverter should not be used. We will only consider inverters whose total harmonic distortion (THD) does not exceed 5%. The users only need single-phase service. As discussed later, the distribution system could be three-phase. In this case each user would be supplied a line-to-neutral connection. However, the mini-grid is small enough that this complexity is not justified. Instead, we will use a single-phase inverter. The mini-grid's architecture is such that power will never flow from the AC bus to the DC bus, so the inverter need not be bi-directional. The battery bank needs protection from deep discharge. Therefore, the inverter must have a low-voltage disconnect (LVD) feature (see Sect. 10.3).

After considering several manufacturers and models of inverters, assume the selected model has the characteristics shown in the third column of Table 12.3.

There are a few other considerations worth noting. Most inverters have a "peak power" or "surge power" rating in addition to their continuous rating. This refers to the power that the inverter can supply for a short duration (often a few seconds to several minutes). Depending on the nature of the load, especially if the peak load is due to motor starting, the peak rating can be used instead of the continuous rating. For Mwase we will use the continuous rating, which is a more conservative approach. We also note the no-load power of the inverter is 25 W. Over the course of the day, the inverter will consume 600 Wh, which represents about 6% of the daily load. The load estimate can be increased to account for this in later iterations of the design. In the case that a single inverter is not capable of supplying the required peak power, multiple inverters can be connected in parallel. The inverters should be from the same manufacturer, and they should be capable of synchronizing with each other. If this is not possible, then separate AC buses can be created, with each inverter supplying power to its own AC bus and the subset of users connected to it.

12.5.6 Battery Bank Design

The battery bank is designed independently of the PV array. The following steps can generally be applied regardless of the energy conversion technology selected. Note the approach presented here is conservative, and it is possible that a smaller-capacity battery bank would be able to produce a similarly high reliability. However, without detailed knowledge of the actual load and given the inherent limitations of intuitive approaches, we cannot know this for certain.

The three most important factors affecting the design of the battery bank are (1) the nominal voltage of the DC bus, (2) the discharge current, and (3) the required reliability. The nominal voltage has already been specified. The discharge current is determined from the inverter input current. The reliability requirement is indirectly expressed as the "Days of Autonomy." The Days of Autonomy is the number of days that the battery bank can supply the average load before being depleted, assuming it is not recharged during this period. This is an extreme scenario. It could occur if the PV array is damaged and the repair was delayed.

A system designed using more Days of Autonomy will have higher reliability than one with fewer. Increasing the Days of Autonomy increases the required battery bank capacity. However, the reliability does not necessarily proportionately scale with the Days of Autonomy. In other words, doubling the Days of Autonomy does not necessarily double the reliability. The specified Days of Autonomy is typically between 2 and 12. We will select just two Days of Autonomy to reduce costs.

It is convenient to specify the battery bank capacity requirement in terms of amphours. The average daily load required from the battery bank, in terms of amphours, is

$$Avg.\ Battery\ Load = \frac{Avg.\ Daily\ Load}{Inverter\ Efficiency \times Nominal\ Battery\ Voltage}. \qquad (12.3)$$

For Mwase, this is

$$Avg.\ Battery\ Load = \frac{10,047}{0.85 \times 48} = 246.25\ Ah.$$

At a minimum, the battery bank must be capable of supplying the average battery load for the specified Days of Autonomy. This must be the case even at the end of the considered 5-year period. However, the battery bank's maximum capacity will naturally decrease over time. Most manufacturers define the maximum capacity at the end of life to be 80% of the initial capacity. The minimum-rated capacity of the battery bank is therefore

$$C_x = Days\ of\ Autonomy \times Avg.\ Battery\ Load \times \frac{1}{End\ of\ Life\ Rating} \qquad (12.4)$$

where x is the discharge current the capacity is associated with, which will be discussed shortly.

The battery bank for Mwase must have a minimum capacity of

$$C_x = 2 \times 246.25 \times \frac{1}{0.80} = 615.63\ Ah.$$

From Sect. 8.5.3, we know that the capacity of a lead–acid battery varies with its discharge current. We must specify the discharge current x the capacity C_x is associated with. The discharge current varies throughout the day, but the capacity is defined at a single discharge current. Two reasonable options are to use the average inverter DC current or use the peak inverter DC current. The latter will yield a larger battery bank. We will take this more conservative approach. The minimum capacity is then 615.63 Ah when discharged at 58 A (C-rate of $58/615.63 = 0.094$, hour rate of $615.63/58 = 10.61$ h). In other words, $C_{58} = 615.63$ Ah. In a very simple design approach, this is the capacity requirement of the battery bank. However, the designer may wish to consider other factors, in which case the following adjustments can be made:

First, recall that deeply discharging a battery can permanently degrade it. To prevent this, the state-of-charge (SoC) should not be zero even after a battery bank has supplied the average load for the specified days of autonomy. We select a maximum depth-of-discharge DoD_{max}, typically 0.50 to 0.80. In other words, we want the battery to still have 20 to 50% of its charge remaining after supplying the average load for the total Days of Autonomy. The capacity requirement is adjusted as

$$C'_x = C_x \times \frac{1}{DoD_{max}}. \qquad (12.5)$$

It is stressed that this is the maximum depth-of-discharge. The battery should not be discharged this deeply on a daily basis. For the Mwase system

$$C'_{58} = C_{58} \times \frac{1}{0.8} = 769.49 \text{ Ah}.$$

We next check to see what daily DoD is consistent with the battery bank lasting the targeted number of years. For the Mwase system, the battery bank will be replaced every 5 years, corresponding to $5 \times 365 = 1825$ cycles.

We consult a chart such as Fig. 8.16 provided by a battery's manufacturer to determine the daily DoD that corresponds to 1825 cycles. This will vary, perhaps appreciably, for each battery considered. For the AGM battery in the Fig. 8.16, the DoD corresponding to 1825 cycles is 40%.

We then check what the average daily DoD is when the capacity is C'_x

$$DoD_{\text{daily}} = 100 \times \frac{Avg. \ Battery \ Load}{C'_x} \tag{12.6}$$

which for the Mwase system is

$$DoD_{\text{daily}} = 100 \times \frac{246.25}{769.49} = 32\%.$$

This is less than the targeted DoD of 40%, and so we expect the battery bank to last at least 1825 cycles. Should we find that DoD_{daily} exceeds the targeted value, then we increase the capacity requirement according to

$$C'_x = \frac{Avg. \ Battery \ Load}{End \ of \ Life \ Rating} \times \frac{100}{DoD \ for \ Required \ Number \ of \ Cycles}. \tag{12.7}$$

The required battery bank capacity can be further adjusted by a design margin to account for the effects of temperature, load estimation error, and losses not modeled elsewhere by applying a design margin:

$$C''_x = C'_x \times (1 + Battery \ Design \ Margin). \tag{12.8}$$

We will apply a design margin of 7.5% bringing the total required capacity to

$$C''_{58} = 769.49 \times (1.0 + 0.075) = 827.21 \text{ Ah}$$

at a discharge rate of 58A (C-rate of 58/827.21 = 0.07C, hour rate of 827.21/58 = 14.25 h). The battery bank requirements are summarized in Table 12.4. A summary of the battery capacity calculations is provided in Table 12.5.

Table 12.4 Battery bank specifications

Parameter	Bank requirement	Bank actual (5 strings, 8/string)	Battery actual
Battery type	AGM	AGM	AGM
Nominal voltage	48 V	48 V	6 V
Capacity	827.21 Ah (0.07C)	≈ 1000 Ah (0.07C)	≈ 200 Ah (0.07C)
Min. cycles at 40% DoD	1825	1825	1825
Max. bulk current	–	220 A	20% of C_{20} (44 A)
Absorption set-point	–	57.60V	7.20
Float set-point	–	54.00 V	6.75 V
Cut-off voltage	–	42 V	1.75 V/cell

Table 12.5 Battery charge capacity design summary

Quantity	Value	Computed from	Equation
Avg. Battery Load	246.25 Ah	Load estimate, inverter efficiency and battery bank nominal voltage	(12.3)
C_x	615.63 Ah	Avg. battery load, days of autonomy and end of life rating	(12.4)
C'_x	769.49 Ah	C_x, maximum depth-of-discharge OR	(12.5)
	–	Avg. battery load, end of life rating, and depth-of-discharge for required number of cycles	(12.7)
C''_x	827.21 Ah	C'_x and design margin	(12.8)

Table 12.6 Battery charge capacity

	10 hr	20 hr	48 hr	72 hr	100 hr
Hour-rate	(19A, 0.1C)	(11A, 0.05C)	(4.8A, 0.021C)	(2.96A, 0.014C)	(2.35A, 0.01C)
Capacity	190Ah	220Ah	228Ah	231Ah	235Ah

12.5.6.1   Battery Bank Configuration

We have determined the requirements of the battery bank and now must select the individual battery and the configuration of the bank. We will assume that the particular battery considered has characteristics in the last column of Table 12.4 and whose capacity is described in Table 12.6. These data are typically found in the battery's specification sheet provided by the manufacturer.

The number of batteries in series is computed from

$$Number\ of\ Series\ Batteries = \frac{Battery\ Bank\ Nominal\ Voltage}{Battery\ Nominal\ Voltage} \qquad (12.9)$$

The given battery has a nominal voltage of 6 V. Therefore, each string will have 48/6 = 8 series-connected batteries.

Next, we determine the number of parallel strings that will be needed to supply the required current. The capacity requirement of the battery bank is 827.21 Ah at a discharge of 58 A. This corresponds to a C-rate of 0.07C and an hour rate of 14.25 h. Although the capacity at the 14.25 hour rate is not provided by the manufacturer, we can approximate it as being somewhat above the 10-hour rate but below the 20-hour rate. We will use 200Ah. This estimate could be somewhat improved using the Peukert equation as discussed in Sect. 8.5.3.6. The number of strings required is found from

$$\text{Number of Battery Strings} = \frac{\text{Required Battery Bank Capacity}}{\text{Battery Capacity}}. \tag{12.10}$$

Using a capacity of 200 Ah, 827.21/200 = 4.13 strings are needed, which we round up to 5 strings. With this design, each string supplies 58/5 = 11.6 A during the peak load.

The battery bank therefore has 5 strings of 8 batteries, for a total of 40 batteries. This arrangement is shown in Fig 12.5. We note that it would also be reasonable to use four strings instead of five. Recall that we conservatively used the maximum discharge current to determine the discharge rate. If we used the average discharge current instead, we would find that four strings would be viable. This would reduce the material cost of the battery bank cost by 20%. It also results in fewer batteries

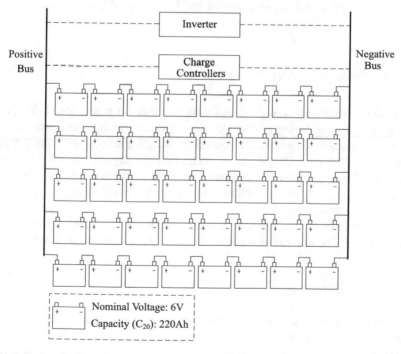

Fig. 12.5 Battery bank configuration for the Mwase system

Table 12.7 Average daily insolation by month

	Month (kWh/m^2/day)												
Tilt	1	2	3	4	5	6	7	8	9	10	11	12	Avg.
Latitude -15°	5.51	5.61	5.69	5.76	5.36	**4.90**	5.15	5.99	6.51	6.78	6.19	5.54	5.79
Latitude	**5.08**	5.39	5.77	6.31	6.40	6.01	6.20	6.87	6.85	6.64	5.76	5.09	6.06
Latitude +15°	4.47	4.94	5.54	6.46	6.89	6.58	6.78	7.23	6.76	6.15	5.08	**4.44**	5.94

placed in parallel, which is desirable from a safety perspective. Nonetheless, because this is a first iteration design, we will proceed using five strings.

From Table 12.4, the battery's cut-off voltage is 1.75 V/cell. Recall that this is the terminal voltage during discharge (not the open-circuit voltage) that corresponds to a zero SoC. Since there are 24 series-connected cells in a 48 V bank, the bank's cut-off voltage is 42 V. The LVD on the inverter should be set to at least this voltage. However, since the DoD_{max} is 80%, not 100%, it is prudent to slightly increase this to, for example, 43 V.

The characteristics of the battery bank are shown in the third column of Table 12.4. To briefly summarize, the average daily battery load 5 years after installation is 246.25 Ah for a 48 V battery bank. The first iteration of the design produced by the intuitive approach calls for a battery bank rated at 1000 Ah at 0.07C. The battery bank has enough capacity to supply the average load for at least 2 days without recharging, to a depth-of-discharge of less than 80%, even after 5 years of operation.

12.5.7 Energy Source Design

The guiding principle of energy source design is that the source should be capable of supplying enough energy to supply the expected average load accounting for generation and storage losses. The system is designed around the month with the lowest capacity factor. Chapter 11 explains how the capacity factor is estimated for different energy conversion technologies. If the load has a seasonal component, then designs are produced for each month or season, and the design with the largest energy source rating is used.

The average daily insolation by month for Mwase was found by consulting a solar database. The data are provided in Table 12.7 for three different array tilts.

12.5.7.1 Tilt Selection

The tilt of PV array is selected by comparing the insolation data in Table 12.7. The month with the lowest insolation for each tilt is highlighted in bold. We select the tilt that corresponds to the largest of these three values. That is, we select the tilt with the greatest minimum monthly insolation. In this case it corresponds to an array tilted at the latitude, whose minimum insolation occurs in January with 5.08 kWh/m^2/day.

12.5.7.2 Capacity Factor

From (11.13), we know that the capacity factor is related to the insolation as

$$Capacity\ Factor = \frac{\bar{I}}{24} \tag{12.11}$$

where $\bar{I}$ is the average insolation. For Mwase during January, the average capacity factor is $5.08/24 = 0.21$. The daily energy production potential for a PV array rated at $P_{PV,rated}$ is

$$E_{PV} = 24 \times Capacity\ Factor \times P_{PV,rated}. \tag{12.12}$$

Therefore, if all losses are ignored, the PV array must be rated at no less than

$$P_{PV,rated} = \frac{Avg.\ Daily\ Load}{24 \times Capacity\ Factor \times Inverter\ Efficiency}. \tag{12.13}$$

For Mwase, this corresponds to

$$P_{PV,rated} = \frac{10.05}{24 \times 0.21 \times 0.85} = 2.33\ kW.$$

This represents an idealized case. A higher rating is needed to account for losses.

12.5.7.3 Generation and Storage Losses

The actual usable energy production from a PV array will be less than E_{PV}. This is due to a number of factors:

- Array shading, including dust (see Sect. 7.12)
- Wire and connection resistive losses
- Parasitic losses (stand-by consumption of controllers, monitors, data acquisition systems, and other devices)
- Module mismatch (caused by PV strings or modules having different maximum power points)
- Array degradation over time (aging)

The production is also affected by coincidence of the load and the irradiance. Recall that during the absorption charging stage, the power to the battery bank is intentionally limited. If the load is also low during this time, the PV array production will be reduced (throttled). Further, any energy stored incurs losses associated with the battery (see Sect. 8.5.4).

It is difficult to estimate the generation and storage losses. Typical ranges are shown in Table 12.8. Whether it is prudent to select a value toward the low end

Table 12.8 Typical
generation and storage losses

Type	Low (%)	High (%)
Shading	0	40
Wire and connection loss	0	10
Parasitic loss	1	10
Module mismatch	0	5
Aging	0	15
Coulombic effect	5	25

or high end of the range, or even exceeding it, depends on the particular local conditions. For example, the soil and wind conditions can inform whether the losses due to dust are closer to 20% or 0%. Use of oversized cables, high-quality batteries, and efficient charge controllers and monitoring equipment can make loss estimates toward the lower end of the range plausible. Note that these losses are associated with the production and storage of energy, not its distribution. Distribution and inverter losses have already been accounted for.

Estimates are made for each type of loss, and the results are summed to determine the total loss K_L. It is obvious that $0 \leq K_L \leq 100$.

The required PV array rating is therefore

$$P'_{PV,rated} = \frac{P_{PV,rated}}{1 - K_L/100}. \tag{12.14}$$

We note that $P'_{PV,rated}$ will always be greater than $P_{PV,rated}$ when losses are considered. For Mwase, we will assume a loss of 22% so that

$$P'_{PV,rated} = \frac{2.33}{1 - 0.22} = 2.98 \text{ kW}.$$

12.5.7.4   Temperature Effects

As discussed in Chap. 7.10, the power output of a PV module decreases with temperature. The temperature of a PV cell during the daytime is usually much higher than the ambient air temperature. A reasonable range of operating temperatures is between 30°C and 60°C. Given a typical temperature power coefficient α_p of -0.5 % per degree Celsius (or Kelvin) above 25°C, the presumed power output of the PV array should be reduced by 2.5% to 17.5%. A value within this range can be selected based on an informal assessment of the climate (warmer locations warrant more severe reduction). However, more accurate estimation can be made using the specification of the PV array and temperature data. The required rating becomes

$$P''_{PV,rated} = 100 \times \frac{P'_{PV,rated}}{100 - Temperature\ Related\ Reduction}. \tag{12.15}$$

Table 12.9 PV array requirements and actual values

Parameter	Array requirement	Array actual (5 strings, 3/string)	Module actual
Maximum power	4.71 kW	5.25 kW	0.350 kW
Optimum operating voltage (V^*)	–	115.62 V	38.54 V
Optimum operating current (I^*)	–	45.40 A	9.08 A
Open-circuit voltage (V_{OC}), STC	≤ 150 V	142.29 V	47.43 V
Short-circuit current (I_{SC}),STC	≤ 60 A	47.45 A	9.49 A
Short-circuit current temp. coeff. (α_i)	–	0.04 %/K	0.04 %/K
Open-circuit temp. coeff. (α_v)	–	-0.29 %/K	-0.29 %/K
Max. power temp. coeff. (α_P)	–	-0.38 %/K	-0.38 %/K
NOCT	–	45°C	45°C

Assume that the PV modules to be used in Mwase have electrical characteristics provided in the last column of Table 12.9. The average ambient daytime temperature in January is 25°C. With this information the temperature de-rating is calculated as described in Sect. 7.10 and is found to be 9.5%. The required rating of the PV array is

$$P''_{PV,rated} = 100 \times \frac{2.98}{100 - 9.5} = 3.29 \text{ kW}.$$

12.5.7.5 Design Margin

The reliability of the mini-grid can be improved by sizing the PV array such that its energy production capability somewhat exceeds the energy supplied by the battery bank to the load each day. The PV array will be able to supply additional energy to the battery bank in case it is deeply discharged or if the insolation is below average for several consecutive days.

The PV array design margin, K_{PV}, generally ranges from 0.1 to 0.2 (10 to 20%) for systems with non-critical loads or with consistent average daily load and consistent insolation. Systems with higher reliability requirements or with inconsistent load or insolation should use higher values, perhaps up to 0.4.

The PV array requirement after accounting for the design margin is

$$P'''_{PV,rated} = \frac{P''_{PV,rated}}{1 - K_{PV}}. \tag{12.16}$$

January is in the middle of Mwase's rainy season. It is commonly overcast for several days in a row. Therefore we use a higher design margin of 0.3, so that

$$P'''_{PV,rated} = \frac{3.29}{1 - 0.3} = 4.71 \text{ kW}.$$

Table 12.10 PV array capacity design summary

Quantity	Value	Computed from	Equation
$P_{PV,rated}$	2.33 kW	Avg. daily load, capacity factor, and inverter efficiency	(12.13)
$P'_{PV,rated}$	2.98 kW	$P_{PV,rated}$, generation and storage losses	(12.14)
$P''_{PV,rated}$	3.29 kW	$P'_{PV,rated}$, temperature and temperature coefficient	(12.15)
$P'''_{PV,rated}$	4.71 kW	$P''_{PV,rated}$, design margin	(12.16)

Table 12.11 Charge controller parameters

Parameter	Required	Actual
Nominal battery voltage	48 V	4 8V
Power	5250 W	3500 W
Max. short-circuit current	>59.31 A	60 A
Max. input open-circuit voltage	>148.48 V	150 V
Max. charge current	–	60 A
MPPT	Yes	Yes
3-stage algorithm	Yes	Yes

This is the final required capacity of the PV array. A summary of the PV array capacity calculations is provided in Table 12.10. We can produce a preliminary design of the PV array, but before it is finalized, it must be checked for compatibility with the charge controller. Assume the charge controller being considered has parameters in the last column of Table 12.11.

The number of PV modules needed is found from

$$Number\ of\ Array\ Strings = \frac{P'''_{PV,rated}}{Module\ Rating} \qquad (12.17)$$

and then rounding up. For the Mwase system

$$Number\ of\ Array\ Strings = \frac{4.71}{0.35} = 13.46$$

meaning that at least 14 PV modules are needed. Now we have to determine how they are arranged. We generally try to minimize the number of strings required so that the string voltage is increased. This reduces losses and wiring complexity. From the last column in Table 12.9, the open-circuit voltage per module is 47.43 V. The maximum number that we can place in series is limited by the charge controller. From Table 12.11, the charge controller is rated at 150 V. This means that the input voltage from the PV array cannot exceed 150 V. Therefore at most three modules ($150/47.43 = 3.16$) are placed in series. When three are placed in series, the open-circuit voltage under Standard Test Conditions is $3 \times 47.43 = 142.29$ V. While this is below the maximum voltage allowed for the charge controller, it is close enough that we should consider the effects of temperature.

Recall that the open-circuit voltage of a PV array increases when the temperature is low. The module's open-circuit voltage temperature coefficient α_v is used to determine the open-circuit voltage at the minimum operating temperature. This likely occurs early in the morning. We will assume this to be 10°C in Mwase. From (7.20) in Sect. 7.8, the open-circuit voltage is

$$V_{OC}(T_C) = V_{OC}(25° \text{ C}) \left(1 + \frac{\alpha_v}{100} \times (T_C - 25)\right)$$
$$= 142.29((1 - 0.0029 \times (10 - 25))$$
$$= 148.48 \text{ V}.$$

This is below the charge controller's maximum voltage of 150 V. If it were found to be above 150 V, then either the number of modules of per string would be limited to two (but the number of strings increased) or different modules (or charge controller) would be necessary.

Each string must have the same number of modules connected in series. Therefore, we can use 15 modules arranged in 5 strings, yielding a rated power of $15 \times 0.35 = 5.25$ kW. Note that if four strings are used, the capacity of the array is 4.20 kW, which is somewhat less than the required 4.71 kW. Using four strings instead of five would lower the cost of the system but would also decrease the reliability.

We must also check that the short-circuit current of the module does not exceed the current rating of the charge controller. Based on the modules' electrical characteristics in Table 12.9, the short-circuit current of the array is $9.49 \times 5 = 47.45$ A. However, remember that the short-circuit current is proportional to the irradiance. In especially sunny areas, the irradiance can exceed the Standard Test Condition value of 1000 W/m^2. To account for this, the short-circuit current is theoretically increased by 25% so that it is 59.31 A. The short-circuit current rating of the charge controller is 60 A. This requirement is nearly at the limit, but it assumes extreme conditions.

12.5.8   Charge Controller Selection

The last major system component to be specified is the charge controller. The major parameters that must be specified are in Table 12.11. The parameters of the selected charge controller are in the last column. The charger satisfies the voltage and current constraints, but not the power constraint. We can either select a larger charge controller or use two in parallel. We will consider two in parallel. One controller will have three strings and the other two. The PV array and charge controllers in this configuration are shown in Fig. 12.6. Note that this also reduces the maximum short-circuit current that each charge controller could be exposed to.

We next check that the current supplied by the charge controller does not exceed the maximum current recommended by the battery's manufacturer. Per the battery specifications, the maximum charging current is 20% of the C_{20} rating, which equals 44 A. The battery bank has five strings, and so the maximum battery bank charging current is $44 \times 5 = 220$ A. The maximum charging current per charge controller

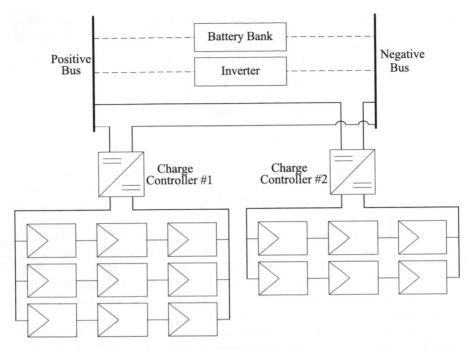

Fig. 12.6 PV array and charge controller configuration for Mwase

is 60 A (120 A total), and so we are not concerned with exceeding the battery bank charging current limit.

There are few other parameters that should be specified. Due to the size of the PV array, the additional cost of using a charge controller with MPPT functionality is justified. To prolong the lifespan of the battery bank, we should select a charge controller that uses a three-stage charging algorithm. The charge controller's absorption and float voltage set-points are set according to the battery manufacturer's specifications.

This completes one iteration of an intuitive design process. This design serves as a starting point for economic and other analyses. Based on the results of these analyses, new designs will be produced using different inputs—for example, serving additional or fewer users—or changing assumptions and specifications such as the Days of Autonomy.

12.5.9 Cost Estimate

We will briefly consider the implementation cost of the proposed design. The material costs will vary widely from one country to the next and from one vendor to the next. Table 12.12 gives an indication of costs that might be considered typical

Table 12.12 Component cost

Item	Each (US$)	No.	Total (US$)
PV module (350 W)	580	15	8700
Battery (220Ah)	204	40	8160
Inverter (2.4kW)	1275	1	1275
Charge controller (150V/60A)	600	2	1200
		Total	US$19,335

Table 12.13 Other costs for Mwase

Item	Each (US$)	No.	Total (US$)
PV mounting rack	1200	1	1200
Battery monitor	290	1	290
Other BoS components	2000	1	2000
Installation (labor and travel)	1800	1	1800
Power house construction	2000	1	2000
		Total	US$7290

in Zambia in 2016, inclusive of the value-added tax. Additional costs are shown Table 12.13. Balance of system (BoS) components typically include power house wiring, outlets, switches, circuit breakers, grounding rods, and surge arrestors. The total estimated cost is US$19, 335 + US$7290 = US$26,625. This excludes the cost of the distribution system and any costs associated with the site assessment and engineering of the system. If this cost is too high, then alterations to the design can be made. For example, the number of battery bank or PV module strings can be reduced. If the cost remains too high, then the system can be designed to serve fewer users or to supply electricity at a lower access tier.

12.6 Numerical Approach

An alternative to an intuitive approach is a numerical approach. Numerical approaches can be applied to simple systems, but they are particularly useful in designing hybrid systems with complex control.

The numerical design approach relies on a computer program to guide the design of the mini-grid. Most are simulation-based. That is, they simulate the operation of a mini-grid over a period of time and provide the user with a summary of the results. The programs do not design the mini-grid. They are better thought of as aides or tools assisting the designer. The programs require engineering judgment at both ends of the process—defining the technical and economic environment in which the mini-grid will operate and interpreting the results provided by the program.

Numerical approaches allow the designer to make better-informed decisions than intuitive approaches. The programs can provide the designer with information regarding the reliability and cost and other technical and economic information. However, the programs have heavier input data requirements than intuitive approaches. For example, a simulation might require the hourly or daily variability of the load to be specified. If the input data provided by the user has a wide range of uncertainty, as it often does, then so does the output of the simulation. There is no guarantee that the actual mini-grid will perform as simulated, especially if the load or resource is substantially higher or lower than input into the program. The designer should always keep this mind. Despite this caveat, computer-aided simulation can be very beneficial when the user understands the limitations and capabilities of the program. Uncertainty in input data can be guarded against by simulating many load and resource scenarios, for example, high, medium, and low projections. The design that best balances cost and risk is selected. Computer-aided design programs are continually improving and will likely play an increasing role in mini-grid design in the future.

There are a few commercial-grade computer programs that can assist in mini-grid design [15]. Among the most popular are HOMER (see Figs. 12.7 and 12.8) and RetScreen [4, 12]. There are at least a dozen other programs developed by universities, but these tend not to be well-supported.

Most computer-aided tools use a time-based simulation of the mini-grid. This means that the program simulates the operation of the mini-grid at discrete time steps over a period of time. The simulation period is usually 1 year but can be longer or shorter. Most programs use a time step of 1 h. A power flow model, like that developed in the previous chapter, is used to simulate the operation and control of the mini-grid at each time step. The different programs might use slightly different models, but the general approach is similar.

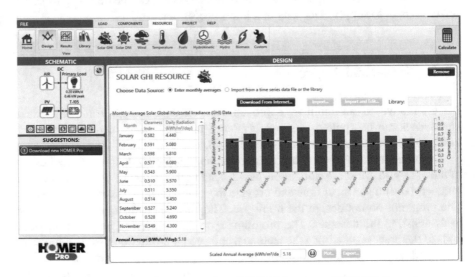

Fig. 12.7 The solar resource input screen from HOMER (courtesy of HOMER Energy)

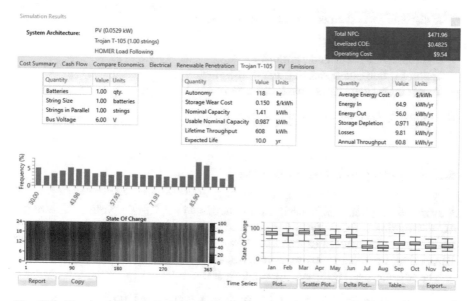

Fig. 12.8 The simulation result screen from HOMER (courtesy of HOMER Energy)

12.7   Distribution System Design

We next consider the distribution system of a mini-grid. The distribution system encompasses the wires, transformers, switches, protection equipment, poles, civil structures, and other components between the AC or DC bus and the users' meter or distribution panel. We will focus on the electrical design of distribution. The reader should consult references such as [16] for information related to other aspects of distribution systems.

The basic purpose of a distribution system is to facilitate the transfer of power from the mini-grid's generators to the user. Its design should be consistent with the targeted electricity access tier (see Sect. 2.7) of the mini-grid as a whole. Especially relevant are the capacity, reliability, safety, and quality (voltage drop) attributes. The distribution system can cost as much or more than the combined cost of the mini-grid's energy conversion, storage, and control components. The distribution system is also known as the reticulation system.

The principles governing the sizing of the conductors, namely, the voltage drop and thermal limits, and the effect of the nominal voltage discussed in Chap. 3 are also relevant to mini-grid distribution design. They are not repeated in this chapter. The economic calculations are also equally applicable, although many mini-grids are not large enough to warrant substations or medium-voltage lines. The reader may wish to review Chap. 3 before continuing. The distribution lines are usually at the same voltage as provided to the user and generally do not exceed 415 V.

A distribution system is characterized by its:

- Voltage type: AC or DC
- Method: overhead or underground
- Topology: hub-and-spoke or trunk-and-branch
- Type: single-phase, split-phase, or three-phase

The following sections discuss these characteristics.

12.7.1   Distribution Voltage Type

A distribution system can be designed to operate with AC or DC voltage. Presently, AC distribution is the de facto approach for larger mini-grid systems, particularly those above 1 kilowatt in capacity. At lower capacities some systems will use AC distribution, and others will use DC distribution.

The advantages of using AC distribution are:

- maturity: large-scale AC systems have been in widespread use for over a century, and engineers, electricians, and inspectors are familiar with their design and use;
- availability of components: AC appliances, generators, protection equipment, meters, and other components are widely available, mass-produced, and inexpensive;
- voltage transformation: voltage levels can be increased for distribution using inexpensive transformers, reducing losses; however, the limited geographic scope of most mini-grids can make this relatively unimportant;
- compatibility with rotating machines: most large motors require three-phase AC supply;
- standards and codes: there are established international quality standards[1] and local codes for the design, installation, and maintenance of AC systems.

The advantages of using DC distribution are:

- compatibility with solar power and batteries: PV modules and batteries are inherently DC;
- compatibility with electronic devices: devices such as mobile phones, laptops, televisions, and LED lights use DC internally; direct supply of DC can eliminate or simplify their power supplies;
- reduced distribution line impedance: the reactive component of a distribution line's series impedance can be ignored, reducing the voltage drop and power loss along the line;

---

[1] At the time of writing, both the IEEE and IEC are working on standards for low-voltage DC mini-grids.

- measurement and metering: measuring the power flow and energy consumption of DC systems is easier than in AC systems because there is no phase angle between the voltage and current.

To improve efficiency, a rule of thumb is to minimize the number of conversions (AC/DC, DC/AC, AC/AC, and DC/DC) between the generation and load. AC loads should be supplied by AC generation using an AC distribution system and vice versa for DC loads. However, it is not difficult to imagine scenarios where this is not possible. When the choice of AC or DC distribution is not obvious, it is reasonable to default to using AC. However, DC should be considered if any of the following conditions are true: the majority of the energy produced is from a DC source (PV module or DC generator); the geographic reach of the distribution system is limited, perhaps no more than 1 kilometer from the point of generation; the load is primarily or entirely LED lighting which can be more efficaciously powered by DC.

12.7.2   Distribution Method

AC or DC distribution lines can be strung overhead or buried underground. Local codes might dictate which method is required, but in many cases it is left to the designer. Burying distribution lines involves digging trenches along the distribution path and then burying the cables, as seen in Fig. 12.9. Overhead lines require installing poles and stringing cables between them. Overhead lines seem to be more common, primarily due to economic reasons.

12.7.2.1   Underground Lines

The advantages of underground lines include:

- protection from sun exposure, rain, vandalism, and aboveground animals;
- discouragement of theft and tampering as it is more difficult to access the lines; however, tampering and theft is less noticeable;
- preservation of community aesthetics;
- elimination of the need for poles, which reduces material, design, and maintenance costs.

The disadvantages include:

- locating faults and troubleshooting the system are more challenging as visual inspection is impossible;
- accessing the cables to connect new users is more difficult;
- the conductors must be protected against burrowing animals, insects, and moisture by using conduit or specialized insulation that prevents moisture intrusion and is resistant to corrosion and fungi;
- burying cables is not practical in areas with rocky terrain.

Fig. 12.9 Trenches being dug for underground distribution (courtesy of PowerGen)

The cables must be buried deep enough to avoid being exposed or disturbed by erosion, people, animals, and vehicles. A minimum of 0.5 m is often recommended. Burying cables underneath earthen roads should be avoided. If this cannot be avoided, then the cables should be buried at a sufficient depth to avoid damage should the road be re-graded. Buried distribution lines cannot be easily moved or re-routed, so the distribution paths must be carefully planned and sized based on anticipated future load growth.

12.7.2.2   Overhead Lines

Overhead lines are strung from pole tops, as seen in Figs. 12.10 and 12.11. The poles add material cost and require some labor, in particular if vegetation needs to be cleared or pruned. The poles must be of sufficient height to discourage tampering and to avoid accidental contact with people or vehicles. This is especially important when the cables span a road. Sometimes the cables can be strung from existing trees, but this requires diligent maintenance and inspection.

The conductors might or might not be insulated. Uninsulated conductors require additional safety precautions to protect against accidental contact. If insulated conductors are used, they must be capable of safe and reliable operation in the exposed outdoor environment. It is important that the insulation is resistant to ultraviolet rays.

Fig. 12.10 An overhead
low-voltage distribution line
being installed (courtesy of
PowerGen)

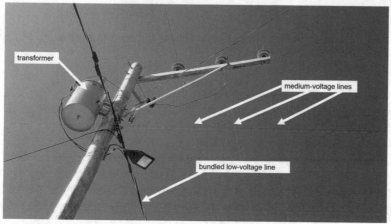

Fig. 12.11 The medium-voltage (22.8 kV) overhead lines in this mini-grid are uninsulated. The
low-voltage, split-phase wires from the transformer are bundled and insulated (courtesy of Sigora
Haiti)

The added weight of the insulation means that the poles must be placed closer
together. Some developers have found success using aerial bundled conductor
(ABC) cable. In ABC cable, each phase is separately insulated and bundled around
a usually bare neutral conductor. Although the ABC cable itself is more expensive
and heavier than uninsulated alternatives, it tends to be safer and easier to install,
and it makes electricity theft more difficult.

12.7.3   Distribution Topology

The distribution topology is concerned with the physical routing of the distribution
lines. Figure 12.12 shows the two types of distribution topology in use in mini-grids:

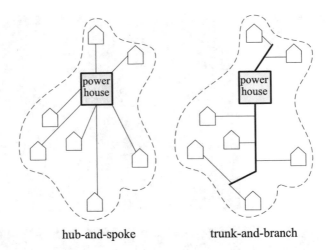

hub-and-spoke   trunk-and-branch

Fig. 12.12 Example distribution system layout of the same users with a hub-and-spoke (left) and a trunk-and-branch (right) topologies

hub-and-spoke and trunk-and-branch (also known as a "radial" topology). In each there is a single path from the power house to the user. This is different than in some grid-connected systems which use "meshed" topologies.

In a hub-and-spoke topology, individual lines are run to each customer. In a trunk-and-branch topology, high ampacity trunk lines are tapped by smaller "branch" lines serving an individual or small group of users. The trunk-and-branch topology tends to be less expensive, especially if the users are located far from the power house along the same direction. In the hub-and-spoke topology, each user has a dedicated line. This can prove to be less expensive if the users are close to and surrounding the power house. The hub-and-spoke approach is less complicated to design. Further, the meters can be centrally and securely located in the power house, rather than in the users' premises. This can reduce meter tampering. It might also encourage users to watch for illegal connections on their lines, knowing that they will be billed for any stolen electricity.

12.7.4   AC Distribution Configuration

There are four basic AC distribution line configurations, as shown in Fig. 12.13. Deciding which configuration to use depends on the user requirements as well as economic considerations. Large industrial, agricultural, or commercial users might require three-phase service to operate motors, pumps, and other equipment. These users must be supplied with either a three-phase delta or wye connection. Most users, particularly households, only require single-phase service. A three-phase distribution configuration can still be used to supply these users, as shown in Fig. 12.13. When split-phase systems are used, they are usually in countries where the nominal voltage is 120 V.

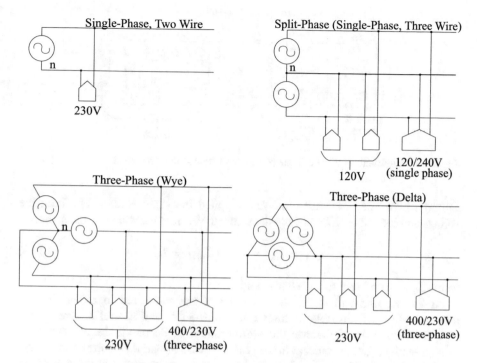

Fig. 12.13 Distribution system types

The generator and inverter might limit the configuration that can be used. For example, single-phase generators and inverters cannot distribute three-phase electricity (but in some cases three single-phase inverters can be arranged to supply three-phase power). Regardless of what configuration is used, at a minimum, each user should have access to the nominal voltage used in that country's national grid, typically 230 V or 120 V.

12.7.4.1   Single-Phase, Two-Wire

This configuration supplies each user with two wires: one phase wire and one neutral wire. This configuration requires the least design work and is conceptually simple, but it is more expensive and results in greater losses when compared to other configurations. It is especially suitable for a hub-and-spoke topology.

A simple circuit model of a single-phase, two-wire configuration is shown in Fig. 12.14. In this model, the users are located at the end of the line. Each wire is modeled as an impedance $Z_w = R_w + jX_w$. The wires are identical in length, material, and cross-sectional area. Therefore, the phase and neutral wires have the same impedance. We will assume that there are H households with identical loads whose impedance is Z_L. The loads are connected in parallel across the phase and

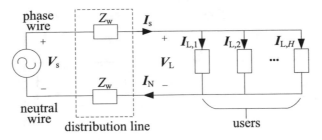

Fig. 12.14 Equivalent circuit of a single-phase, two-wire distribution system

neutral wires. They each therefore are supplied the same voltage V_L. With the voltage and load of each household the same, the current must also be the same

$$I_{L,1} = I_{L,2} = \cdots = I_{L,H} = I_L \qquad (12.18)$$

where I_L is the current to each household.

At the sending end of the line, the wires are connected to either the AC bus, a single-phase distribution transformer or the line and neutral of three-phase wye distribution line. Whatever the configuration, we designate the voltage at the sending-end V_s. Let the current in the phase wire be I_s and the neutral wire be I_N. Applying Kirchhoff's Current Law to the top and bottom nodes at the end of the line in Fig. 12.14

$$I_s = \sum_{h=1}^{H} I_{L,h} = I_L H = I_N. \qquad (12.19)$$

The current through the phase wire is equal to the sum of the currents to each household. This current returns to the source on the neutral wire. Applying Kirchhoff's Voltage Law to this circuit shows

$$V_s = I_s Z_w + V_L + I_N Z_w = 2 I_s Z_w + V_L. \qquad (12.20)$$

Recall from Chap. 3 that the voltage drop along a line is the magnitude of the difference between sending-end and receiving-end voltage. In this circuit, the receiving-end voltage is at the end of the line (V_L) where the households are located. To ensure that the houses will receive suitable voltage, the drop should be no more than 5 to 10% of the sending-end voltage. For a single-phase, two-wire configuration, the voltage drop is therefore

$$V_{drop} = |V_s - V_L| = 2|I_s Z_w| = 2|I_L|H|Z_w|. \qquad (12.21)$$

We see that the voltage drop increases in proportion to number of households served as well the current supplied to each household. Half of the voltage drop is caused by the phase wire and half by the neutral wire. Equivalently, the total voltage drop is twice of that along the phase or neutral wire. There is also power loss along each wire. The total power loss is

$$P_{\text{Loss}} = |I_s|^2 R_w + |I_N|^2 R_w = 2|I_s|^2 R_w = 2|I_L|^2 H^2 R_w \qquad (12.22)$$

showing that the losses increase with the square of the number of households and current supplied to each household.

12.7.4.2   Split-Phase

Split-phase is also known as "single-phase, three-wire." It is more efficient and has a lower-voltage drop and losses than the single-phase configuration. The simplified circuit model of the split-phase circuit serving H households is shown in Fig. 12.15. There are two phase wires and one neutral. Each wire has the same impedance. The circuit is supplied by two voltage sources with equal magnitude but are out of phase by 180° so that

$$V_{s,\text{even}} = -V_{s,\text{odd}}. \qquad (12.23)$$

Some gen sets are internally wired for split-phase distribution. Split-phase inverters are also commercially available, but they most often are rated at 120 V line-to-neutral (240 line-to-line). Single-phase generators and inverters can supply a split-phase distribution system by using a "center-tapped" transformer.

We will assume that there are an even number of households H, each with identical power consumption. Although users can be connected between both phase wires, we omit these connections for clarity. It does not change the general

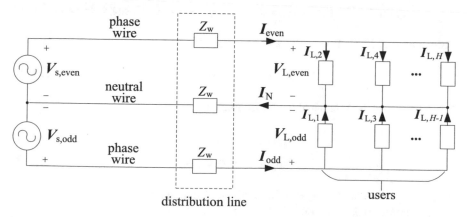

Fig. 12.15 Equivalent circuit of a split-phase (single-phase three-wire) distribution system

results. These connections supply twice the phase-to-neutral voltage and are used in countries where the line-to-neutral voltage is 120 V. They provide access to 240 V supply which is used in high-power appliances.

The households with even indices are connected in parallel and so have the same voltage and current; similarly, the households with odd indices are in parallel and have the same voltage and current.

Let the current to and the voltage at even-indexed households be $I_{L,even}$ and $V_{L,even}$, and $I_{L,odd}$ and $V_{L,odd}$ for odd-indexed households. The current through each phase wire is the sum of the current to the households it serves

$$I_{even} = \sum_{h:even} I_{L,h} = \frac{H}{2} I_{L,even} \tag{12.24}$$

$$I_{odd} = \sum_{h:odd} I_{L,h} = \frac{H}{2} I_{L,odd}. \tag{12.25}$$

By Kirchhoff's Current Law, the neutral current is the sum of the current to the even- and the odd-indexed households:

$$I_N = I_{even} + I_{odd}. \tag{12.26}$$

Importantly, this shows us that the neutral current is the sum of the phase wire currents.

Applying Kirchhoff's Voltage Law to this circuit shows

$$V_{s,even} = I_{even} Z_w + V_{L,even} + I_N Z_w \tag{12.27}$$

$$V_{s,odd} = I_{odd} Z_w + V_{L,odd} + I_N Z_w. \tag{12.28}$$

Adding (12.28) and (12.27) yields

$$V_{s,even} + V_{s,odd} = (I_{even} + I_{odd}) Z_w + V_{L,even} + V_{L,odd} + 2 I_N Z_w. \tag{12.29}$$

Substituting (12.23) and (12.26) into (12.29) yields

$$0 = 3 I_N Z_w + V_{L,even} + V_{L,odd}. \tag{12.30}$$

We can express the voltage at the households as

$$V_{L,even} = I_{even} Z_{eq,even} \tag{12.31}$$

$$V_{L,odd} = I_{odd} Z_{eq,odd} \tag{12.32}$$

where $Z_{eq,even}$ and $Z_{eq,odd}$ are the equivalent impedances of the even- and odd-indexed households. These impedances are equal because each house consumes the same power, are supplied with current of the same magnitude, and there are an equal number of even- and odd-indexed houses.

Substituting (12.31) and (12.32) into (12.30) yields

$$0 = 3I_N Z_w + I_{even} Z_{eq,even} + I_{odd} Z_{eq,odd} \tag{12.33}$$

$$0 = 3I_N Z_w + 2I_N Z_{eq,even} \tag{12.34}$$

$$0 = I_N \left(3Z_w + 2Z_{eq,even}\right) \tag{12.35}$$

which is only practically possible if $I_N = 0$. In other words, when the loads are balanced in a split-phase system, the current in the neutral wire is zero. From (12.26), the current through each phase is equal in magnitude but opposite in phase:

$$I_{odd} = -I_{even}. \tag{12.36}$$

We generically express the magnitude of the current and voltage to each household as $|I_L|$ and $|V_L|$.

The voltage drop along the line for the odd-indexed households is

$$V_{drop} = |V_{s,odd} - V_{L,odd}| \tag{12.37}$$

Using (12.28) and (12.25)

$$V_{drop} = |\left(I_{odd} Z_w + V_{L,odd} + I_N Z_w\right) - V_{L,odd}| \tag{12.38}$$

$$= |I_{odd} Z_w + I_N Z_w| \tag{12.39}$$

$$= |I_{odd} Z_w| = \frac{H}{2}|I_L||Z_w|. \tag{12.40}$$

This assumes that the system is balanced so that $I_N = 0$ A as shown previously.

It can be shown that the voltage drop for the even-indexed households is same. The total power loss is calculated considering only the phase wires since the neutral current is zero:

$$P_{loss} = |I_{s,odd}|^2 R_w + |I_{s,even}|^2 R_w = 2\left(\frac{H}{2}|I_L|\right)^2 R_w. \tag{12.41}$$

Comparing (12.21) and (12.22) with (12.40) and (12.41), we see the advantage of using split-phase: the voltage drop and power losses are one-fourth that of a single-phase configuration. This is caused by each phase wire carrying half the current of the phase wire in the single-phase configuration, and, because there is no neutral current, there is no voltage drop or power loss associated with the neutral wire.

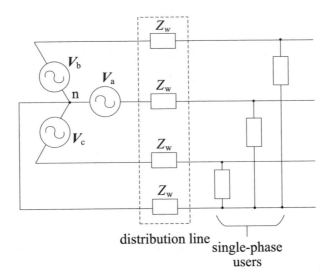

Fig. 12.16 Three-phase wye-connected system

12.7.4.3 Three-Phase Configuration

Higher-capacity systems generally use three-phase distribution. There are two common ways that a three-phase distribution system can be configured: as a wye or delta. Three-phase systems are well covered in other texts [2, 3], and so we only present a short overview here.

Three-phase distribution requires three-phase generators or inverters. In a wye-connected system, a total of four wires are used: three-phase wires and a neutral wire. A circuit model for a three-phase, four-wire configuration is shown in Fig. 12.16. Users can be supplied three-phase or single-phase service. Single-phase users are connected between the phase and neutral wires as in a single-phase two-wire configuration. If the loads are balanced, then there is no neutral current. Center-tapped transformers can be used to supply some users with a split-phase service, if needed.

Delta-connected systems are also known as "three-wire" systems because they do not use a neutral conductor. A circuit model for the three-phase, three-wire configuration is show in Fig. 12.17. Single-phase users are connected phase-to-phase; three-phase users are connected to all three phases.

A wye configuration is the most efficient of the four distribution configurations. Less current is needed than in a delta connection to provide same power because the line–line voltage is $\sqrt{3}$ times greater. This reduces the voltage drop and losses along the conductors. Users should be connected to balance the load as nearly as possible on all phases. When this is not done, the voltage drops and power loss increases.

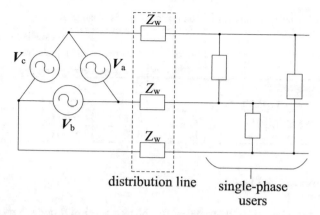

Fig. 12.17   Three-phase delta-connected system

Table 12.14   Distribution configuration summary

Configuration	Voltage drop	Power loss
Single-phase, two-wire	$2H\lvert I_L\rvert\lvert Z_w\rvert$	$2\,(H\lvert I_L\rvert)^2\,R_w$
Split-phase (single-phase, three-wire)	$\frac{1}{2}H\lvert I_L\rvert\lvert Z_w\rvert$	$\frac{1}{2}\,(H\lvert I_L\rvert)^2\,R_w$
Three-phase, wye (four-wire)	$\frac{1}{3}H\lvert I_L\rvert\lvert Z_w\rvert$	$\frac{1}{3}\,(H\lvert I_L\rvert)^2\,R_w$
Three-phase, delta (three-wire)	$H\lvert I_L\rvert\lvert Z_w\rvert$	$(H\lvert I_L\rvert)^2\,R_w$

12.7.4.4   Distribution Configuration Comparison

For a given conductor size and total user load, the voltage drop and power loss
will be different depending on the distribution configuration used, as shown in
Table 12.14. From this table we see that, for example, the power loss associated
with a three-phase delta configuration is twice that of a split-phase configuration
but half that of a single-phase configuration. The split-phase configuration has one-
fourth the voltage drop and power loss of the single-phase configuration.

Interpreted in another way, the split-phase system can use conductors with four
times the resistance and have the same voltage drop and power loss as a single-phase
system. Since resistance is inversely proportional to cross-sectional area, the split-
phase configuration can use conductors whose cross-sectional area is one-fourth that
of the single-phase configuration. However, the split-phase configuration requires
three conductors, not two. The total cross-sectional area relative to a single-phase
configuration is $(0.25 \times 3)/2 = 0.375$. The material costs of the conductor should
therefore be approximately 37.5% of the single-phase configuration. Table 12.15
shows the conductor area and cost for each configuration relative to the single-phase
configuration. We see that the three-phase wye configuration requires the least total
area of conductors—just 33% of the single-phase configuration.

Table 12.15 Distribution configuration conductor comparison summary

Configuration	No. of conductors	Area per conductor	Total area	Conductor cost relative to single-phase, two-wire
Single-phase, two-wire	2	1.0	2.0	1.0
Single-phase, three-wire (split-phase)	3	0.250	0.75	0.375
Three-phase, four-wire (wye)	4	0.167	0.667	0.333
Three-phase, three-wire (delta)	3	0.50	1.5	0.75

12.7.4.5 Imbalanced Load

The results in the last section are based on the assumption that each user consumes the exact same power. That is, the load is balanced. In practice this is almost never the case, especially in smaller-capacity mini-grids with a limited number of users. An imbalance erodes some of the benefits of split-phase and three-phase (delta and wye) configurations. When the load is imbalanced, the neutral current is no longer zero in the split-phase and three-phase wye configurations. There is now a voltage drop and power loss associated with the neutral current. The severity of the effect depends on the magnitude of the imbalance. The voltage drop and power loss for the unbalanced loads can be calculated through AC circuit analysis but can be tedious to do by hand. The voltage drop and power loss in split-phase and wye-connected configurations might increase by 50%, but this still is an improvement of a single-phase configuration. Delta-connected systems are less affected by imbalance.

12.7.4.6 Location of Load

The circuit models in Figs. 12.15–12.17 assumed the users were lumped at the end of the line. More than likely, the users are spread along the line. When this is the case, the voltage drop and power losses are reduced. This scenario is explored in the following example.

Example 12.1 Consider the split-phase circuit in Fig. 12.18 where $H = 40$. Twenty users are located at the midpoint of the line and 20 are at the end. The impedance of each wire is $Z_w = 0.14 + j0.01 \ \Omega$. The sending-end line-neutral voltage is 120 V, and the magnitude of the current to each house is 3A. Compute the voltage drop for the users at the midpoint and end of the line and the total power loss.

(continued)

Solution Due to symmetry, we only need to consider the users connected to one of the phases. We will arbitrarily consider the odd-indexed customers. The current magnitude in the segment of the wire between the source and the midpoint is

$$|\boldsymbol{I}_{\text{odd}}| = |\boldsymbol{I}_{\text{L}}| \frac{H}{2} = 3 \frac{40}{2} = 60 \text{ A}.$$

The impedance of this segment is half the total impedance of the phase wire. The voltage drop across the segment 1 is

$$V_{\text{drop,mid}} = V_{\text{drop,1}} = |\boldsymbol{I}_{\text{odd}}| \frac{|Z_{\text{w}}|}{2} = 60 \frac{0.14}{2} = 4.21 \text{ V}.$$

This corresponds to a $4.21/120 = 3.51\%$ voltage drop. The associated power loss is the magnitude of the current squared multiplied by the resistance:

$$P_{\text{loss,1}} = (|\boldsymbol{I}_{\text{odd}}|)^2 \frac{R_{\text{w}}}{2} = (60)^2 \frac{0.14}{2} = 252.0 \text{ W}.$$

Only half of the current continues to segment 2. The voltage drop is therefore

$$V_{\text{drop,2}} = \frac{|\boldsymbol{I}_{\text{odd}}|}{2} \frac{|Z_{\text{w}}|}{2} = \left(\frac{60}{2}\right)\left(\frac{0.14}{2}\right) = 2.11 \text{ V}$$

The total voltage drop at the end of the line is

$$V_{\text{drop,end}} = V_{\text{drop,1}} + V_{\text{drop,2}} = 6.32 \text{ V}$$

which corresponds to a $6.32/120 = 5.26\%$ voltage drop. The power loss of segment 2 is

$$P_{\text{loss,2}} = \left(\frac{|\boldsymbol{I}_{\text{odd}}|}{2}\right)^2 \frac{R_{\text{w}}}{2} = 30^2 \times \frac{0.14}{2} = 63 \text{ W}.$$

The total power loss, considering both the even and odd phase wires, is twice that of the one of the phases:

$$P_{\text{loss,total}} = 2 \times \left(P_{\text{loss,1}} + V_{\text{loss,2}}\right) = 630 \text{ W}.$$

We note that the voltage drop associated with the segment between the midpoint and end of the line is one-half of those associated with the first segment of the line, and the power loss is one-fourth.

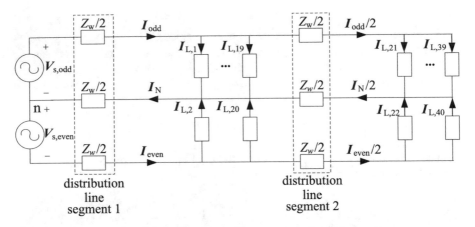

Fig. 12.18 A split-phase system with loads at the midpoint and the end of the line

In general, as the load is distributed across the line, the voltage drop at the end of the line and the total power loss decrease. It can be shown that if the load is uniformly spread along the line, then the voltage drop is one-half of the case when the loads are located at the end, and the power loss is reduced to one-third. It turns out that these reductions are independent of the configuration, as long as the loads are balanced. Keep in mind, it is always more advantageous to have the load closer to the power house than farther away.

12.7.5 Estimating Voltage Drop and Power Losses

We now return to Mwase and consider the design of the distribution system. Since the location of the users and the power house is known, the path for the distribution system can be determined. The users are all located on one side of the power house, and so a radial (trunk-and-branch) topology is selected. It is good practice for the path of the distribution line to be as short and as straight as possible. The lines will also be overhead in order to reduce costs. The planned distribution system for Mwase is shown in Fig. 12.19. The nodes correspond to certain landmarks in Mwase. The landmarks are used to measure the length of the different line segments. The segment lengths are given in the second column of Table 12.16.

As previously described, a single-phase distribution system will be used. The supply voltage is 230 V, and so we will consider a single-phase two-wire design. We must decide what size and type of wire to use. For mechanical reasons, we will use an uninsulated ASCR conductor with spacing of 0.30 m. We will consider two different sizes with characteristics shown in Table 12.17. We will start with the 13 mm^2 wire, the smaller of the two. Its ampacity is far above the peak current of the mini-grid, and so we continue in estimating the voltage drop and power loss.

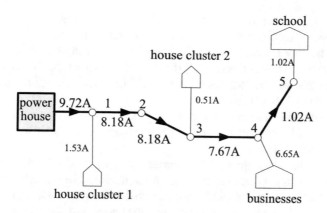

Fig. 12.19 Distribution line path and segment and node current for Mwase

Table 12.16 Distribution system calculation table

Node	Length (km)	Estimated coincident peak load (W)	Estimated load current $|I_{L,n}|$ (A)	Segment $|Z_{w,n}|$ (Ω)	Segment $|I_s|$ (A)	Segment voltage drop (V)	Segment power loss (W)
5	0.15	200	1.02	0.34	1.02	0.70	0.71
4	0.10	1300	6.65	0.23	7.67	3.49	26.49
3	0.08	100	0.51	0.18	8.18	2.97	24.11
2	0.05	0	0	0.11	8.18	1.86	15.07
1	0.05	300	1.53	0.11	9.72	2.21	21.25
					Totals	11.22	87.63

Table 12.17 Distribution wire characteristics

Cross-section (mm²)	Resistance (Ohms/km)	Reactance (Ohms/km)	Ampacity (A)
13	2.25	0.31	105
21	1.40	0.31	140

A rough estimate of the voltage drop and losses can be made by assuming the entire load is located at the end of the distribution line and applying the equations in Table 12.14 for a single-phase, two-wire system. However, since we have rough plan of the distribution line, we will use a more accurate method.

We begin by grouping all users near each node together. The surveys are used to estimate each node's contribution to the aggregate peak load, which we have rounded to 1.9 kW. These values are shown in the third column of Table 12.16. We make the simplifying assumptions that the current supplied to the users at each node n can be approximated as

$$|I_{L,n}| = \frac{P_{L,n}}{PF_{L,n} \times |V_s|} \qquad (12.42)$$

where $P_{L,n}$ is the real power associated with all users connected to the node n and $PF_{L,n}$ is the power factor of the users at node n. The PF is assumed to be 0.85 for all nodes. The node current computed from (12.42) is an approximation because the voltage at each node is not exactly V_s due to the voltage drop. This approximation usually does not significantly alter the result. The calculated values are shown in fourth column of Table 12.16.

The magnitude of the impedance for each segment $|Z_{w,n}|$ is computed, taking into account each segment's length. The resulting values are shown in the fifth column of Table 12.16.

The current in each segment is calculated by applying Kirchhoff's Current Law at each node, starting with the node at the end of the line. For example, the current in the segment between node 4 and 5 must be equal to the load current of the school at node 5 (1.02 A). The current in the segment from node 3 to node 4 must be the sum of the current in the segment between node 4 and 5 and the current to businesses at node 4 (6.65 A), for a total of 1.02 + 6.65 = 7.67 A. The calculation proceeds in this manner until the total current that supplied the line (9.72 A) is determined.

Once again, starting at the node 5, the voltage drop and power loss are computed. The voltage drop and power loss equation is found from Table 12.14 for the distribution configuration considered. In this case, the equations for the single-phase configuration are used. The results are shown in the last two columns of Table 12.16. After this is done for all segments, the voltage drop and power losses are totaled. In this case, the percent voltage drop at the end of the line (node 5) is $11.22/230 = 0.0488$ or 4.88%. Note that if the approximation was made that the entire load was lumped at the end of the line, then the voltage drop would be 22.08 V (9.6%). The total power loss is 4.6% of the peak load, which is somewhat less than our initial estimate of 6.5% used in the design of the battery and PV array. This is close enough to not warrant revising the design. If the voltage drop or power loss are too large, then the process is repeated using the next larger-size wire. It is left to the reader to compute the voltage drop and losses if the 21 mm^2 wire is used.

12.8   Economic Considerations

Associated with each proposed technical design is an economic analysis. This is required to compare different designs and to understand trade-offs during the design phase. The economic analysis can be presented in several ways: levelized cost of energy (LCOE), yearly cash flow, total life cycle costs, and annualized maintenance, operating, and replacement costs. The last three are covered in most business economics textbooks. We will briefly discuss the LCOE.

Energy economists and power system planners use the (LCOE) to express the cost of supplying each unit of energy. A system with a lower LCOE is able to supply energy at a lower cost than one with a higher LCOE. The basic premise of computing the LCOE is straightforward: divide the lifetime cost of a system by its lifetime energy production, yielding a figure whose units are cost per unit energy:

In practice, the LCOE is computed before the system is operational, so its lifetime costs and energy production must be estimated. The LCOE is particularly useful in comparing the economics of systems that have unequal lifetimes, capacities, and costs. As you might suspect, the LCOE depends heavily on the energy conversion technology used. For context, the LCOE for utility-scale power plants in the United States ranges from about US$0.05 to US$0.15 per kilowatthour. It is usually much larger for small-scale off-grid systems.

12.8.1 Simplified LCOE

The lifetime cost of an off-grid system includes jurisdiction-dependent factors such as insurance and taxes paid on revenue and real estate. We will skip over these factors and consider the *simplified LCOE* (sLCOE), which approximates the LCOE while using a conceptually simple cost model. The sLCOE includes many of the key elements of LCOE:

- capital costs
- fuel costs
- operation and maintenance cost.

Any complete cost estimation would have to include the localized factors mentioned above.

12.8.1.1 Capital Cost

The overnight capital cost is the cost of bringing the system to a commercially operable state, assuming the cost is incurred (and the system was commercially operable) overnight. The overnight capital costs typically include costs of not only equipment but also land, buildings, construction labor, permitting, and other nonrecurring costs.

The overnight capital cost can be expressed as K dollars per kilowatt capacity of the system. The overnight capital cost is converted to an annuity extending over the lifetime Y years of the system as

$$Cap = \frac{K}{\frac{(1+i)^Y - 1}{i(1+i)^Y}} \tag{12.43}$$

where i is the interest rate, so that Cap is the total capital cost per kilowatt of capacity per year. Note the similarity between (12.43) and (3.31) from Chap. 3.

12.8.1.2   Fuel Cost

The fuel cost per kilowatthour of energy produced are computed as

$$Fuel = \frac{p_f \times h}{1 \times 10^6} \tag{12.44}$$

where p_f is the price of fuel in \$/MMBTU and h is the heat rate of the power plant in BTU per kilowatthour. Fuel prices are commonly quoted in MMBTU, where each "M" is "thousand" in Roman numerals. The "MM" is interpreted as one thousand thousands or one million. Hence, the scaling by 10^6 in the denominator of (12.44) is needed. For renewable mini-grids, the fuel costs are zero (an exception is biomass, where the feedstock might have an associated cost).

12.8.1.3   Operations and Maintenance Costs

Operations and maintenance costs are split into fixed and variable costs. Fixed costs O_f are incurred regardless of the energy produced and are associated with the capacity of the system. It is expressed as dollars per kilowatt of capacity per year. Variable costs O_v are based on the amount of energy produced and expressed as dollars per kilowatthour.

12.8.1.4   sLCOE Calculation

The sLCOE is computed as

$$sLCOE = \frac{Cap + O_f}{E_{annual}} + Fuel + O_v \tag{12.45}$$

where E_{annual} is the average annual energy production per kilowatt of capacity. The annual production, rather than lifetime production, is used to be consistent with the units of the capital and fixed operation costs.

All else being equal, a design that results in the lower sLCOE is the more economically favorable. The sLCOE also has very useful commercial interpretation: it is the minimum cost that can be charged for electricity per kilowatthour without losing money. The sLCOE for off-grid systems is typically much higher than for the grid, which is often subsidized. It is common for the sLOCE to exceed US\$1/kWh for mini-grids.

Table 12.18 Mwase sLCOE parameters

Parameter	Value
Lifespan	15 (years)
Interest rate	5%
Overnight capital cost	18465 (US$); 3920 (US$/kW)
Annual production	3668.3 (kWh); 778.8 (kWh/kW)
Fuel price	0 ($/MMBTU)
Heat rate	N/A (BTU/kWh)
Fixed OM	2132 (US$/year); 452.7 (US$/kW/year)
Variable OM	0 (US$/kWh)

12.8.1.5 sLOCE Example

We now compute the sLCOE of the Mwase system using the costs in Tables 12.12 and 12.13. We will assume an interest rate of 5%. The lifespan of the system is taken as 15 years. However, the system was designed for the batteries to be replaced every 5 years. To represent this, we will model the battery replacement as a fixed operating cost whose annual value is one-fifth of the cost of the battery (US$8160/5 = US$1632/year). We will assume that other fixed operating costs include routine maintenance every 6 months, for an annual cost of US$500. The capital cost excluding the battery is US$18,465. There are no fuel costs or variable operation and maintenance costs. We do not know if the load will continue to grow at 5% after the fifth year. We will assume that the average annual production during the 15-year lifespan of the system is equal to the consumption at during the fifth year (10.05 kWh/day = 3668.3 kWh/year). These parameters are summarized in Table 12.18. Recall that the capacity of the system is 4.71 kW, which is used in computing the cost per capacity values.

The sLCOE is found by first computing the overnight capital cost in the form of an annuity from (12.43):

$$Cap = \frac{3920}{\frac{(1+0.05)^{15}-1}{0.05(1+0.05)^{15}}} = US\$411.1/kW.$$

Now applying (12.45)

$$sLCOE = \frac{411.1 + 452.7}{778.8} + 0 + 0 = US\$0.973/kWh.$$

This sets the break-even price for the Mwase system considering the energy production system. The analysis could be further expanded to include distribution cost, which would further increase the break-even price. If this price is deemed too high, it can be lowered by decreasing the size of the system components, but this may reduce the reliability of the system.

12.9   Cost and Reliability Trade-Off

As is often the case in engineering, there is a trade-off between the reliability of an off-grid system and its cost. There are several ways to improve reliability: increasing the Days of Autonomy or design margin when specifying components, adding redundancy, or using a hybrid architecture to diversify the energy sources. These tend to increase the capital and perhaps the operating cost of the system. This trade-off is perhaps the most important to understand of mini-grid design. An example of a cost-versus-reliability curve is shown in Fig. 12.20. Here the cost refers to the capital cost of the energy production system (conversion technology, storage, and controllers), not the distribution system.

The curve, which will be somewhat different for each mini-grid, shows that it becomes increasingly more expensive to increase reliability. For example, increasing the reliability from 98.5% to 99.5% might increase the energy production capital cost by 40%. Whether or not this additional outlay is worth it depends on the context of a particular project.

There a tendency to over-design off-grid systems. That is, making them more reliable than needed. However, it must be acknowledged that the goal of many off-grid systems is to provide life-improving electricity access, and this can be accomplished at somewhat low reliability. A target of 95% is typical. The budget saved by not installing a system with higher reliability can be used to install additional systems, having a wider impact.

The designer should also be mindful of the impact that low reliability has on the LCOE. Certain costs are fixed. When the reliability is low, less energy is supplied and the revenue decreases. This causes the LCOE to increase because the fixed cost are spread over a smaller amount of production [18].

Fig. 12.20 Example relationship between the capital cost of the energy production system and reliability of an off-grid system

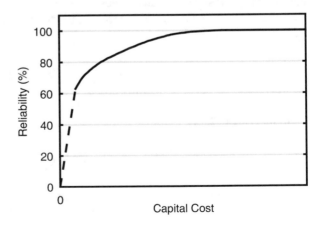

12.10   Example Systems

We conclude this chapter by briefly discussing three real-world off-grid systems.

12.10.1   Môle-Saint-Nicolas Mini-Grid: Sigora Haiti

In 2015, Sigora Haiti implemented a hybrid solar–diesel mini-grid in Môle-Saint-Nicolas, Haiti. Môle-Saint-Nicolas is located on Haiti's northwestern coast. The grid presently supplies a mix of 1,100 households, businesses, and social institutions. Household users are supplied with a 120 V, 60 Hz connection rated at 15 A. Images from Môle-Saint-Nicolas are shown in Figs. 12.21, 12.22, 12.23. Additional technical details are summarized in Table 12.19.

Sigora Haiti uses a proprietary multifunctional meter. The meter allows users to prepay for electricity. Credits can be purchased in small amount through the local

Fig. 12.21 Two 100 kW gen sets used in the Môle-Saint-Nicolas mini-grid (courtesy of Sigora Haiti)

Fig. 12.22 A 200 kW PV array is used in the Môle-Saint-Nicolas mini-grid (courtesy of Sigora Haiti)

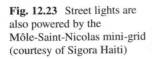

Fig. 12.23 Street lights are also powered by the Môle-Saint-Nicolas mini-grid (courtesy of Sigora Haiti)

Table 12.19 Môle-Saint-Nicolas system details

Year commissioned	2015
Users (present)	1,100 (5,500 people)
Users (future)	Eight additional towns
Energy sources	Solar (200 kW), Diesel (200 kW)
Peak load	70 kW (evening)
Energy storage (future)	130 kW, 630 kWh, lithium–ion
Low-voltage distribution	AC, 120/240 V, split-phase
Medium-voltage distribution	24 km (total), 22.8 kV, #2 ASCR, overhead, three-phase, radial

network of Sigora vendors, online and directly by mobile money payments. The maximum power that can be drawn by the user can be set via the meter's software.

Although many mini-grids are designed to serve a single community, the Môle-Saint-Nicolas mini-grid was designed at the outset to rapidly scale. In total, nine towns will be served, representing a significant increase over the present load. Sigora Haiti has designed the energy production system to evolve and scale as additional towns are added.

At present, there are two identical 100 kW diesel gen sets and a 200 kW PV array. Although this may seem like excessive capacity given the peak load of 70 kW, it allows Sigora Haiti to serve additional towns as the mini-grid is extended. A 630 kWh lithium–ion battery bank with 130 kW bi-directional converter will be added in the near future. In the interim, the gen sets are operated using a load-following scheme. The PV array is AC coupled using grid-tied string inverters. The inverters are not capable of forming the AC bus voltage, and so one gen set must be online at all times. Once installed, the battery bank will allow the utilization of the PV array to increase and dramatically reduce the use of the gen sets.

The Môle-Saint-Nicolas mini-grid has achieved a high level of reliability, in excess of 99.5%. This is remarkable given the prevalence of tropic storms and hurricanes in the region.

Table 12.20 Ighombwe system details

Year commissioned	2016
Users	50
Energy source	Solar (3.18 kW, 12 × 265 W panels)
Energy storage	24 V, 18 kWh lead–acid (12 batteries at 2 V, 750 Ah)
Low-voltage distribution	AC, 230 V, underground, single-phase, hub-and-spoke

Fig. 12.24 The PV array of the Ighombwe mini-grid with the batteries, inverters and meters securely located in the housing below (courtesy of PowerGen)

12.10.2 Ighombwe Mini-Grid: PowerGen

The Ighombwe mini-grid was implemented by PowerGen in 2016. Ighombwe is a farming community in central Tanzania. The Ighombwe site was chosen because it is one of the largest villages in the district with no national grid power, and the businesses served by the mini-grid are tightly clustered. Existing infrastructure was another consideration. There is a well-maintained road nearby. The grid was entirely self-funded by PowerGen, a Kenyan company. The technical details are provided in Table 12.20, and an image of the energy production equipment is in Fig. 12.24.

12.10.3 Filibaba–LiChi's Community Solutions and KiloWatts for Humanity

The Filibaba energy kiosk was implemented in 2015 by the nonprofit organizations KiloWatts for Humanity (USA) and LiChi's Community Solutions (Zambia) [10]. The energy kiosk is funded by a grant from IEEE Smart Village. Filibaba is a farming community located in the Copperbelt province of Zambia. Filibaba was selected in particular because LiChi's Community Solutions had previously been active in the community. The kiosk is shown in Figs. 12.25 and 12.26 with technical specifications in Table 12.21.

Fig. 12.25 The energy kiosk in Filibaba, Zambia (courtesy of KiloWatts for Humanity)

Fig. 12.26 The Filibaba energy kiosk sells groceries including refrigerated food and beverages (courtesy of KiloWatts for Humanity)

The kiosk is operated as a social enterprise. The kiosk recharges mobile phones and batteries and sells groceries and sundries. Frozen food and cold beverages are sold from a refrigerator inside the kiosk. Filibaba is a sparsely populated community. It was deemed not cost-effective to install a distribution system. Instead, small solar home systems were made available to the community for purchase. Financing was provided so that the solar home systems, each of which cost about US$35, could be paid off over several months. In addition, two nearby houses and a church were provided a wired connection.

Table 12.21 Filibaba system details

Year commissioned	2016
Users	4
Energy source	Solar (1.8 kW, 6 × 300 W panels)
Energy storage	24 V, 10.56 kWh lead–acid (4 batteries at 12 V, 220 Ah)

12.11 Summary

This chapter discussed the implementation and technical design of mini-grids. The mini-grid life cycle consists of eight steps, starting with prospecting and screening and ending with expansion or retirement. Early in the design phase, the different energy conversion technologies are compared against several objectives, constraints, and other considerations.

The technical design can proceed in two ways: using an intuitive approach, guided perhaps by rules of thumb, standards, and best practices. This approach can be used with limited information. Becoming more popular is the numerical or computer-aided approach. This approach uses computer simulations to evaluate and possibly identify feasible designs that are optimized in some sense—typically economic.

The design of the distribution system was discussed. There are several options that should be considered: AC or DC, overhead or underground, and hub-and-spoke or trunk-and-branch topology. The configuration can be single-phase, split-phase, three-phase wye, and three-phase delta. Hybrid configurations are also possible. The design should consider the targeted access tier, the user requirements, the load, the location of the users, and the cost of implementation.

The economics of each design are important. There are several ways of evaluating a mini-grid for economic viability. One important metric is the levelized cost of energy (LCOE), which shows the cost of providing each unit of energy over the lifetime of the project considering operation and capital costs. Ultimately the design of a mini-grid requires balancing the economic considerations with the technical. With the input data often very uncertain, there is almost never a single "best" design.

Problems

12.1 Design the battery bank for the users in Table 12.22. The characteristics of the available batteries are provided in Tables 12.23 and 12.24. Provide details of the battery selected, and the number of strings and batteries per string needed, if applicable.

12.2 Design the PV array for the users in Table 12.22. The characteristics of the available PV modules are provided in Table 12.25. Provide details of the module

Table 12.22 Standalone system details

User	Community center	Maternity ward	Restaurant	School
Avg. daily load (kWh)	0.68	0.40	0.56	2.36
Inverter efficiency (%)	0.85	0.83	0.78	0.82
Peak DC current @ 12 V	13.0	12.5	20.8	11.0
End of life rating	0.8	0.8	0.9	0.8
Days of autonomy	1.5	6	2	2
Maximum DoD	0.90	0.85	0.80	0.85
Cycles required	2000	1000	1500	2000
Battery design margin	0.08	0.10	0.15	0.05
Insolation kWh/m^2/day	4.3	5.2	5.5	5.1
Generation and storage loss(%)	18	18	25	32
Temperature reduction (%)	8	10	14	17
PV design margin	0.09	0.20	0.12	0.05
Minimum operating temp. (Celsius)	5	10	10	10

Table 12.23 Battery characteristics

Parameter	Battery A	Battery B	Battery C	Battery D	Battery E
Type	Flooded	Flooded	Flooded	AGM	AGM
Nominal voltage (V)	2	6	12	2	12
DoD for 1000 cycles	0.70	0.70	0.50	0.70	0.70
DoD for 1500 cycles	0.45	0.45	0.25	0.48	0.48
DoD for 2000 cycles	0.25	0.25	–	0.35	0.35

Table 12.24 Battery charge capacity

Hour-rate	10 hr	20 hr	50 hr	100 hr
Battery A	955 Ah	1124 Ah	1338 Ah	1452 Ah
Battery B	268 Ah	315 Ah	375 Ah	419 Ah
Battery C	72 Ah	85 Ah	95 Ah	106 Ah
Battery D	1040 Ah	1150 Ah	1200 Ah	1275 Ah
Battery E	189 Ah	210 Ah	218 Ah	230 Ah

Table 12.25 PV module characteristics

Parameter	Module A	Module B	Module C	Module D
Maximum power (W)	50	80	190	300
Open-circuit voltage, STC (V)	22.5	22.3	43.2	45.5
Short-circuit current, STC (A)	2	4.96	5.98	8.56
Open-circuit voltage temp. coeff.	−0.34%/K	−0.34%/K	−0.34%/K	−0.34%/K

selected, and the number of strings and modules per string needed, if applicable. Assume the maximum input voltage and current of the charge controller is 75 V and 15 A, respectively.

Table 12.26 Community parameters

Parameter	Initial
Avg. daily load (kWh/Day)	25
Peak load (kW)	38
Battery bank voltage (V)	48
Inverter efficiency (%)	81
Inverter design margin	1.05
End of life rating	0.80
Days of autonomy	3
Maximum DoD	0.90
Cycles required	2000
Battery design margin	0.05
Generation & storage loss (%)	30
Temperature reduction (%)	8.75
PV design margin	0.05
Minimum operating temp.	10

Table 12.27 WECS capacity factor by month

	Month (kWh/m²/day)											
	1	2	3	4	5	6	7	8	9	10	11	12
Capacity factor	0.25	0.29	0.40	0.18	0.34	0.36	0.38	0.10	0.24	0.39	0.22	0.34

12.3 Repeat the design for Mwase using a coincidence factor of 0.80. Specify the requirements for the inverter, battery bank, and charge controller. Compare this to the design when the coincidence factor is 0.37.

12.4 Repeat the design of the battery bank for Mwase using Battery D whose characteristics are in Tables 12.23 and 12.24.

12.5 Repeat the design for Mwase, but assume the average daily worst-month insolation is 5.5 kWh/m²/day. Specify the required PV array capacity. Provide details of the number of strings and modules per string needed, if applicable.

12.6 Design an off-grid system for a community whose characteristics are in Table 12.26. Assume the average daily worst-month insolation is 5.2 kWh/m²/day. Select the PV modules and batteries from Tables 12.25 and 12.23. Assume the maximum input voltage and current of the charge controller is 250 V and 60 A, respectively. Assume the selected inverter has a maximum rating of 5000 W and is able to be synchronized with other generators.

12.7 Design a DC-coupled wind-powered mini-grid for Mwase using the monthly capacity factors in Table 12.27. Determine the required rating of the WECS and rectifier. Assume each WEC is rated at 1 kW. Use a design margin of 0.5; assume the generation and storage losses are 15%. There is no temperature-related reduction.

12.8 Assume that there is viable hydro resource near Mwase. The effective head is 32 m and the maximum flow rate that can be used is 10 l/s. The combined efficiency of the turbine and generator is 82%. Determine the required power rating of the MHP system and the required flow rate if the system is AC coupled without batteries. Draw the architecture of the system. Include the rating of the electronic load controller.

12.9 Consider a 1 km single-phase, two-wire distribution line. The line serves 30 households. Ten households are located at 0.33 km down the distribution line; another ten are located at 0.67 km down the line, and the rest are located at the end of the line. Compute the voltage drop at each location along the line and the total power loss. Assume the impedance of the line is $Z_w = 0.4 + j0.3\Omega$, the current to each customer is 1.25 A, and the sending-end voltage is 230 V.

12.10 Compute the voltage drop and power loss for the Mwase system if a split-phase distribution system is used. Assume the loads are evenly balanced.

12.11 Compute the voltage drop and power loss for the Mwase system if the 21 mm^2 wire is used, as described in Table 12.17.

12.12 Compute the sLCOE of an diesel-based internal combustion engine used in an off-grid system with the parameters found in Table 12.28.

12.13 Consider a biomass gasification system serving 400 houses. Each house consumes 300 Wh of electricity per day. The gasification system is 60% efficient, and the combined efficiency of the engine and generator is 33.3%. The energy content of the dry husk feedstock is 12.6 MJ/kg, which costs US$25 per metric ton. The peak power demand is 20 kW. The overnight capital cost is US$1200/kW, and the distribution infrastructure and user premise costs total US$5000. The interest rate is 10%. The fixed operation and maintenance costs are $500 per month. There are no variable operation and maintenance costs. The lifespan of the system is 15 years. Compute the sLCOE associated with this system.

Table 12.28 System parameters

Parameter	Value
Lifespan (years)	8
Interest rate (%)	5%
K ($/kW)	400
Annual production (kWh/kW)	7884
Fuel price ($/MMBTU)	29
Heat rate (BTU/kWh)	11,500
Fixed OM ($/kW-year)	15
Variable OM ($/kWh)	0.007

References

1. Africa-EU Energy Partnership: AEEP energy access best practices (2016). URL http://www.euei-pdf.org/en/aeep/thematic-work-streams/aeep-energy-access-best-practices-2016
2. Berge, A., Vittal, V.: Power Systems Analysis, 2nd edn. Prentice-Hall (2000)
3. Gönen, T.: Electrical Power Transmission System Engineering, 3rd edn. CRC Press (2014)
4. HOMER Energy: (2018). URL https://www.homerenergy.com/
5. IEC: Recommendations for small renewable energy and hybrid systems for rural electrification. IEC Std. TS 62257-1:2015, International Electrotechnical Commission, Geneva, Switzerland (2007)
6. IEEE: Recommended practice for testing the performance of stand alone photovoltaic systems. IEEE Std. 1526–2003, Institute for Electrical and Electronic Engineers, Piscataway, NJ (2003)
7. IEEE: Recommended practice for sizing lead-acid batteries for stand-alone photovoltaic (pv) systems. IEEE Std. 1013–2007, Institute for Electrical and Electronic Engineers, Piscataway, NJ (2007)
8. International Renewable Energy Agency: Off-grid renewable energy systems: Status and methodological issues (2015). URL http://www.irena.org/publications/2015/Feb/Off-grid-renewable-energy-systems-Status-and-methodological-issues
9. Inversin, A.R.: Mini-grid design manual. World Bank (ESMAP technical paper no. 007.) (2000). URL "http://documents.worldbank.org/curated/en/730361468739284428/Mini-grid-design-manual"
10. Louie, H., Shields, M., Szablya, S.J., Makai, L., Shields, K.: Design of an off-grid energy kiosk in rural Zambia. In: 2015 IEEE Global Humanitarian Technology Conference (GHTC), pp. 1–6 (2015). doi:http://dx.doi.org/10.1109/GHTC.2015.7343946
11. Mentis, D., Howells, M., Rogner, H., Korkovelos, A., Arderne, C., Zepeda, E., Siyal, S., Taliotis, C., Bazilian, M., de Roo, A., Tanvez, Y., Oudalov, A., Scholtz, E.: Lighting the world: the first application of an open source, spatial electrification tool (OnSSET) on sub-saharan africa. Environ. Res. Lett. **12**(8), 085,003 (2017). URL http://stacks.iop.org/1748-9326/12/i=8/a=085003
12. Natural Resources Canada: RETScreen (2018). URL http://www.nrcan.gc.ca/energy/software-tools/7465
13. Odarno, L., Sawe, E., Sai, M., Katyega, M., Lee, A.: Accelerating mini-grid deployment in Sub-Saharan Africa: Lessons from Tanzania (2017)
14. Schnitzer, D., Lounsbury, D.S., Carvallo, J.P., Deshmukh, R., Apt, J., Kammen, D.M.: Microgrids for rural electrification: A critical review of best practices based on seven case studies (2014)
15. Sinha, S., Chandel, S.: Review of software tools for hybrid renewable energy systems. Renew. Sustain. Energy Rev. **32**, 192–205 (2014). doi:https://doi.org/10.1016/j.rser.2014.01.035. URL http://www.sciencedirect.com/science/article/pii/S136403211400046X
16. Willis, H.L.: Power Distribution Planning Reference Book, 2nd edn. Marcel Dekker (2004)
17. Zalengera, C., Blanchard, R.E., Eames, P.C., Juma, A.M., Chitawo, M.L., Gondwe, K.T.: Overview of the Malawi energy situation and a PESTLE analysis for sustainable development of renewable energy. Renew. Sustain. Energy Rev. **38**, 335–347 (2014). doi:https://doi.org/10.1016/j.rser.2014.05.050. URL http://www.sciencedirect.com/science/article/pii/S136403211400375X
18. Zimmerle, D., Manning, D.T.: Optimizing rural village microgrids to provide affordable and reliable renewable electricity in developing countries. In: 2017 IEEE Global Humanitarian Technology Conference (GHTC), pp. 1–6 (2017). doi:http://dx.doi.org/10.1109/GHTC.2017.8239318

Chapter 13
Solar Lanterns and Solar Home Systems

13.1 Introduction

We now shift our focus to solar lanterns (SLs) and solar home systems (SHSs). SLs and SHSs are small-scale, stand-alone solar-powered systems. Examples are shown in Figs. 13.1 and 13.2. Some SLs provide just enough electricity for a single light and perhaps to recharge a mobile phone, as shown in Fig. 13.3. Larger SHS can power small appliances such as radios, fans, and televisions as shown in Fig. 13.4. The exact distinction between a SL and SHS has become blurred. However, a useful definition is that solar lanterns have less than 10 W of PV capacity. They are also known as "pico solar." SHSs typically, but not always, have capacities exceeding 10 W. Both of these technologies also require a battery to properly function.

SLs and SHSs provide lower-tier (less than tier 3) electricity access than mini-grids as defined by the multi-tier framework in Sect. 2.7. However, the electricity they provide is meaningful as evidenced by their robust and rapidly growing sales in recent years (see Fig. 13.5). SHSs are more expensive than SLs, and so fewer SHSs have been sold. Some estimates place the number of SHS sold per year to be around one million units. The leading markets for SLs and SHS are India, Kenya, Tanzania, and Ethiopia [2]. Some projections predict SLs or SHSs will be in one out of every three off-grid households. The contribution of SLs and SHSs to improving access to electricity should not be overlooked.

© Springer International Publishing AG, part of Springer Nature 2018
H. Louie, *Off-Grid Electrical Systems in Developing Countries*,
https://doi.org/10.1007/978-3-319-91890-7_13

Fig. 13.1 A solar lantern
includes a small solar panel
and LED light (courtesy of
d.Light)

Fig. 13.2 Solar home
systems are often capable of
powering several LED lights
and small appliances
(courtesy of BBOXX Ltd.)

Fig. 13.3 Solar lanterns
provide a modest amount of
electricity, usually enough for
a single lamp and perhaps
enough to recharge a mobile
phone each day

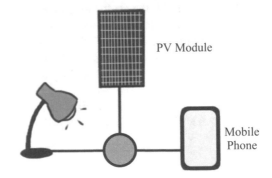

Fig. 13.4 Solar home systems are usually able to supply several lights and even small appliances

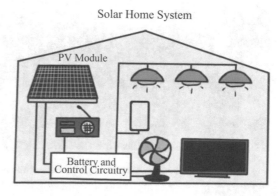

Fig. 13.5 Cumulative sales of pico solar products [2]

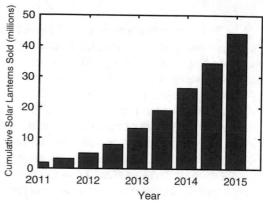

Example 13.1 How long would it take for SLs or SHSs to be in every presently unelectrified household? Assume a combined 20 million units are sold each year and an unelectrified population of 1.1 billion people. Assume that each SHS or SL can serve a single household with five people.

Solution The total number of households requiring a SLS or SL is found from

$$\frac{1.1 \text{ billion people}}{5 \text{ people per household}} = 220 \text{ million households.}$$

If 20 million units are installed each year, then it will take 11 years. However, this does not account for the limited life span of the units. This aspect is explored in Problem 13.1.

Fig. 13.6 Mobile phones can be recharged at businesses with electricity access for a fee (courtesy of the author)

SLs and SHSs are conceptually different from mini-grids in that they are consumer products, rather than industrial assets. Customers typically purchase or lease SLs and SHSs, whereas they purchase energy from mini-grids. Consequentially, the associated business models are very different, as discussed later.

The main drivers for the popularity of SL and SHS are rapidly reducing PV prices, improvements in the efficacy and reduction of costs of LED lighting, and the relatively high cost of traditional lighting methods, kerosene in particular. This makes SHSs and SLs cost-competitive. Mobile phones are another driver. In Sub-Saharan Africa, there are 74 mobile phone subscriptions for every 100 people [3]. This far exceeds the electrification rate. In rural areas, even those without electricity access, mobile phones are common. As you can imagine, recharging mobile phones in these areas is problematic. The owner must either travel to the nearest electrified town, perhaps tens of kilometers away, or pay a merchant with an off-grid system to recharge. See, for example, Fig. 13.6. In either case, the fee is often US$0.20 to US$0.50—a large sum given that many in these communities live off less than US$2.00 per day. The energy required to recharge a mobile phone is very low, often less than 10 Wh, making the effective rate for the energy several orders of magnitude greater than what grid-connected customers pay.

Recently, improvements in batteries—more specifically, lithium iron phosphate (LiFePO$_4$) chemistries—and cloud connectivity have enabled scalable business models for SHSs in particular.

SLs can be purchased from US$10 to about US$40. They are often available in or near rural communities. SHSs are more expensive. Some cost more than US$350, depending on the capacity and appliances included. It should be kept in mind that grid connection fees are often several hundred dollars. Like other off-grid systems, the levelized cost of energy (LCOE) of a SL or SHS is usually several times greater than the tariff for grid-connected electricity, especially if the tariff is subsidized. Sellers of SL and SHS usually tout the services the SL and SHS can provide—clean, modern lighting, for example—instead of the LCOE. This makes sense because with a SL or SHS, the customer is buying a product that provides services, not electricity.

13.2   Solar Lanterns

The first solar lanterns were designed to replace kerosene lamps, candles, and battery-powered torches (flashlights) commonly used in rural settings. They were designed to provide basic task lighting from a single light source and did not require permanent installation. Instead, they were designed to be portable. Portability is particularly useful when walking outside at night. They can also be hung, mounted, or placed on a flat surface. They were typically designed so that a typical day's insolation would be sufficient for the PV module to charge to the battery, allowing the light to be run for several hours in the evening.

LED prices have since dropped considerably, and the efficacy of LED bulbs—the amount of light output per unit of electric power input—increased. The result was a halving of LED costs, in terms of cost per lumen, between 2009 and 2015. With less energy needed for lighting, manufacturers began offering SLs that included additional functionality. Most important is the ability to recharge mobile phones through dedicated charging ports. SLs are now able to save the customer lighting and mobile phone recharging-related expenses. This can be significant. However, the market for SLs providing single-purpose lighting remains strong.

SLs are designed to be economical entry-points for electricity access. As such, reliability is often sacrificed for lower prices. SLs are typically designed to provide electricity only for 3 to 5 h in the evening if charged for a full day. Their batteries are usually smaller for a given PV size than a SHS. The PV module rating versus battery capacity for several SLs and SHSs is shown in Fig. 13.7. Note the difference in axes scaling. It is worthwhile to note that in 2009, nearly 40% of the SLs sold would last between 9 and 24 h on a single day of charging [2]. Most newer models

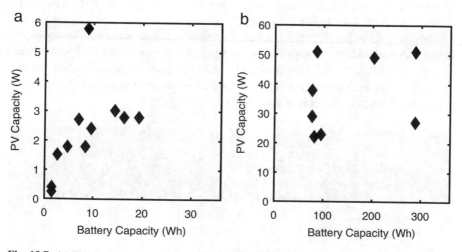

Fig. 13.7 (a) PV size versus battery capacity for SL and (b) SHS

do not offer this level of autonomy. This suggests that the increased reliability was not important to the customers—a less expensive unit with less reliability was more desirable.

13.3   Solar Home Systems

SHSs offer more functionality than SLs and greater solar and battery capacities. SHSs are typically rated below 100 W, but even at this size, appliances such as televisions and fans can be powered. Like SLs, SHSs include PV modules, a battery, control, and protection circuitry. They typically feature several ports for multiple LED lights, USB charging of devices, and plugs for DC appliances. Some SHSs feature inverters so that AC appliances can be powered.

SHSs are not intended to be conveniently portable. The PV array is roof-mounted and the lights are designed to be hung or wall mounted. The lights are connected to cables that are long enough to reach different rooms or even be hung outside a home. The batteries are often lithium ion, which offers longer cycle life and reduces disposal concerns when compared to lead–acid batteries. SHSs are designed so that their battery is completely charged by a typical day's insolation.

13.4   Design

The architecture of a SL and SHS mirrors that of a DC-coupled mini-grid with DC load as discussed in Sect. 4.6. The basic components are PV module, charge controller, battery, and DC load. Most have battery discharge protection in the form of a LVD. The sophistication of the charge controller varies: many do not include maximum power point tracking functionality, and some can be recharged by an external AC power supply. An example of a generic SHS design is shown in Fig. 13.8. SHSs and SLs are designed to be user-friendly. Many include battery state-of-charge indicator lights and have special protections to prevent the external sources and loads from being plugged into the wrong ports.

Example 13.2 A solar lantern provides light through an LED that consumes 1.25 W when on. Determine the required capacity of the battery, in amp-hour, and solar module for the LED to be powered for 6 h each day with an insolation of 5.0 kWh/m^2/day. Assume the battery is LiNCM and the daily depth-of-discharge is 80%.

Solution The energy required each day is

(continued)

$$6 \, h/day \times 1.25 \, W = 7.5 \, Wh/day.$$

The required battery size, in watthours, accounting for the depth-of-discharge is at least

$$\frac{7.5 \, Wh}{0.8} = 9.375 \, Wh.$$

The battery is LiNCM, so we use the nominal voltage of 3.7 V to determine the amp-hour rating:

$$\frac{9.375 \, Wh}{3.7 \, V} = 2.534 \, Ah.$$

The discharge current is

$$\frac{1.25 \, W}{3.7 \, V} = 0.338 \, A.$$

Therefore, the battery should have a rating of $C_{0.334}=2.58Ah$. Recall from Chap. 11 that, for example, a PV module rated a 1 W will produce 5 Wh with an insolation of 5.0 kWh/m^2/day. Therefore, the required PV capacity rating for the solar lantern is

$$\frac{7.5}{5} = 1.5 \, W.$$

This simple design does not include a design margin or other losses, but nonetheless, it generally agrees with the PV and battery capacities for solar lanterns in Fig. 13.7.

13.5 Light Output

Providing light is one of the most beneficial uses of electricity. It is the primary reason why SLs and SHSs are in demand. The brightness of a light is associated with its luminous flux, measured in "lumens" (lm). Luminous flux is the power of the electromagnetic radiation emitted by a light source that is within the visible spectrum as perceived by the human eye.

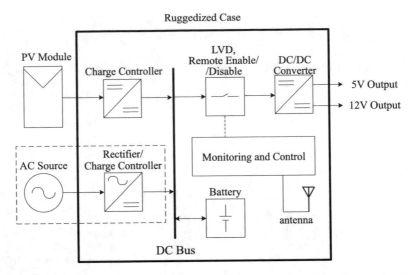

Fig. 13.8 A generic design for a SHS system cable of PV module and AC source inputs and multiple DC voltage output. External charging from an AC source is becoming less common in SHS design

Table 13.1 Typical lumen output

Technology	lumens (lm)
White LED (1 W)	25–120
Kerosene lamp	10–100
Incandescent (40 W)	325
White LED (7 W)	450
Fluorescent (18 W)	1250
Incandescent (100 W)	1750

Table 13.1 shows the luminous flux associated with different light sources. The luminous efficacy η_L of a source, for example, an LED bulb or kerosene lamp, is defined as the luminous flux F divided by the input power to the lamp P_{lamp}

$$\eta_L = \frac{F}{P_{lamp}}. \tag{13.1}$$

A lamp whose emitted radiation is largely outside the visible spectrum has low efficacy. The efficacy of kerosene lamps is low, around 0.1 lm/W. The efficacy of an LED is very high by comparison, about 50 to 100 lm/W. Most SLs and SHSs use LED lights, with each light fixture consisting of one or more LEDs, providing between 100 and 200 lm total. The lumen output of an LED slowly fades over time. A quality target for SL and SHS is that their lights maintain at least 85% of the rated lumen output after 2000 h of use.

How large of a space can be meaningfully lit by a solar lantern? The answer depends on type of space being lit and the activities being done in the space. For example, classrooms and offices should have an "illuminance" of about 250 lm/m^2. Using this target, two SLs each outputting 125 lm are needed to illuminate a 1 square meter surface, about the size of a large desk. In practice, not all of the light produced is cast upon the surface and so more SLs are needed. Many rural homes are lit by just one SL or SHS. While this is inadequate by illumination standards in developed countries, the light is still meaningful and much safer, cheaper, and more convenient than alternatives such as kerosene lamps or candles.

13.6 Remote Data Capture and Control

A recent trend is to equip SHSs with remote data capture and control capability. SHSs with this capability are outfitted with sensors measuring quantities such as battery terminal voltage, charging/discharging current, and temperature. The data are locally stored on a memory device and perhaps compressed or preprocessed. An internal radio periodically transmits the data to a remote server over the cellular network, where it is included in a centralized database. It is possible to transmit the data using a low-speed technology such as GPRS (General Packet Radio Service), which is widespread in Africa, even in remote areas.

There are technological and commercial uses of this capability. In-field measurements of voltage, current, and temperature provide insight into how the products are used. These data are valuable in improving designs of next-generation SHS and can be used to monitor battery degradation and predict failure. This allows for planned maintenance and replacement of the SHS, which can be done more efficiently and cheaply than waiting for failure to occur. Understanding the typical use of SHS can help researchers and engineers select appropriate battery technologies and chemistries and design charge controllers that are better-suited to how the SHS is actually used. It is also possible to track the location of SHSs using GPS (if included) or through triangularization from the cellular network.

Some SHSs have remote control capability. Commands can be sent to the SHS using a mobile network to enable or disable it through an internal solid-state switch or adjust the LVD set-point. This effectively disconnects the ports on the SHS from the internal battery, rendering the SHS inoperable. There is also a commercial reason for this functionality, as discussed next.

13.7 Business Models

The upfront cost of a SHS is a barrier to ownership. A large SHS can cost several hundred dollars, equivalent to several month's wages in many parts of the world. There are multiple strategies to lowering this barrier.

13.7.1   Rent-to-Own

In a "rent-to-own" business model, instead of paying the full price of the SHS upfront, periodic payments are made by the customer over an extended but definite period, for example, 6 to 24 months. At the end of the rental period, the SHS is owned outright by the customer and further payments are not necessary. The customer is now in sole ownership of an asset, which they can continue to use, resell, or offer as collateral. However, should the SHS fail after the rental period, the system will likely not be replaced by the seller.

13.7.2   Perpetual Lease

In a "perpetual lease" model, the customer never owns the SHS outright. They make payments as long as they wish to use the SHS. The advantage to the customer is that the payments are smaller compared to those under a rent-to-own model, and the risk of malfunction or failure of the SHS is transferred to the manufacturer. The seller repairs or replaces the SHS as needed as long as payments are being made. The customer payments are usually pegged to a period of service—for example, US$0.75 per day, not energy consumed, as is done with grid or mini-grid electricity. Conceptually then, the perpetual lease model is similar to selling lighting as a service, rather than a SHS as a product. Some perpetual lease agreements also require an upfront payment by the customer. Typically the user prepays, meaning they pay before they can use the SHS. This arrangement is commonly referred to as "pay-as-you-go" (PAYGO).

From the perspective of the seller, a rent-to-own arrangement reduces their financial risk as the money is repaid faster than under a perpetual lease scheme. However, some sellers see value in maintaining an ongoing relationship with customers through the perpetual lease model. Over time, customers can be "up-sold" to more expensive and higher-quality SHSs, as well as appliances.

13.7.3   Payment Models

There are several considerations in designing a payment model for SHSs which apply to the rent-to-own and perpetual lease models. These considerations also apply to mini-grids.

Payment method: the transfer of money from the customer to the SHS seller can be done in several ways. The payment can be made in cash to the seller or an agent of the seller. However, this requires the seller to have a presence in rural locations and places an administrative burden on collecting and tracking cash payments.

In countries where mobile money[1] is available, it is possible for the customer to transfer money to the seller via a mobile phone. This is often more convenient for the customer and seller than in-person cash payments.

Frequency of payment: the frequency of payment is linked to the payment method. There is a burden placed on the seller and customer with each transaction, especially for in-person cash payments. This consideration favors weekly or monthly payments (typically, the customer prepays rather than post-pays). However, as the frequency of payments decreases, the amount per payment increases, which requires more financial planning and saving by the customer. Mobile money on the other hand facilitates more frequent payments. Daily payment requirements are common in mobile money schemes.

Amount of payment: the amount charged by the seller per payment depends on whether or not the customer is on a rent-to-own or perpetual lease arrangement. The pricing is often set to be cost-competitive with other fuel sources, typically kerosene. Payments ranging from US$0.30 to US$2.00 per day, depending on the size of the system, are common. Of course, the payment amounts and frequency and duration of payment must be sufficient for the seller to profit in the long term.

Example 13.3 An off-grid family has two mobile phones and uses three kerosene lamps for lighting. Each mobile phone requires 5.0 Wh of energy to recharge and must be recharged every 3 days. Each kerosene lamp produces 20 lm and consumes 0.02 liters of kerosene each hour. The lamps are used for 4 h each night. The family pays US$0.20 to recharge their mobile phone in the nearest electrified town. Kerosene costs US$1.4/liter. What is the maximum daily amount that can be charged for a SL or SHS that is capable of recharging the mobile phones and replacing the kerosene use without increasing the family's expenditure?

Solution The average daily kerosene expense is

$$3 \text{ lamps} \times 0.02 \text{ liters/h} \times 4 \text{ h/day} \times US\$1.4/\text{liter} = US\$0.336/\text{day}$$

The average daily expense for mobile phone charging is

$$2 \text{ phones} \times US\$0.2/\text{charge} \times \frac{1}{3} \text{ charges/day} = US\$0.133/\text{day}$$

(continued)

---

[1]Mobile money is common in developing countries. It is a payment service platform that enables individuals to transfer and receive money using mobile phones.

The total expenditure is US$0.336 + US$0.133 = US$0.469/day. This is the maximum amount that can be paid by the family each day. While it might not seem like much money, a SL providing this functionality would likely cost less than US$40. If the family paid for it under a rent-to-own plan at US$0.50/day, they would own it outright in less than 3 months.

13.7.4   Payment Technologies

There are also technological considerations in choosing a SHS payment scheme. The primary consideration is how payment terms are enforced. There must be a consequence of nonpayment. This is typically removal or disablement of the SHS until payments are made. A low-technology way of enforcing payment is by physically repossessing a SHS if payments are not made. This method is time-consuming and could incite conflict between the seller and the customer as the SHS is being repossessed.

The repossession method of payment enforcement is falling out of favor as SHS sellers adopt more technologically advanced methods. One method is analogous to the prepaid method used in some grid- and mini-grid-connected systems. Upon payment, the customer is issued a unique code. The code is entered into a key pad on the SHS, which enables it until the next payment is due.

In systems that have remote control capability, the seller can remotely enable and disable the SHS. If a payment is missed, then the SHS is disabled.

There is another benefit of remote payment aside from the direct commercial and technological benefits of remote monitoring. The consumption patterns and payment profiles of the SHS can be stored and analyzed over time. This can lead to the construction of credit profiles of households, which can unlock and de-risk loans to rural individuals. It is too early to see if this application will be used, but the potential is at least interesting.

13.7.5   Practical Considerations

SLs and SHSs are products sold to individual customers. Like most products, branding and marketing are important. The leading companies strive to be associated with reliable, high-quality products. To this end, they strive to meet the quality standards set forth by independent organizations such as the Global Off-Grid Lighting Association (GOGLA) [1]. The GOGLA quality standards include targets for lumen maintenance, battery protection, physical quality (drop test), water protection, and warranty, among others. With so many fake and imitation products in the marketplace, it is good practice to check which, if any, quality standards are met before purchasing a SL or SHS.

13.8 Summary

Solar lanterns (SLs) and solar home systems (SHSs) offer lower-tier electricity access on an individual or household basis. SLs provide sufficient light for basic tasks, and some can recharge mobile phones. SHSs are more expensive but can power multiple lights and in some cases small appliances. Over 40 million units have been sold, with sales now exceeding 10 million per year. The popularity of SLs in particular can be attributed to their low cost, coupled with the high price and low quality of competing lighting sources, especially kerosene.

There are several business models that can be used to lower the financial barrier to solar home system ownership. Rent-to-own and perpetual lease with pay-as-you-go payment plans are popular. These payment plans are enabled by remotely controllable SHSs, which can enable and disable SHSs depending on if payments have been made.

Most experts expect the markets for SLs and SHSs to continue to grow for the foreseeable future. The capabilities should increase as appliances designed for the off-grid market increase in availability and decrease in cost. Although the electricity access tier is modest, some see them as a reasonable transitional solution until higher-tier mini-grid or national grid access is available.

Problems

13.1 Repeat Example 13.1 but assume the SLs and SHSs require replacement every 6 years.

13.2 The LED used in a SL has a luminous efficacy of 130 lm/W. The total lumen output is 90 lm. Compute the power required by the LED.

13.3 A certain SHS includes four 5 W lights and one 7.5 W television. The battery is rated at 17 Ah at 12 V. Compute the number of hours the SHS can operate for under the following scenarios: (1) the four lights only, (2) the television only, and (3) the four lights and the television. Assume the daily depth-of-discharge is 60%.

13.4 Determine the required capacity of the PV module of the SHS in the previous example assuming an average daily insolation of 5.0 kWh/m^2/day. Assume the generation and storage losses are 20%. Do not include a design margin for the PV array.

13.5 Determine the required charge capacity of the battery in a SHS designed to supply three 1.5 W LEDs for 8 h per night and a 15 W television for 3 h per night. The battery should be sized so that the daily depth-of-discharge is 40%. The nominal voltage is 14.4 V.

13.6 Determine the required capacity of the PV module of the SHS in the previous example assuming an average daily insolation of 5.0 kWh/m^2/day. Assume the generation and storage losses are 20%. Do not include a design margin for the PV array.

13.7 Describe, in your own words, the difference between a perpetual lease and rent-to-own payment models.

References

1. GOGLA: Global off-grid lighting association (2018). URL https://www.gogla.org/
2. Sturm, R., Njagi, A., Blyth, L., Bruck, N., Slaibi, A., Alstone, P., Jacobson, A., Murphy, D., Elahi, R., Hasselsten, J., Melnyk, M., Peters, K., Appleyard, E., Orlandi, I., Tyabji, N., Chase, J., Wilshire, M., Vickers, B.: Off-grid solar market trends (2016). URL http://documents. worldbank.org/curated/en/197271494913864880/Off-grid-solar-market-trends-report-2016
3. World Bank: World bank open data (2017). URL https://data.worldbank.org/

Chapter 14
Other Considerations

14.1 Introduction

We have focused so far on the technical and some economic aspects of off-grid electrification. However, there are a host of other considerations that a practitioner should be mindful of. In this chapter, we discuss several of these considerations, drawing largely upon firsthand experiences of the author and other organizations. When appropriate, advice and generally accepted best practices are given.

14.2 Electricity Theft

Theft of electricity is a problem that utilities face worldwide. It is also a concern for mini-grids. Electricity theft, also referred to as "nontechnical losses," is generally low, at 1 to 2% of generation. However in many countries struggling with electricity access, the nontechnical losses can exceed 15% per year.

Precautions should be taken to guard against electricity theft. One way of stealing electricity is by tampering with or bypassing the meter. There are a few ways of making it more difficult to tamper with a meter. One way is to locate the meter in a place where it is difficult or impossible for the customer to access the meters. For example, it can be placed at the top of the distribution pole as seen in Fig. 14.1 or at the sending end of the customer's line, rather than the end. Tamper-resistant meters can also be installed.

Low-voltage distribution lines can be susceptible to unauthorized connections by tapping or splicing them. Aside from the danger, this is problematic because it increases the load on lines, transformers, and inverters, perhaps exceeding their rated values. This can cause premature failure and increased voltage drop and losses along the line.

© Springer International Publishing AG, part of Springer Nature 2018
H. Louie, *Off-Grid Electrical Systems in Developing Countries*,
https://doi.org/10.1007/978-3-319-91890-7_14

Fig. 14.1 Placing customer meters on distribution pole tops, like these in Nigeria, can reduce meter tampering (courtesy GVE Projects)

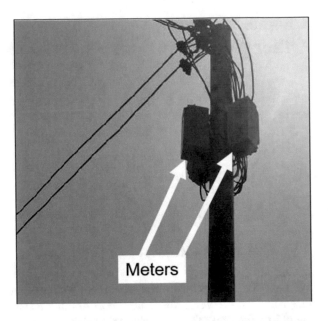

Employees of the mini-grid can be bribed to inaccurately bill the customer or record payment when a payment was never received. Customers might refuse to pay all or a portion of the bill or chronically make late payments. Mini-grid operators should have checks and procedures in place to detect improprieties by its employees and to disconnect customers in arrears. Pay-as-you-go billing with remote disconnection functionality is a popular method to discourage late payment.

14.3   Use of Donated Equipment

Nonprofit organizations abroad often find that manufacturers of electrical equipment are willing to donate their products. These "in-kind" donations can offset several thousand dollars of project costs. However, an organization should carefully assess whether or not to accept these donations.

The most important disadvantage of using donated equipment is that it is often up to the recipient organization to manage the shipping logistics and importation. This task should not be taken lightly, as arranging for shipment by sea or air is expensive and can add weeks of uncertainty to the project schedule. Additionally, import taxes and duties—perhaps 25% of the declared value—might need to be paid. In many countries, solar equipment can be imported without duty; however, the organization bears the administrative burden of completing and submitting the necessary forms. The equipment must clear customs—a process that in some countries can take weeks or months. In some cases, unscrupulous customs officers have expected a bribe to expedite the process. Holding fees at the port of entry might need to be paid while the components wait to clear customs. Batteries will self-discharge during this period, perhaps leading to permanent damage.

Additionally, many manufacturers will not provide warranties for donated equipment or will only provide equipment that did not make their quality standards (this is somewhat common with PV modules). When the donated equipment fails, the organization is on their own to find a replacement. For these reasons some organizations will refuse to use donated equipment, seeing it as well-intentioned but more trouble than its worth.

14.4 Capacity-Building

The goal of many organizations, particularly nonprofits and social enterprises, is to positively impact a community in a more broad way than just providing access to electricity. One way is to build capacity—the skills, experience, and infrastructure—of the community. Mini-grids in particular offer an avenue for this.

In many countries, use of in-country components is not possible. However, using in-country equipment distributors, logistics companies, and installers *is* possible. The advantage of this beyond building capacity is that the equipment supply chain and regular and warranty-related service can be managed in-country.

Many volunteer organizations from abroad rely on volunteer or student labor to implement the project. Although this offers a meaningful, if not transformative, experience for the participants, it must be done carefully, respectfully, and in a way that preserves the dignity of the community. Many tasks can be done by locally available labor—digging trenches for the distribution system, clearing land for poles, or making bricks. Reverse outsourcing of these tasks by volunteers should be avoided. There may be an opportunity to train and hire locals to manage the system once installed.

14.5 Substandard and Counterfeit Components

Many consumer products found in Sub-Saharan Africa are substandard or counterfeit. Unfortunately, this is also true for off-grid systems. Substandard components are those that do not meet applicable performance and quality standards set by manufacturers and local or international organizations. Counterfeit components are those that appear to be made by one manufacturer (typically a widely recognized, brand name manufacturer) but are made by another manufacturer. The counterfeit components are almost always of lower quality than the authentic components. Counterfeiting also occurs when a component is mislabeled and a lower-quality or lower-rated component is substituted. Used or damaged authentic products might be sold as new, sometimes after a superficial cosmetic treatment. Although hard numbers on the extent of problem are difficult to come by, one study estimated that approximately 50% of the solar home systems and solar lanterns worldwide or counterfeit [3]. Counterfeit and substandard components are appealing because

Fig. 14.2 A "Ciemans" charge controller (courtesy of P. Dauenhauer)

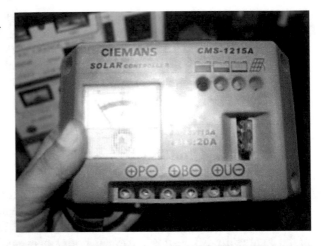

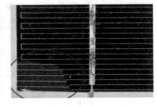

Fig. 14.3 Examples of PV modules with "dummy" material replacing active material (courtesy K. Sinclair, M. Sinclair, Zayed Energy and Ecology Centre)

they are often offered at a discounted price, sometimes accompanied by a reasonable story for the discount. Substandard components tend to reduce the reliability of the system due to premature failure or malfunction.

Practitioners in off-grid electrification frequently encounter substandard or counterfeit components. An example of a misleading label of a charge controller is shown in Fig. 14.2. Here, the controller brand name is "Ciemans" a simple play on the recognized brand name of "Siemens." Figure 14.3 shows a PV module where instead of silicon, some of the cells are simply pictures printed on paper. Another example is when a fake nameplate indicates that a component, for example, an inverter, is rated a higher value than it actually is.

Practitioners can and should take precautions against using substandard or counterfeit components. Use of established, reputable, and official distributors is the best way to reduce the risk. Even still, diligence is recommended. All components should be at least visually inspected for signs of counterfeiting or substandard quality. Inspection of PV modules is the subject of IEC 62257-10 [1]. All components should have labels and nameplates specifying the manufacturer, ratings, and model and serial number. If inspection of each component is not possible, batch-checking should be done, in which one or more components from every batch or pallet are inspected.

Contracts with suppliers should clearly state the ratings, manufacturers, and models to be used and should be written to include warranties—typically 1 or 2 years. A portion of the payment should be withheld until the materials have been verified to be of acceptable quality. If the system is installed by a contractor, it should be tested to ensure proper installation and operation.

14.6 Standards

In engineering, a "standard" refers to a document that specifies the technical details and characteristics of a product, system, or process. Standards are used to ensure consistency, safety, compatibility, and quality. However, adherence to a particular standard is not mandatory. Manufacturers, installers, and design engineers may or may not decide to follow a standard.

There are many standards that apply to certain aspects of off-grid electrical system components, including inverters, PV arrays, and lead–acid batteries. For example, IEC 61215 and IEC 61730 pertain to quality and safety of crystalline PV modules. However, standards relating to the off-grid systems as a whole are limited. At the time of writing, there are two major international standards relating to mini-grid systems:

- IEC TS 62257 Series—Recommendations for renewable energy and hybrid systems for rural electrification.
- IEEE 1546.4—Guide for Design, Operation, and Integration of Distributed Resource Island Systems with Electric Power Systems

Additionally, IEEE 1526 and IEEE 1013 provide a method for sizing stand-alone PV systems as discussed in Chap. 12. Some practitioners feel that the presently available standards require updating—indeed several are being revised—or that additional standards are needed for developing country context. Both the IEEE and IEC are developing independent standards for low-voltage DC mini-grids.

Adherence to any standard is optional. However, new practitioners should be familiar with practices described in the standards as a starting point. Lighting Global also has standards for solar lanterns and solar home systems, considering characteristics such as lumen maintenance, durability, health and safety, and truth-in-advertising [2]. It is recommended that before purchasing a product, it is checked to see if it meets the appropriate standard.

14.7 Applications Requiring Heightened Reliability

Losing power for an everyday application such as lighting can be annoying; loss of power to more critical applications, however, can result in economic loss or even danger to the community. Health care is one such critical use of electricity.

Access to adequate medical services in remote, unelectrified communities is often limited. Among the challenges is storing vaccines at the proper temperature. Vaccines must be stored at low temperatures in order to prolong their shelf life and to prevent the vaccines from degrading. However, if vaccines are exposed to too low of a temperature, they risk freezing, which causes an irreversible loss of potency. Solar-powered refrigerators offer a solution to store and regulate the temperature of vaccines at off-grid health-care facilities. The vaccine refrigerator should be designed to be extremely robust and reliable and meet performance quality standards set by the World Health Organization WHO [4]. To improve reliability, the vaccine refrigerator electrical system should not supply other loads, including lights and other medical equipment. If the off-grid system is capable of serving an AC load, WHO-approved refrigerators can be used. Consumer refrigerators intended for home or business use generally do not meet the WHO standards and should not be used for vaccines.

14.8 Human-Powered Generation

Using human motion to power generators for electricity access has been proposed by various researchers and organizations. A few commercial products exist, particularly those with applications in emergency or disaster situations, such as a hand-cranked radio or shake-style flashlight. Human-powered generation has found limited success in providing higher-tier electricity access. To many, it is treated as a novelty rather than a practicable, scalable solution. Human-powered generation does offer several advantages: no resource assessment is needed; the systems can be cheaply manufactured using locally available components such as bicycles and automobile alternators; and it can create jobs locally in the manufacturing but also powering the devices.

The disadvantages tend to outweigh the advantages. Chief among them are that the end-to-end efficiency of human-powered generation is prohibitively low. Unlike most renewable resources, the input energy to a human-powered generator is not free—the person powering the device must ingest additional calories to compensate for the added production and be compensated for their time. Average humans are not capable of sustained high-power output, perhaps only tens of watts. To supply 1 kWh requires several hours of effort and about 10 kWh of input (food) energy, roughly four times an average daily intake. This is likely not practical. In addition, time spent generating electricity is time that cannot be used for other, probably more productive, activities.

14.9 Summary

No textbook can replace field experience in off-grid system implementation in developing countries. Practitioners usually have many stories about ways that off-grid projects did not go as anticipated. Some of these are amusing, and some resulted in delayed, over-budgeted, or aborted projects. Volunteering for organizations such as IEEE Smart Village and KiloWatts for Humanity is a good way to gain experience prior to pursuing a career in off-grid electrification. New practitioners and organizations should remember that off-grid communities are underserved and vulnerable in many ways and should not be used as testing grounds simply to gain experience or to implement hobby projects and products.

References

1. IEC: Recommendations for renewable energy and hybrid systems for rural electrification – part 10: Silicon solar module visual inspection guide. IEC Std. TS 62257-10: ED1, International Electrotechnical Commission, Geneva, Switzerland (2017)
2. Lighting Global: Quality assurance program (2018).   URL https://www.lightingglobal.org/quality-assurance-program/
3. Sturm, R., Njagi, A., Blyth, L., Bruck, N., Slaibi, A., Alstone, P., Jacobson, A., Murphy, D., Elahi, R., Hasselsten, J., Melnyk, M., Peters, K., Appleyard, E., Orlandi, I., Tyabji, N., Chase, J., Wilshire, M., Vickers, B.: Off-grid solar market trends (2016).   URL http://documents.worldbank.org/curated/en/197271494913864880/Off-grid-solar-market-trends-report-2016
4. World Health Organization: A guideline for manufacturers of solar power systems. Tech. Rep. WHO_PQS_E003_GUIDE 2 3.doc, WHO (2010)

Appendix A

A.1 Review of AC Power Fundamentals

Steady-state AC circuit analysis is conveniently performed in the phasor domain. A sinusoidal voltage in the time domain can be written as

$$v(t) = v_{max} \cos(\omega t + \theta_v) \tag{A.1}$$

where v_{max} is the voltage amplitude; ω is the frequency, in radians per second; and θ_v is the phase angle in radians. When $v(t)$ is transformed into the phasor domain, it becomes

$$v(t) \leftrightarrow \mathbf{V} = |\mathbf{V}|\angle\theta_v = \frac{v_{max}}{\sqrt{2}}\angle\theta_v = V_{RMS}\angle\theta_v \tag{A.2}$$

where V_{RMS} is the RMS value of the voltage. The phasor transform can also be applied to a sinusoidal current $i(t)$ so that

$$i(t) = i_{max} \cos(\omega t + \theta_i) \leftrightarrow \mathbf{I} = |\mathbf{I}|\angle\theta_i = \frac{i_{max}}{\sqrt{2}}\angle\theta_i = I_{RMS}\angle\theta_i. \tag{A.3}$$

Note that the frequency of the voltage or current is not explicitly shown in the phasor representation. A circuit can only be analyzed in the phasor domain if the voltages and currents are sinusoidal and have the same frequency. Let the angle difference between the voltage and current be

$$\theta = \theta_v - \theta_i. \tag{A.4}$$

The angle θ is known as the *power factor angle*. The power factor PF is defined as

$$PF = \cos(\theta). \tag{A.5}$$

© Springer International Publishing AG, part of Springer Nature 2018
H. Louie, *Off-Grid Electrical Systems in Developing Countries*,
https://doi.org/10.1007/978-3-319-91890-7

When $\theta < 0$, the power factor is "leading" and when $\theta > 0$ the power factor is "lagging." Lagging power factor is associated with inductive loads, such as motors, whereas leading power factor is associated with capacitive loads. The voltage and current are related to the impedance Z as

$$V = IZ \tag{A.6}$$

where $Z = R + jX\ \Omega$.

The instantaneous power of a circuit element at time t is the product of the time-domain voltage and current at time t.

$$p(t) = v(t)i(t) = v_{max}i_{max}\cos(\omega t + \theta_v)\cos(\omega t + \theta_i) \tag{A.7}$$

The average power, also called the "real" power P, is the average of the instantaneous power. It can be shown that

$$P = \frac{v_{max}i_{max}}{2}\cos\theta = |V||I|\cos\theta. \tag{A.8}$$

The apparent (complex) power S is defined as

$$S = VI^* = P + jQ \tag{A.9}$$

where $*$ is the complex conjugate operator, j is $\sqrt{-1}$ and Q is the reactive or imaginary power. The complex power and reactive power are commonly expressed with units of volt-amperes (VA) and volt-amperes reactive (VAR), respectively. The relationship between S, P, Q, and θ are

$$P = \text{Re}\{VI^*\} = \text{Re}\{S\} = |S|\cos(\theta) = |V||I|\cos(\theta) = |S| \times PF \tag{A.10}$$

$$Q = \text{Im}\{VI^*\} = \text{Im}\{S\} = |S|\sin(\theta) = |V||I|\sin(\theta) \tag{A.11}$$

A.2   Three-Phase Analysis

A balanced three-phase source consists of three voltage sources whose RMS values V_ϕ are all the same and have the following relationships:

$$v_a(t) = \sqrt{2}V_\phi\cos(\omega t) \leftrightarrow V_a = V_\phi\angle 0° \tag{A.12}$$

$$v_b(t) = \sqrt{2}V_\phi\cos(\omega t - 120°) \leftrightarrow V_b = V_\phi\angle -120° \tag{A.13}$$

$$v_c(t) = \sqrt{2}V_\phi\cos(\omega t + 120°) \leftrightarrow V_c = V_\phi\angle 120° \tag{A.14}$$

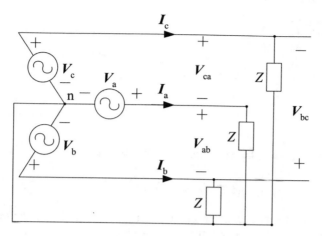

Fig. A.1 A balanced three-phase wye-circuit

That is, the sources are of the same magnitude and are displaced in phase by 120°. The voltage sources can be arranged as a wye or a delta. We will consider a wye-connected source; the fundamental results also apply to delta-connected circuits. A wye-connected circuit is shown in Fig. A.1. Note that common point of connection is known as the "neutral" point and is marked with an "n." The voltages V_a, V_b, and V_c are known as the "line-to-neutral" voltages. The voltages V_{ab}, V_{bc}, and V_{ca} are known as the "line-to-line" voltages.

In this book, we are mostly concerned with converting between the magnitude of the line-to-line voltage $V_{\ell\ell}$ and the magnitude of the line-to-neutral voltage V_ϕ. The line-to-line voltage is the voltage between any two lines. For a wye-connected source

$$V_{ab} = V_{an} - V_{bn} = \sqrt{3}V_\phi\angle 30° = V_{\ell\ell}\angle 30° \tag{A.15}$$

$$V_{bc} = V_{bn} - V_{cn} = \sqrt{3}V_\phi\angle -90° = V_{\ell\ell}\angle -90° \tag{A.16}$$

$$V_{ca} = V_{cn} - V_{an} = \sqrt{3}V_\phi\angle 150° = V_{\ell\ell}\angle 150°. \tag{A.17}$$

From this we see that the line-to-line voltage magnitude $V_{\ell\ell}$ is equal to the line-to-neutral voltage V_ϕ multiplied by a factor of $\sqrt{3}$. That is

$$V_{\ell\ell} = \sqrt{3}V_\phi. \tag{A.18}$$

A three-phase load is said to be balanced if the line-to-neutral impedances are equal. In this case, the phase currents have the same magnitude I_ϕ and are displaced from one another by 120°:

$$I_a\angle\theta_{i,a} = I_b\angle\theta_{i,a} - 120° = I_c\angle\theta_{i,a} + 120° \tag{A.19}$$

$$I_\phi = |I_a| = |I_b| = |I_c|. \tag{A.20}$$

Fig. A.2 Equivalent
per-phase circuit of a
three-phase wye-connected
circuit

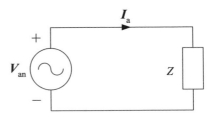

By symmetry, each phase supplies the same power:

$$S_\phi = V_{an}I_a^* = V_{bn}I_b^* = V_{cn}I_c^* \tag{A.21}$$

where S_ϕ is the power supplied or consumed by any single phase. The total apparent power is the sum of the power delivered by each phase:

$$S_{total} = 3V_{an}I_a^* = V_\phi I_\phi^* \tag{A.22}$$

and

$$P_{total} = 3\text{Re}\{V_{an}I_a^*\} \tag{A.23}$$

$$Q_{total} = 3\text{Im}\{V_{an}I_a^*\}. \tag{A.24}$$

We can therefore model any balanced three-phase circuit as a single-phase or "per-phase" equivalent circuit as shown in Fig. A.2. We use the line-to-neutral voltage in this circuit, and we must remember to apply a factor of three if we are interested in the total power. It is customary to use a-phase quantities, but this is arbitrary.

Solutions

Problems of Chap. 1

1.1 1 TOE; 1 MWh; 100 kWh; 10,000 calories; 5000 J; 1 Calorie; 1 BTU

1.3 3,4 days; US$0.294

1.7 18 panels (rounding up)

1.8 SSA: 454,666 GWh; USA: 4,069,400 GWh.

Problems of Chap. 2

2.1 851.67 l/yr

2.2 1481.4 Wh/day; 540.7 kWh/year

2.11 Tier 3

Problems of Chap. 3

3.1 69.47 A

3.4 Ababju: infrastructure cost US$600,500; cost per connection US$1201.0

3.5 Ababju: annual cost per connection US$63.97

© Springer International Publishing AG, part of Springer Nature 2018
H. Louie, *Off-Grid Electrical Systems in Developing Countries*,
https://doi.org/10.1007/978-3-319-91890-7

Problems of Chap. 4

4.3 $101.34\angle - 76.5°$A

Problems of Chap. 5

5.2 $112.68\angle - 23.44°$V; 11.8 kVA

5.4 One 50 kW gen set: US$148,263

5.7 9900 MJ/day

Problems of Chap. 6

6.1 3.90 m

6.5 45 m: for 1 kW 2.67 l/s, for 5 kW 13.3 l/s, for 10 kW 26.7 l/s

6.6 Location A: specific speed 4.37, Kaplan Turbine

6.7 PCD: 0.185 m; jet diameter: 0.011 m

Problems of Chap. 7

7.1 Cell Temperature: 56°C. short-circuit current: 9.61 A; open-circuit voltage: 43.17 V, maximum power: 308.77 W. short-circuit current at 500 W/m^2: 4.745 A. power produced: 217.77 W

7.4 Array short-circuit current: 30.28 A; array open-circuit voltage: 171.98 V; array power: 3685.3 W

Problems of Chap. 8

8.1 5 moles: 49.952 V

8.4 Current and C-rate at 72 hour rate: 5.47 A, 0.0139C

8.6 47.24 h

8.9 11 h, 63.36 kWh

Problems of Chap. 9

9.1 0.414 (41.4%)

9.3 Duty cycle: 0.6055; PV current: 9.081 A; MPPT current: 5.916 A

9.6 Power to battery: 0.487 kW; average DC voltage: 58.07 V; average DC current 47.80 A

9.10 $I = 4.06\angle - 2.93°$A; real power 1.266 kW; reactive power: 67.79 VAR

Problems of Chap. 10

10.3 Charge controller current: 40.83 A; PV power: 610.4 W

10.7 Battery voltage: 58.40 V; diversion load current: 2 A; charge controller current: 0 A

10.8 Gen set 1 power:57.2 kW; gen set 2 power: 52.8 kW; frequency: 59.91 Hz (both gen sets)

Problems of Chap. 11

11.3 User A: 140 Wh/day, load factor: 0.194

11.8 Capacity factor at 20 l/s: 1.00 (summer), 1.00 (winter); capacity factor at 30 l/s: 0.70 (summer), 1.00 (winter)

Problems of Chap. 12

12.9 Voltage drop segment 1: 12.375 V; segment 2: 8.500 V; segment 3: 4.125 V. power loss segment 1: 371.25 W; segment 2: 170 W; segment 3: 41.25 W

12.12 US$0.350/kWh

Problems of Chap. 13

13.5 14.06 Ah

Index

© Springer International Publishing AG, part of Springer Nature 2018
H. Louie, *Off-Grid Electrical Systems in Developing Countries*,
https://doi.org/10.1007/978-3-319-91890-7

Printed in the United States
By Bookmasters